Information Fusion and Data Science

Series editor
Henry Leung, University of Calgary, Calgary, AB, Canada

This book series provides a forum to systematically summarize recent developments, discoveries and progress on multi-sensor, multi-source/multi-level data and information fusion along with its connection to data-enabled science. Emphasis is also placed on fundamental theories, algorithms and real-world applications of massive data as well as information processing, analysis, fusion and knowledge generation. The aim of this book series is to provide the most up-to-date research results and tutorial materials on current topics in this growing field as well as to stimulate further research interest by transmitting the knowledge to the next generation of scientists and engineers in the corresponding fields. The target audiences are graduate students, academic scientists as well as researchers in industry and government, related to computational sciences and engineering, complex systems and artificial intelligence. Formats suitable for the series are contributed volumes, monographs and lecture notes.

More information about this series at http://www.springer.com/series/15462

Basel Solaiman • Éloi Bossé

Possibility Theory for the Design of Information Fusion Systems

Springer

Basel Solaiman
Image and Information Processing
Department
IMT Atlantique
Brest, France

Éloi Bossé
Image and Information Processing
Department
IMT Atlantique
Brest, France

ISSN 2510-1528 ISSN 2510-1536 (electronic)
ISBN 978-3-030-32852-8 ISBN 978-3-030-32853-5 (eBook)
https://doi.org/10.1007/978-3-030-32853-5

This Springer imprint is published by the registered company Springer Nature Switzerland AG
The registered company address is: Gewerbestrasse 11, 6330 Cham, Switzerland

Preface

Possibility theory is a mathematical theory, coined by L.A. Zadeh in the late 1970s (1978) to deal with vague pieces of information described by means of fuzzy sets and fuzzy logic. Thereafter, Didier Dubois and Henri Prade have been the main founders of this theory to the extent that we have today, a credible alternative to probability theory. A considerable body of literature has flourished around fuzzy sets and possibility theory concepts in a very wide range of applications, from mathematics and logics to advanced engineering methodologies, from medical domain to finance, from human factors to consumer products, and so on. There is a plethora of books and papers describing this rich domain of applications.

The ambition of this book is to address a niche still uncovered by the existing available books: a comprehensive assemblage of the basic concepts, the mathematical developments, and the engineering methodologies to position and exploit possibility theory for the design of computer-based decision-support systems. Usually, decision-support systems comprise three main parts: analysis (analytics), synthesis (information fusion), and prescription (decide and act). Literature shows that possibility theory can be applied to the three parts.

This book consists of nine chapters. The first three chapters discuss the fundamental possibilistic concepts: distribution, necessity measure, possibility measure, joint distribution, and the important concept of conditioning. Chapter 4 examines the concept of similarity that plays an essential role in a wide range of application fields like pattern recognition, reasoning, data and knowledge mining but with respect to what can possibility theory bring to implement that complicated concept. Chapter 5 addresses the links and transformations between the interrelated uncertainty modeling theories. The following next two chapters treat aspects of decision-making through possibilistic and fuzzy integrals, fusion operators, and decision-making criteria in the framework of possibility theory. Chapter 8 presents three low-level complexity applications of possibilistic concepts: (1) on pixel-based image classification, (2) on spatial unmixing, and (3) on image segmentation.

The book is concluded by Chapter 9 on the use of possibility theory in the design of information fusion systems in today's ever-increasing complexity of our real world. Information overload and complexity are core problems to most organizations of today. The advances in networking capabilities have created the conditions of complexity by enabling richer, real-time interactions between and among individuals, objects, systems, and organizations. Fusion of Information and Analytics Technologies (FIAT) are key enablers for the design of current and future decision-support systems to support prognosis, diagnosis, and prescriptive tasks in such complex environments. Hundreds of methods and technologies exist, and several books have been dedicated to either analytics or information fusion so far. This book presents the overall picture in which possibility theory can be of any use.

Brest, France Basel Solaiman
Brest, France Éloi Bossé

Contents

Chapter 1
Introduction to Possibility Theory

1.1 Introduction

The problem of decision-making, arising from everyday practice, belongs to the oldest problem types studied from the seventeenth century, and since then the best reference decision-maker is certainly the human brain. For this reason, modeling of the human brain reasoning operation in decision-making has been a topic of intense studies in many application domains leading to creative methodologies, algorithms, and deductive approaches giving, thus, way to sustainable researches and developments. The major difficulty faced by systems supporting decision-making is due to the fact that we have to deal with imperfect decision-relevant information. At this level, we have to admit a basic assumption that human knowledge, reasoning, and exchanged evidences and information are intrinsically, for the most part, characterized and expressed by a degree of ambiguity and uncertainty rather than in a probabilistic uncertainty manner. In fact, uncertainty and ambiguity capture two rather different types of information imperfections.

Uncertainty is the main cognitive process that makes human free to choose. Its presence (due to lack of knowledge, imperfect, or insufficient information) is the price affecting experts' decision when handling complex systems. It is derived by the nondeterministic membership of a point from the set of decisions containing all possible elementary decisions (called *singletons*, *states of the world*, *basic events*, *decisions*, etc.). The framework of uncertainty modeling has been rooted in probability theory in which the analyst's uncertainty about the integrity of the model is expressed in probabilistic terms. For a long time, probability theory has been considered as the unique normative model to cope with imperfection by presenting a classical well-founded framework manipulating uncertain but precise information. Nevertheless, probability theory, as good as it is, does not remain the best alternative where imprecision is inherent in the studied domain, where available information is simply preferences or ambiguous. In fact, ambiguity is derived from the partial

B. Solaiman, É. Bossé, *Possibility Theory for the Design of Information Fusion Systems*, Information Fusion and Data Science,
https://doi.org/10.1007/978-3-030-32853-5_1

membership of one or several points from the set of decisions. Thereby, over the last five decades, a lot of effort has been put into developing new nonclassical uncertainty theories (fuzzy sets theory, belief functions or evidence theory, imprecise probability theory, and possibility theory). Among the new theories, possibility theory is said to be amenable to the framework for representation of human perceptive uncertainty. This point has been suggested by prominent systems scientists such as Shackle [1] and Cohen [2]. They argue that the traditional approaches for choice modeling using probability theory do not completely represent the true level of uncertainty in people's behavior. Possibility theory deals with uncertainty when the evidence points to a nested set of propositions; and hence, it can deal with propositions that refer to an interval as well as a single value. Possibility theory deals with uncertain information where the available knowledge is expressed in an ambiguous framework and representing our inability to distinguish which of several alternatives is the true one in a particular situation. The word "possibility" can be interpreted in several ways: physical, epistemic, and logical [3].

In the physical sense, "possible" refers to feasibility or easiness. For example, in the sentence "it is possible for a vehicle to hold six passengers," possibility means the physical capacity of the vehicle. In the epistemic sense, "possible" means plausible. Finally, in the logical point of view, possibility provides a means to deal with incompleteness. With a given piece of incomplete information about an event, logical interpretation of possibility gives a degree of confidence in the occurrence of the event as in *it is possible that it will rain tomorrow*.

The concept of possibilities was, first, mooted by the economist G. L.S. Shackle [1, 3] who, unhappy with the use of subjective probability for handling uncertainty, proposed an alternative formalism. This formalism was the calculus of *potential surprise* where uncertainty about an event is characterized by a subjective measure of the degree to which the observer in question would be surprised by its occurrence. Potential surprise is clearly linked to the intuitive notion of possibility. If an event is entirely possible, then there is no surprise attached to its occurrence. If an event is wholly impossible, or is believed to be so, then if it occurs, it will be accompanied by the maximum degree of surprise.

Nevertheless, possibility theory was later introduced again by L. Zadeh [4], who related possibility theory to fuzzy sets theory. Consider the following example to show the scope of application of possibility theory. Let Ω denote a collection of individuals within which we are looking for precisely one and only one person but we don't know the person who we are searching for looks like. This uncertain environment, or situation, constitutes the global framework of application of possibility theory (which is exactly the same as the framework of other uncertain information processing theories like probability theory and belief functions theory). Now, imagine the available knowledge, called *evidence*, to identify our individual is given as a *fuzzy evidence* such as the individual we are looking for is *young*. Faced with this situation, our human reasoning will:

(i) "Indirectly" attribute, to each individual in Ω, a membership degree, or a belongingness value, on the scale from 0 to 1, to the fuzzy evidence *young*

(ii) Project this membership degree into a kind of mental possibilistic space of representation, where each individual in Ω has a *possibility degree of being* the one we are looking for.

This operation is referred to as the projection of the available evidences into a *possibilistic knowledge representation* form. In possibility theory, the result of this projection is called a *possibility distribution* defined on the set Ω.

In the case where the available evidence is not fully reliable, the obtained possibility distribution has to be "adjusted" taking into account this important knowledge of what is called information sources reliability. This possibility distribution adjustment is called possibility distributions *discounting*. If several sources of knowledge are available (the individual we are looking is *young*, *tall*, etc.), then and after the first step of transforming these sources into possibilistic knowledge representation form, the resulting possibility distributions have to be "merged" together, using adequate *possibilistic fusion operations*, in order to "resume" our global state of knowledge into a single possibility distribution over Ω. An important question is related to the decision-making process. In fact, as possibility theory operates in uncertain environments, the ultimate expected output is, therefore, to identify one and only one individual within the set Ω. At this level, possibility theory offers some interesting decision-making tools like the *possibility measure* (producing a degree of possibility that the individual we are looking for lies within a subset $A \subseteq \Omega$ of individuals) and the *certainty measure* (producing a *degree of trust* that the individual we are looking for lies within a subset $A \subseteq \Omega$ of individuals).

This simple example of searching for an individual within a set can be extended to extremely important categories of engineering problems: pattern recognition, investment risk evaluation, classification, estimation, automatic target recognition and tracking, etc. Before going in-depth through different concepts of possibility theory, it is important to precisely define the concept of "information," hereafter called *information element*, as well as different forms of information imperfections and make a brief visit to some existing theories allowing to process imperfect information.

1.2 Information Concept

1.2.1 Information Element Definition

One of the historical barriers to technology transfer in information processing systems has been the lack of a unifying terminology. *Information*, by contrast, is a most heterogeneous term. What is information? While there exists a conventional, commonsense hierarchy of information ranging from data (usually raw) to information (processed data) to knowledge (synthesized information), even these are not precise distinctions. The dividing lines are blurry and partly subjective. Recall that the success of an information processing system is strongly related to the way its basic components are defined and to the quality of their associated knowledge as well as to the knowledge produced by the processing system.

Nevertheless, there is a relatively small body of literature in the information processing community that addresses this topic; moreover, there is no precise definition of what *information* is and what defines information characteristics. A clear statement of *what is information* and *what is informative* can lead to a strong qualitative understanding of the fundamental nature of information. A definition of information should capture the essential nature of the information and should allow frameworks, theories, and results to be transferred across disciplinary boundaries [5, 6]. According to Losee [5], information can be defined in terms of a process, or a function, in the following way: "Information is the value currently attached or instantiated to a characteristic or a variable returned by a function, or a process. The value returned by a function is informative about the function's argument, or about the function, or about both." This definition has the merit of positioning information as a relational concept linking data sets [7]. Asking for the meaning, the word "information" seems trivial since we use this word so frequently. Consider a simple example. What does it mean if we are given the value $x = 39$? From a mathematical point of view, it is just a positive integer; it is simply *data*. If $x = 39$ °C, then this becomes more "informative" since x certainly refers to the temperature (i.e., x denotes an element from a set of meaningful semantics). Nevertheless, this is still not enough informative because we still don't know the temperature of what! Now, if we are given that this concerns the temperature of Paul, then we possess a real information element concerning the temperature of Paul, and we can even start "reasoning" on this information by saying that Paul has a fever and must be examined by a physician. In other words, *observing a data value from a given set isn't enough to make the observed data an informative act* [7]. We need the context in which that information has been obtained. Having real information means that we know the concern of the information and how the content outcome is obtained. This leads to the following pragmatic definition of information [7, 8] which extends the definition proposed by Losee [5] by including a formal structure associated with important information characteristics:

Definition An information element is *a functional relation between two data sets: definition and content sets, through an informative function*[1]. Here, the "informative function" has to be understood in its cyber-physical sense.[2]

Therefore, the main components of an information element are (Fig. 1.1):

1. A ***definition set***: representing the potential information input elements

[1]Here, "function" has to be taken not in its formal mathematical sense but rather as *an activity or purpose natural to or intended for a person or thing.*

[2]Cyber-physical means a mechanism (or a machine) that is controlled or monitored by computer-based algorithms. See https://en.wikipedia.org/wiki/Cyber-physical_system for a definition of a cyber-physical system.

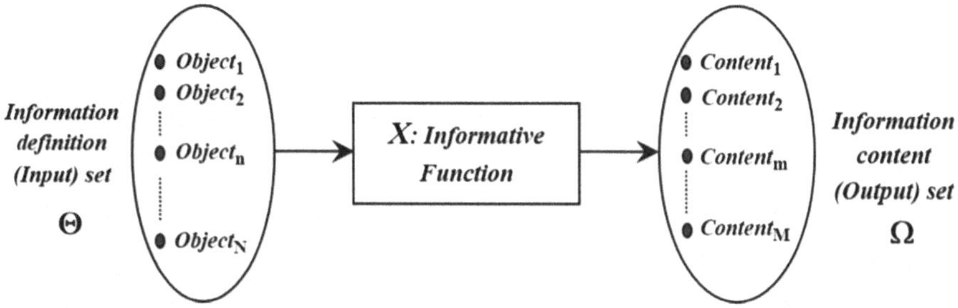

Fig. 1.1 Basic information element structure

2. A ***content set***: encoding the possible knowledge produced by the information such as measurements or estimations of physical parameters, decisions, hypothesis, etc.
3. An ***input-output relational function***: producing the mathematical or physical model representation associating the input elements with the produced information contents

An information element is always informative about something, being a component of the output or result of the informative relation.[3] In fact, having the sole information content isn't enough to make an input object/event as an informative act; the information meaning as an entity (*definition set*, *informative relation*, *content set*) must be perceived to make the "information" informative. This aspect has already been pointed by Stonier [9]: "we must not confuse the content reading and/or interpretation of information with information itself."

An information element is called ***exhaustive*** if and only if the content set contains all possible outcomes produced by the informative relation. This property, related to the information content set, is also called the ***closed world assumption***. Otherwise, the information is said to be operating under the ***open world assumption***.

Another important property is also used to characterize an information element: exclusivity. In fact, an information element is called ***exclusive*** if and only if two different information contents cannot be simultaneously produced as the outcome of the informative function.

Consider, for instance, the case of a digital image. In this case, if we consider the digital image as a 2D array representing gray levels, then what is considered is *not* information but just a kind of *abstract data* (i.e., observed pixel gray levels). This abstract data becomes a full information element when it is associated with the basic objects we are imaging as well as with the physical model leading to obtain the

[3]A *relation* between two sets is a collection of ordered pairs containing one object from each set. If the object is from the first set and the object is from the second set, then the objects are said to be related if the ordered pair is in the relation. A *function* is a type of relation. But, a relation is allowed to have the object in the first set to be related to more than one object in the second set. So a relation may not be represented by a function machine, because, given the object to the input of the machine, the machine couldn't spit out a unique output object that is paired to.

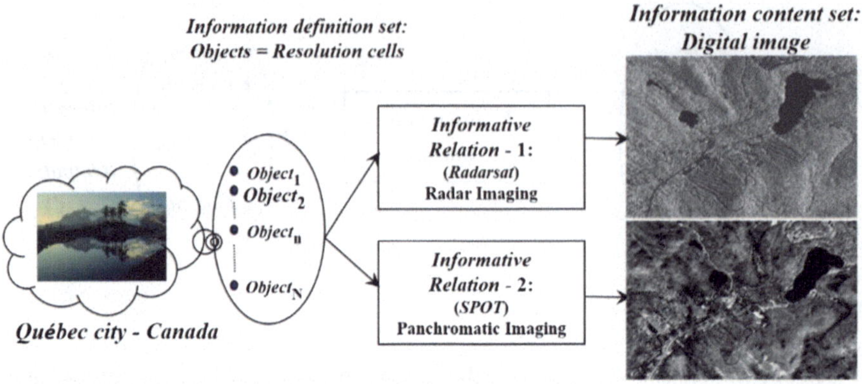

Fig. 1.2 Remote sensing examples of information using two informative functions (active radar imaging and passive panchromatic imaging functions)

digital image (i.e., the physical process used by the imaging sensor). In Fig. 1.2, image information elements of the Québec City are given using two imaging modalities (i.e., two distinct informative physical functions: *radar imaging* and *optical panchromatic imaging*).

The knowledge of the informative function is crucial for all information processing tasks that some call as "intelligent" tasks like scene interpretation, information fusion, data/knowledge mining, etc. For example, in remote sensing, let us imagine that a resolution cell (i.e., input object) is "observed" as having a zero gray level (i.e., information content). The interpretation of this cell in terms of its thematic contents (i.e., giving a semantic meaning to the observed gray level) cannot be conducted if the *physical model* of the sensor is not known. If the used sensor is an imaging radar, then, the resolution cell contents can be interpreted as being a flat surface or may correspond to a shadow area. On the other hand, if the imaging sensor corresponds to a given spectral band in multispectral imaging, then the resolution cell thematic content corresponds to the content absorbing the emitted electromagnetic signals in the considered spectral band. From this example we can easily understand that the adjunction of the physical model (i.e., informative function) and the resolution (i.e., input element) makes the essential difference between data and information.

If we go further in this explanation, we can easily understand the difference between "generic" data processing techniques and what can be called "knowledge-based" processing techniques. To illustrate this idea, we can simply imagine the difference between classical image filtering techniques and filtering techniques adapted to speckled images (radar, sonar, ultrasound, etc.). In fact, the knowledge incorporated in speckled images filtering techniques corresponds to the mathematical modeling of the physical model exploited by the imaging sensors (i.e., a multiplicative Rayleigh noise). Moreover, the explicit positioning of the informative function within the basic information element structure clarifies the concept of ***information partiality*** or ***incompleteness*** (i.e., the information does not capture all relevant aspects of a phenomenon, an entity, or an input object). In fact, considering a sensor-based information element and since a sensor exploits a specific physical process in order to extract one or several "facets" of the observed objects, then the huge informative aspects of an observed object will be restricted to those acquired through the "physical" window of the considered sensor. For instance, the spectral signature of objects is only measured in few small spectral bands in a multispectral imaging system. In radar imaging systems, this partial sensor vision is related to the frequency, polarization, and geometrical acquisition configuration used by the sensor. As a direct consequence, and to overcome information incompleteness, the use and the development of information fusion systems are very desirable and even become crucial.

It is important to notice that this information element definition and structure are extremely general and can be applied to all types of encountered information elements: sensor issued information, data transformation information, feature extraction information, decisional information, etc.

1.2.2 Intrinsic Information Imperfection Types

Information imperfection usually arises at the early stages of the development of information processing systems, since it pervades the description of the domain or the real-world situation. It can be considered as the multifaceted concept characterizing the fact that a considered information element lacks fulfilling a predefined targeted objective. Most efforts for handling imperfect information have focused on "modeling" imperfections and on "processing" imperfect information elements through mature mathematical theories and approaches.

Quality of information (QoI) [10] provides the foundation and the reasoning framework for the conception, design development, operation of information processing, and fusion systems. Considerable research on studying and classifying various quality aspects into broad categories has been conducted. Wang and Strong [11] have classified QoI into four major categories: *intrinsic*, *contextual*, *representational*, and *accessibility*.

In the framework of information processing systems, QoI is addressed at only two levels: intrinsic and contextual levels. The intrinsic level concerns information characterization in terms of imperfection nature, interpretation, and modeling while considering the information element as an "independent" entity out of the global fusion context, whereas the contextual level concerns the information characterization in terms of its impact, completeness, relevance, conflict, redundancy, etc. within the global fusion context.

From an intrinsic point of view, various sources of imperfections are encountered ranging from the early input definition to the content outcome of the information element (including the nature of the informative function, or relation, as well as the available external sources of knowledge used by the information element).

Consider the information element $I = (\Theta, X, \Omega)$ of Fig. 1.1, where Θ (resp. Ω) denotes the definition (resp. content) set and X the informative relation. Imperfection modeling is studied in terms of impact of different information imperfection sources on the information content outcome. Three major intrinsic imperfection types are considered: *uncertainty*, *imprecision*, and *ambiguity*.

I. ***Uncertainty*:** Most information processing systems efforts are concerned with adequate modeling of information uncertainty, which is the result of noisy, imprecise, erroneous or ill-suited to the problem data, ambiguous observations, and incomplete and poorly defined *prior* knowledge [12].

   Assume the two following conditions hold:

1. The informative relation, X, is a *punctual outcome relation* (i.e., one information content, x_{True} from Ω is produced by X).
2. The set of information contents Ω is exhaustive and exclusive (i.e., the information content produced by X is unique and certainly belongs to Ω).

The information is said to be affected by uncertainty, if and only if the true information content x_{True} is *unknown* with certainty. The main objective of uncertainty imperfection modeling and processing is to represent and to deal with this lack of knowledge and to "determine" with total certainty the unknown true content outcome of the considered information (the class of an observed object, the decision to consider, etc.). Two major approaches are used to model and to process uncertainty type of imperfection: *probabilistic* and *evidential* approaches. Given that uncertainty affects punctual outcome informative functions and relations, both probabilistic and evidential approaches consider the *total certainty* as having a global measure of unity. Depending on the available knowledge concerning the true information content, x_{True}, both approaches differ in the way this global measure is distributed into different elements from Ω.

The probabilistic approach makes a "punctual certainty distribution" of the unity total certainty on different information contents (i.e., each information content, $x \in \Omega$, called a singleton, captures a partial amount of certainty $\Pr\{x\}$ where different $\Pr\{x\}$ add to one). $\Pr\{x\}$ is interpreted as the probability that x is the true information content. An information element $I = (\Theta, X, \Omega)$ affected by uncertainty type of imperfection and for which a probability distribution of uncertainty is available is called a *probabilistic information*.

The evidential approach (based on belief functions theory [13]) constitutes an excellent alternative to the probabilistic approach when the available knowledge does not allow making a punctual certainty distribution. In fact, evidential approach is based on making a "subset certainty distribution" of the unit total certainty on different subsets of Ω (i.e., each information content subset, $A \subseteq \Omega$, captures a partial amount of certainty $m(A)$ with different $m(A)$ add to one). $m(A)$ is interpreted as the mass of belief that the true information formation content x_{True} is in A. An information element $I = (\Theta, X, \Omega)$ affected by uncertainty type of imperfection and for which a mass distribution of uncertainty is available is called an *evidential information*.

Notice that the probabilistic approach can be considered as a special case of the evidential approach. Relaxing the punctual certainty distribution constraint, by the evidential approach, gives a "practical" dimension and a "facility" of knowledge representation. Nevertheless, this relaxation reduces the precision and the quality of information processing results when compared to the probabilistic approach.

II. ***Imprecision***: Imprecision type of imperfection is an issue pertaining to the intrinsic quality of information. It refers to the case where the available knowledge about the true information content is available as a subset Ω_1 of the information content set Ω (i.e., $x_{True} \in \Omega_1 \subseteq \Omega$). In this case, the information $I = (\Theta, X, \Omega)$ is called an *imprecise information*. Therefore, imprecise information involves the lack of precise knowledge of the information content and, thus, should not be considered as erroneous. In the decision-making domain, imprecision represents the uncertainty as "a state of mind" of an agent which does not possess the needed information or knowledge to make a precise decision; the agent is in a state of uncertainty: "I'm not sure that this object is a table" [14]. Special kinds of imprecise information include:

- *Disjunctive* information content subset (e.g., John's age is either 31 or 32, the class of the object is either C_1 or C_2, etc.).
- *Negative* information content subset (e.g., John's age is not 30, etc.).
- *Range* information content subset (e.g., John's age is between 30 and 35, or John's age is over 30).
- *Error Margins* content subset (e.g., measured missile range is 100 ± 5 Km).

The two "boundary" kinds of imprecision are *precise* information (i.e., $\Omega_1 = \{x_{True}\}$), and *null* (also called missing data or total ignorance) information (i.e., Ω_1 encompasses the entire set of possible information content Ω). Notice that imprecise information is generally associated with an accuracy measurement quantifying the closeness of agreement between the information produced outcome (i.e., Ω_1) and the true information content, x_{True}. From an information processing point of view, an imprecise information is considered as a special case of evidential information where the total certainty is attributed to the subset Ω_1 (i.e., $m(\Omega_1) = 1$, and $m(A) = 0$ for all $A \neq \Omega_1$).

Remarks

- The aim of information processing approaches when dealing with both forms of imperfection (imprecision and/or uncertainty) is mainly to determine the unique true information content x_{True} with the highest precision and certainty degrees.
- Imprecision is often confused with uncertainty because both imperfection types are related to the same root (i.e., originated by punctual informative function where the "unique" true content is unknown: precisely, case of imprecision; or certainly, case of uncertainty). Also, both imprecision and uncertainty can be present at the same time, and one can cause the other. It is important to be able to tell the difference between these two antagonistic concepts, even if they can be included in a broader meaning for uncertainty (knowing that $x_{True} \in \Omega_1 \subseteq \Omega$ does not imply the precise and certain knowledge of x_{True}). To illustrate the

difference and potential "mixture" between imprecision and uncertainty, consider the following two situations:

1. Paul has at least two children and I'm sure about it.
2. Paul has three children but I'm not sure about it.

In the first information, the number of children is imprecise but certain; whereas, in the second information, the number of children is precise but uncertain [15].

III. ***Ambiguity*:** Literally, information is said to be ambiguous if it is unclear what the information refers to the fact that it can be interpreted in several ways or that its truth or validity is not totally verified. From an informational point of view and assuming that the information content set Ω is exhaustive, then two types of ambiguity are encountered:

   - *Non-specificity*: i.e., multiple content outcomes are produced simultaneously by the informative function.
   - *Partial truth*: i.e., the information content is partially produced.

L. Zadeh [16] has proposed to model this imperfection type as a fuzzy set defined on the information content set, where each content outcome x is associated with a *membership value* $\mu(x)\in[0,1]$ representing the "strength" or the "truth" of production of the outcome x by the informative function: $\mu(x) = 0$ means that x is not produced (obtained or concerned); and $\mu(x) = 1$ means that x is fully produced.

An information element $I = (\Theta, X, \Omega)$ affected by the ambiguity type of imperfection and for which a membership function $\mu(.)$ is available is called an *ambiguous (or fuzzy) information*. Notice that in this case, it is nonsense to determine single information content. Therefore, the major objective of the application of fuzzy concepts is to combine and to conduct the fusion of multiple ambiguous information elements.

1.3 Possibilistic Information Concept

A particular situation, of high importance, where "hybrid" forms of imperfections is frequently encountered. This concerns the case where the information element is affected by the uncertainty imperfection type (i.e., having a punctual informative function where the true output x_{True} is unknown with certainty), but the available knowledge about x_{True} is "weaker" than probabilities (subjective knowledge, ambiguous, etc.). In this case, each content outcome x is associated with a *possibility value* $\pi(x)\in[0,1]$ representing the possibility strength that the outcome x to be the unique true information content. This type of information imperfection is called *epistemic uncertainty*, and the associated information element is referred as a *possibilistic information* [4]. A practical feature of possibility theory that is worth emphasizing is its interest for modeling uncertainty as well as preferences. This theory is detailed in the next chapters.

References

1. G.L.S. Shackle, *Decision Order and Time in Human Affairs* (Cambridge University Press, New York, 2010)
2. L.J. Cohen, *The Probable and the Provable* (Clarendon Press, Oxford, 1977)
3. E.N. Zalta, U. Nodelman, C. Allen, J. Perry, *Stanford Encyclopedia of Philosophy* (Stanford University. The Metaphysics Research Lab, Stanford, 2003)
4. L. Zadeh, Fuzzy sets as the basis for a theory of possibility. Fuzzy Sets Syst. **1**, 3–28 (1978)
5. R.M. Losee, A discipline independent definition of information. J. Am. Soc. Inf. Sci. **48**, 254–269 (1997)
6. R.M. Losee, *The Science of Information: Measurement and Applications* (Academic Press, San Diego, 1990)
7. B. Solaiman, Information fusion concepts: From information elements definition to the application of fusion approaches, in *Proceedings SPIE Volume 4385 - Sensor Fusion: Architectures, Algorithms, and Applications*, (2001), pp. 205–212
8. É. Bossé, A. Guitouni, P. Valin, An essay to characterize information fusion systems, in *Proceedings of the 9th International Conference on Information Fusion (FUSION2006)*, (Firenze, 2006)
9. T. Stonier, *Information and the Internal Structure of the Universe* (Springer, London, 1990)
10. É. Bossé, G. Rogova (eds.), *Information Quality in Information Fusion and Decision Making*Information Fusion and Data Science (Springer Nature, Basel, 2019)
11. R.Y. Wang, D.M. Strong, Beyond accuracy: What data quality means to data consumers. J. Manag. Inf. Syst. **12**, 5–34 (1996)
12. A. Appriou, Situation assessment based on spatially ambiguous multisensor measurements. Int. J. Intell. Syst. **16**, 1135–1166 (2001)
13. G. Shafer, *A Mathematical Theory of Evidence* (Princeton University Press, London, 1976)
14. A.-L. Jousselme, P. Maupin, É. Bossé, Uncertainty in a situation analysis perspective, in *Proceedings of the Sixth International Conference of Information Fusion*, (2003)
15. P. Smets, Imperfect information: Imprecision and uncertainty, in *Uncertainty Management in Information Systems*, (Springer, New York, 1997), pp. 225–254
16. L.A. Zadeh, Fuzzy sets. Inf. Control. **8**, 338–353 (1965)

Chapter 2
Fundamental Possibilistic Concepts

2.1 Introduction

In the real-world problems, it is impossible to avoid uncertainties. Uncertainty sources are diverse: incomplete domain knowledge, noisy and conflicting data, incomplete information, linguistic imprecise and ambiguous knowledge, etc. Up to middle of the twentieth century, most theoretical advances were devoted to the theory of probabilities. The second half of the twentieth century was very prolific for the development of new theories dealing with uncertainties [1, 2]. One of these theories is possibility theory [3, 4] that was introduced to allow reasoning to be carried out in the framework of a vague knowledge about the uncertainties. In fact, possibility theory can be described as a collection of techniques centered on the concept of a possibility distribution used for the representation and manipulation of the ambiguous or vague knowledge about the encountered uncertainty. In this chapter, the fundamental concept of possibility distributions is detailed in terms of its definition, its informative facets, and its different distribution models. Two important concepts are also detailed: the discounting concept (allowing to adjust a possibility distribution in order to take into consideration some external reliability knowledge) and the extension principle that allows to compute the new possibility distribution resulting from the projection of the set of alternatives using a deterministic extension projection function. Different operators allowing the merging of several possibility distributions are then detailed. Two set measures allowing to characterize subsets occurrence (i.e., possibility and necessity measures) are defined and their different characteristics are expressed. An important issue detailed within this chapter is related to subnormal possibility distributions where the available ambiguous knowledge is inconsistent.

B. Solaiman, É. Bossé, *Possibility Theory for the Design of Information Fusion Systems*, Information Fusion and Data Science,
https://doi.org/10.1007/978-3-030-32853-5_2

2.2 Possibility Distributions Concept

In a similar way to all theories dealing with uncertainty, through possibility theory, the estimation of the tendency, or likelihood, of the occurrence of an elementary event (i.e., singleton) is represented by a *possibility distribution* ($\pi - d$) depicting our state of knowledge of what is plausible from what is less plausible. The $\pi - d$ shows the graded partial belief of the occurrence of an elementary event ranging from "0" (for impossible ones) to "1" (for absolutely possible ones). Namely, let Ω denote a finite set of *mutually exclusive* alternatives that are of concern to us (diagnosis, hypothesis, classes, decisions, etc.). This means that in any situation, one and only one of these alternatives, called the *true alternative*, occurs at a time, and that Ω contains all potential alternatives that may occur. Ω is called *reference set, universe of discourses, frame of discernment, decision set, set of alternatives, set of states of the world*, etc. In this document, Ω will be called the *set of alternatives.* Each element x from Ω is called: *elementary alternative, basic hypothesis, elementary decision, state of the world, singleton*, etc. In this document, x will be simply called an *alternative.*

2.2.1 Defining a Possibility Distribution

Let Ω denote a finite set of *mutually exclusive* alternatives, where the unique occurring alternative, i.e., the true alternative, is unknown. This, in fact, resumes the uncertainty type of information imperfection where we face the problem of "discovering" or identifying the identity of the true occurring alternative. A *possibility distribution* π, defined on the set of alternatives Ω, is a point-wise mapping from the set Ω into the unit interval, i.e.,

$$\begin{aligned} \pi : \Omega &\rightarrow [0,1] \\ x &\rightarrow Poss\{x\} = \pi(x) \end{aligned}$$

The value $\pi(x)$ is interpreted as being our degree of belief or as representing a *flexible restriction* (i.e., a constraint) of the value of x on the set of alternatives Ω. It shows the graded partial belief of the occurrence of different alternatives with the following conventions:

- $\pi(x) = 1$ means that the alternative x is believed to be fully possible (i.e., the occurrence of x is totally compatible with the knowledge available about Ω).
- $\pi(x) = 0$ means that the alternative x is believed to be fully impossible to be the true alternative (i.e., the occurrence of x is totally incompatible, or in a total contradiction, with the knowledge available about Ω).
- $\pi(x) = p \in]0,1[$ indicates that the alternative x is considered as having a partial possibility to degree p of being the true occurring alternative.

- $\pi(x_1) > \pi(x_2)$ means that x_1 is a *preferred alternative* to x_2 for being the true alternative.

  Note that:

- Most of the authors "impose" a normalization condition to the possibility distribution meaning that at least one alternative should be fully possible (i.e., $\exists x_0 \in \Omega$: $\pi(x_0) = 1$). In this case, the possibility distribution is referred to as being *normal*. Otherwise, the possibility distribution is called *subnormal*, i.e., $\forall x \in \Omega$: $\pi(x) < 1$ (subnormal distributions will be discussed in detail later in this chapter).
- The unit interval [0, 1] of π may be replaced by any linearly ordered, possibly finite scale.
- As possibility distributions are defined over basic alternatives (also called *singletons*) and not on the events (i.e., not on subsets of Ω), they are called *point functions*.
- A possibility distribution, π, could be viewed as describing possible values that could be assigned to some unknown variable X taking values in the finite set of alternatives $\Omega = \{x_1, \ldots, x_N\}$ and assuming that X represents a possibilistic information. The unique true alternative is known through the possibility distribution π which acts as an elastic constraint on the alternatives that can be assigned to X. Therefore, $\pi(x)$ represents to what extent it is possible that x is the true alternative: $\pi(x) = Poss\{X = x_{True}\}$.
- (X, π) is called a *possibilistic variable*.

The two extreme forms of knowledge, i.e., the complete knowledge and the total ignorance, are simply modeled by the two following possibility distributions:

- Complete knowledge, CK, (i.e., the occurring true alternative x_{True} is known):

$$\pi_{\mathrm{CK}}(x_{True}) = 1 \text{ and } \pi_{\mathrm{CK}}(x) = 0, \quad \forall x \neq x_{True}$$

- Total ignorance, TI, (i.e., total lake of knowledge concerning the occurring alternative):

$$\pi_{\mathrm{TI}}(x) = 1, \quad \forall x \in \Omega.$$

A possibility distribution π defined on the set of alternatives Ω such that $\pi(x) > 0$ for all $x \in \Omega$ is called *nondogmatic* as it does not definitely exclude any alternative x from Ω (all $x \in \Omega$ are considered as being possible alternatives). The *height* of a possibility distribution π, denoted by $h(\pi)$, is the highest possibility value taken by different alternatives:

$$h(\pi) = max_{x \in \Omega}\{\pi(x)\}$$

The *core* of a possibility distribution π, denoted by $Core(\pi)$, is defined as the subset of fully possible alternatives:

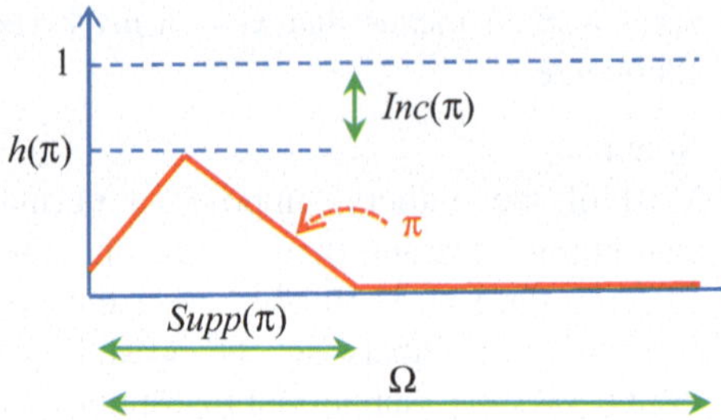

Fig. 2.1 Informative facets embedded in a possibility distribution

$$Core(\pi) = \{x : x \in \Omega, \pi(x) = 1\}$$

In general, a possibility distribution π defined on a set of alternatives Ω vehicles three main informative knowledge facets (Fig. 2.1).

1. *The domain of possible*, $Supp(\pi)$, is also called the support of the possibility distribution. In fact, the set of alternatives Ω is partitioned into two domains: the domain of possible containing alternatives having a possibility degree different from zero (i.e., $Supp(\pi) = \{x \in \Omega$ such that $\pi(x) > 0\}$) and the domain of impossible $\Omega / Supp(\pi) = \{x \in \Omega$ such that $\pi(x) = 0\}$.
2. *Possibilistic ordering*: The second informative source of knowledge is related to relative possibility degrees attributed to different alternatives. In fact, considering two alternatives x_1, $x_2 \in \Omega$ for which the possibilistic source of information has attributed the two possibility degrees $\pi(x_1)$ and $\pi(x_2)$. Besides the importance of the values of the possibility degrees attributed to both alternatives, the fact that $\pi(x_1) > \pi(x_2)$, for instance, encapsulates a relevant informative aspect for which the possibilistic source of knowledge "considers" the occurrence of the alternative x_1 as more credible than the alternative x_2.
3. *Inconsistency*: The third informative source of knowledge encapsulated in a possibility distribution is its degree of inconsistency defined as:

$$Inc(\pi) = 1 - h(\pi)$$

($Inc(\pi) \in [0,1]$), where $h(\pi)$ indicates the height of the possibility distribution (i.e., the highest possibility degree). In fact, this element reflects the degree to which the possibility distribution can confirm, or not, if at least one of the alternatives is fully possible to occur. Unfortunately, this important informative knowledge source element simply "disappears" when a possibility distribution is normalized (forcing, thus, to have at least one alternative to be fully possible).

2.2.2 Possibility Distribution Models

Different types of encountered possibility distributions are detailed in this section. Each type is assumed to model a given form of the available knowledge about the identity of the true, unique but unknown alternative from Ω.

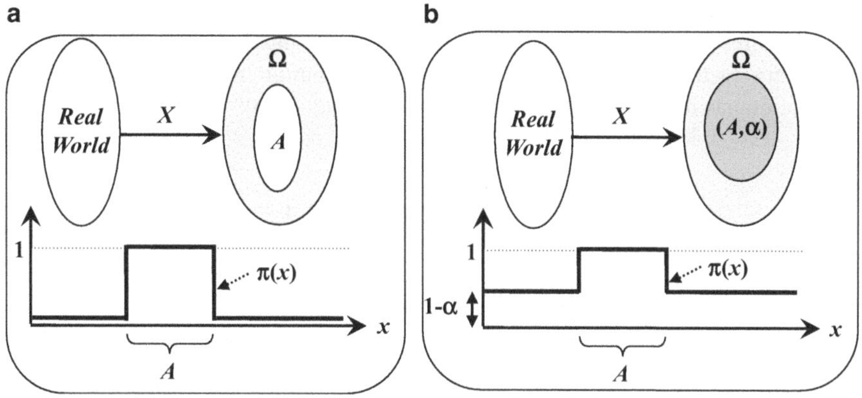

Fig. 2.2 Some types of imprecise type possibility distribution functions. (**a**) Imprecise information, (**b**) α-Certain imprecise information

(A) Imprecise type possibility distribution *Imprecise information* on the set of alternatives Ω, assumed to be exhaustive and the alternatives are mutually exclusive (i.e., one and only one alternative occurs at time), is defined as an uncertain information for which the knowledge about the true alternative is expressed as a subset $A \subseteq \Omega$. This type of information is frequently encountered and is more natural than giving a "point" alternative. Some examples of imprecise information are encountered when an expert claims that the true value x lies within the interval $x_0 \pm \Delta x$ or when a physician declares that the patient's illness is certainly one out of a subset of mutually exclusive illnesses.

The following distribution, called *imprecise information possibility distribution*, allows representing easily this type of information (Fig. 2.2a):

$$\begin{aligned} \pi : \Omega \quad &\rightarrow \quad [0,1] \\ x \quad &\rightarrow \quad \pi(x) = \mathbf{1I}(A) = \begin{cases} 1 & \text{if} \quad x \in A \\ 0 & \text{if} \quad x \notin A \end{cases} \end{aligned}$$

where **1I** (A) denotes the classical algebraic characteristic function of the subset A. Nevertheless, imprecise information possibility distributions are too restrictive. In fact, claiming that $\pi(x) = 0$ for some $x \notin A$ means that x is fully impossible to occur. This is too strong for the expert who is then tempted to give a wide uninformative interval support of π. It is worthwhile to notice that this type of possibility distributions assumes binary values and doesn't offer the possibility to express partial degrees of possibility.

(B) α-Certain imprecise type possibility distribution An *imprecise information with a certainty factor α*, (also called α-Certain imprecise information) is an imprecise information where the available knowledge about the true alternative is expressed as a subset $A \subseteq \Omega$ associated with a certainty level of trust $\alpha \in [0,1]$ concerning the occurrence of A, for instance, the information delivered by a physician declaring that the patient's illness is certainly one out of the two mutually exclusive illnesses $\{H_1, H_2\}$ and that he is 90% sure. In this case, $A = \{H_1, H_2\}$ with a certainty factor $\alpha = 0.9$. The possibility distribution modeling this type of information has been suggested in [5] as follows (Fig. 2.2b):

$$\begin{aligned} \pi : \Omega &\rightarrow [0, 1] \\ x &\rightarrow \pi(x) = max\{\mathbf{1I}(A), 1 - \alpha\} \end{aligned}$$

which expresses that the "remaining trust," i.e., $1 - \alpha$, will be considered as the possibility degree that the true alternative lies outside A.

(C) Epistemic type possibility distribution The objective of a possibilistic assessment is to produce a measure of the degree to which the available/acquired knowledge supports each alternative in the set of alternatives Ω. The result of this assessment is a possibility distribution. In several cases, the available knowledge about the true alternative is given as a constraint defined in terms of a "fuzzy concept" defined on Ω. It is important to notice that the concept of possibility distributions is closely related to that of fuzzy sets. Let Ω denote an exhaustive set of *mutually exclusive* alternatives, on which a fuzzy evidence A is defined, and denote $\mu_A(x)$ the corresponding membership function:

$$\begin{aligned} \mu_A : \Omega &\rightarrow [0, 1] \\ x &\rightarrow \mu_A(x) \end{aligned}$$

Assuming that A is the available knowledge about the true occurring alternative, the question is then: "How can we obtain a possibility distribution on the set of alternatives?"

L. Zadeh [3] has formulated the so called *possibility postulate* which may be considered as the basis for a possibilistic interpretation of the fuzzy evidence:

Possibility postulate In the absence of any information regarding the true alternative from Ω than that conveyed by the fuzzy evidence A, then, $\mu_A(.)$ induces a possibility distribution $\pi_A(x)$ which equates the possibility for an alternative $x \in \Omega$, be the true one, to the grade of membership $\mu_A(.)$.

Example Assume that we are to assess the possibility of the occurrence of a number from 0 to 10 given that the available evidence is that the number is *small* (Fig. 2.3).

Since we know that the compatibility of the concept *Small* with the number "4" is 0.6, $\mu_{Small}(4) = 0.6$, we conclude that the occurrence possibility degree of 4 is considered as being 0.6. By a similar argument, the possibility of 0, 1, 2, and 3 is 1 and of the numbers greater than 5 is 0. Notice that the membership function, $\mu_A(x)$, is viewed as computing a degree of assurance, certainty, or possibility that an alternative $x \in \Omega$ satisfies the property of being a member of the fuzzy set (or the ambiguous piece of information) defined by μ_A. Therefore, if we were to select a

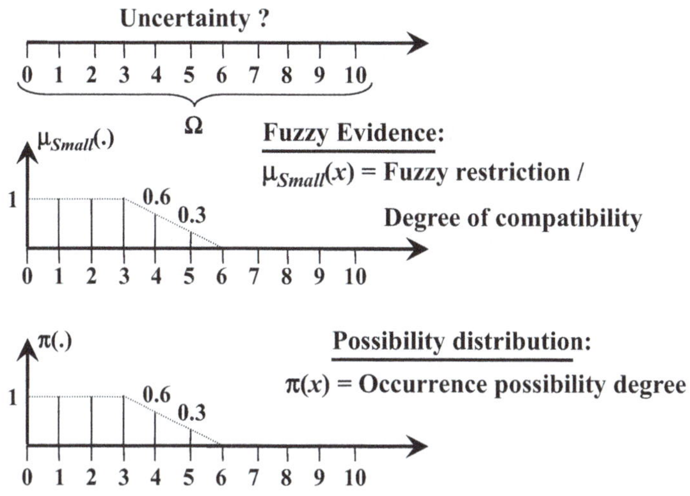

Fig. 2.3 Epistemic possibility distribution assessment

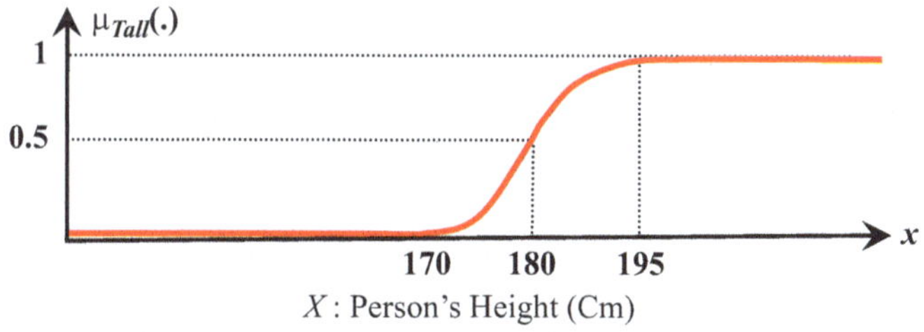

Fig. 2.4 Possibility distribution function associated with *Tall*

given $x \in \Omega$ as being the true alternative, we would be only certain to a degree $\mu_A(x)$ that x satisfies the aforementioned ambiguous information. For instance, if our goal is to know the exact height of Peter, and we only know that Peter is *Tall*, then the possible values of Peter's height are restricted by the possibility distribution associated with the fuzzy set *Tall* and defined by (Fig. 2.4):

$$\pi_{Peter's\ height}(x) =_{\mathrm{def}} \mu_{Tall}(x)$$

It is crucial to understand that the values duplication (i.e., $\pi_{Peter's\ height}(x) = \mu_{Tall}(x)$) is to be positioned at the numerical and not at the semantical level. In fact, $\mu_{Tall}(x)$ reflects the compatibility of one "feature" of persons (here, the size) with the fuzzy property *Tall*. This compatibility value is extended (or projected) to become representing the possibility degree for a given size to be the true one for a given person. An interesting example in mammographic image interpretation can be

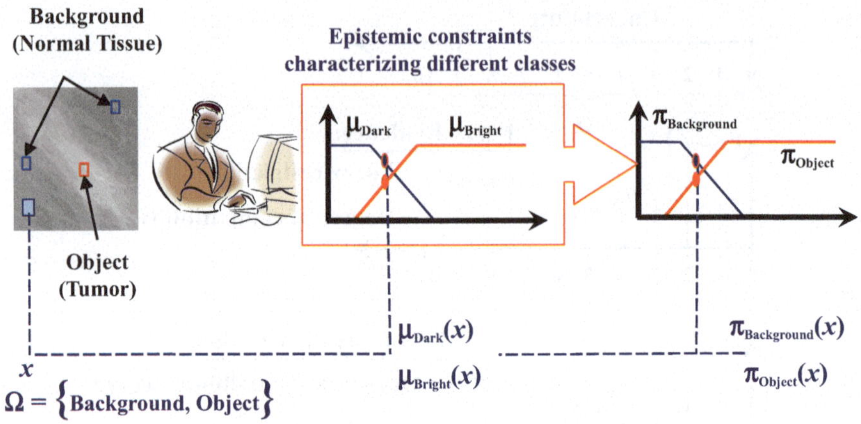

Fig. 2.5 Possibility distribution functions associated with normal tissue and tumor classes in mammography

considered (Fig. 2.5). In this example, physicians express their knowledge describing normal and tumor tissues by characterizing these tissues as being "observed" as *dark* and *bright* pixels in mammographic images.

Notice that this description represents two fuzzy sets (*dark* and *bright*) defined on the observed gray levels definition set. The membership functions of these fuzzy sets constitute the epistemic constraints allowing to define the possibility degrees for an observed pixel to belong to a normal or to a tumor tissue.

It is worthwhile to mention that all standard types defining membership function and representing the fuzzy constraints (i.e., triangular, trapezoidal, Gaussian, singleton-based, piecewise linear, etc.) can be applied for the definition of the epistemic type possibility distributions.

(D) Qualitative possibility distributions Experts frequently meet difficulties in providing precise numerical values of possibility degrees. It seems, thus, more natural for them to give an "order relation" between different alternatives of the universe Ω.

Consider a finite universe of alternatives $\Omega = \{x_1, \ldots, x_N\}$ and an ordered scale $\mathcal{L} = \{a_0 = 1, a_1, \ldots, a_L, a_{L+1} = 0\}$ such that $a_0 = 1 > a_1 > \cdots > a_L > a_{L+1} = 0$. A qualitative possibility distribution is defined as a function associating to each alternative (from the universe Ω) an element $a \in \mathcal{L}$ (from the ordered scale $\mathcal{L}$) enabling, thus, to express that some alternatives are "more possible" than others (without referring to any numerical value, and this, makes the difference with other quantitative possibility distributions setting. In other words, assigning "qualitative" values a_k as possibility values (i.e., $\pi(x_n) = a_k$) implies alternatives ranking importance representation rather than pure numerical possibilistic degrees. However, we can derive an infinity of quantitative possibility distributions from a qualitative one.

2.2.3 Possibility Distributions Discounting

Information reliability is an important issue in information processing. In real-world applications, full reliability of knowledge sources, or experts, could rarely be guaranteed, and thus, we have to deal with unreliable information where the reliability is described using some certainty, or trust, qualifiers. These qualifiers are expressed, most of the time, by a quantitative certainty degree (for instance, 80% sure or 70% certain), but sometimes the qualifiers are expressed using qualitative descriptors (like very sure, more or less certain, fully reliable). Discounting refers to the process of impacting a given certainty degree on the available knowledge. Once discounted, a possibilistic source is assumed to be reliable. Assume that a possibility distribution $\pi(.)$ is provided by a given source of information and that the *degree of certainty* (also called *reliability factor*) that this source is reliable is given by a numerical index $\beta \in [0,1]$ ($\beta = 0$ means a fully unreliable source, and $\beta = 1$ means a fully reliable source). The question is "How can we take into consideration this degree of certainty?"

D. Dubois [6] has proposed the most widely applied discounting method allowing to update the possibility distribution as follows (Fig. 2.6a):

$$\pi'(x) = max\,(\pi(x), 1 - \beta), \quad \forall x \in \Omega$$

This discounting method considers possibility degrees less than the reliability loss (i.e., $x \in \Omega$ for which $\pi(x) < 1 - \beta$) as "meaningless" and should be substituted by $1 - \beta$. Alternatives for which the possibility degrees are greater than the reliability loss remain unchanged. Nevertheless, the mathematical foundation of Dubois's discounting rule was proposed by R. Yager [7]. In fact, R. Yager has proposed an analytical possibilistic discounting method, where the updated possibility distribution is given by:

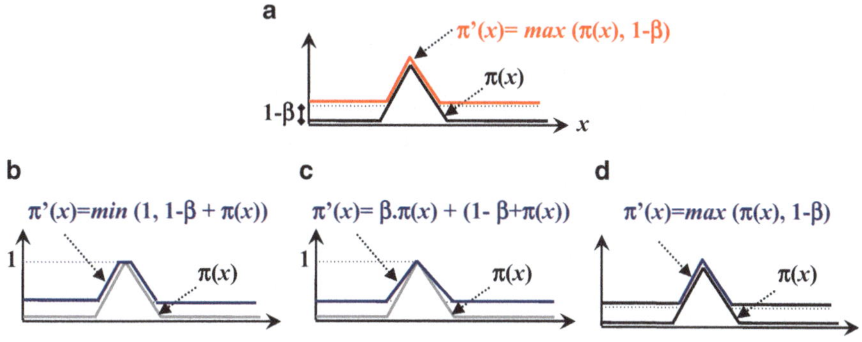

Fig. 2.6 Dubois and Yager possibilistic discounting methods. (**a**) Dubois's Possibilistic discounting method, (**b**) Yager discounting method (Zadeh's *min* T-Norm), (**c**) Yager discounting method (Probabilistic product T-Norm), (**d**) Yager discounting method (Lukasziewicz's T-Norm)

$$\pi'(x) = T[\beta, \pi(x)] + (1 - \beta), \quad \forall x \in \Omega$$

where $T(.,.)$ is a T-Norm operator and $1 - \beta$ refers to the reliability loss amount.

For $T =$ Zadeh *min* T-Norm, we get (Fig. 2.6b):

$$\begin{aligned} \pi'(x) &= min\,[\beta, \pi(x)] + (1 - \beta) \\ &= min\,[1, 1 - \beta + \pi(x)], \quad \forall x \in \Omega \end{aligned}$$

For $T =$ probabilistic product T-Norm, we get (Fig. 2.6c):

$$\pi'(x) = \beta \cdot \pi(x) + (1 - \beta), \quad \forall x \in \Omega$$

and for $T =$ Lukasziewicz's T-Norm (i.e., $T\,(a,b) = max[0, a + b - 1)]$, we get (Fig. 2.6d):

$$\pi'(x) = max\,[\pi(x), 1 - \beta], \quad \forall x \in \Omega$$

Notice that:

- Yager's discounting method, using Lukasziewicz's T-Norm, leads to the same Dubois' discounting method.
- When the available evidence represents a precise information (i.e., complete knowledge possibility distribution: $\exists x_0 \in \Omega$: $\pi(x_0) = 1$ and $\pi(x) = 0, \forall x \neq x_0$) or an imprecise information (i.e., $\exists A \subseteq \Omega$: $\pi(x) = 1\ \forall x \in A$ and $\pi(x) = 0, \forall x \notin A$), then all the proposed discounting methods will lead to the same discounted possibility distribution.

2.2.4 Possibilistic Extension Principle

The *extension principle*, first proposed in the context of fuzzy sets theory [8], allows to determine the possibility distribution π_Y of a possibilistic variable (Y, π_Y) defined over Ω_1 having a functional relation with a possibilistic variable (X, π_X) defined over Ω (i.e., $Y = f(X)$) whose possibility distribution π_X is known (Fig. 2.7).

By virtue of the extension principle, the possibility distribution π_Y is given by [9]:

$$\pi_Y(y) = \begin{cases} \max\limits_{x \in \Omega, f(x) = y} \pi_X(x) & \\ 0 & \text{if } f(x) \neq y \text{ for all } x \in \Omega \end{cases}$$

where $y \in \Omega_1$ is a possible value of the variable $Y = f(X)$.

This definition shows that the possibility value assigned to $y \in \Omega_1$ is the maximum possibility value encountered among all the values $x \in \Omega$ such that $y = f(x)$. If no value $x \in \Omega$ exists such that $y = f(x)$, then the possibility value $\pi_Y\,(y)$ is set equal to zero.

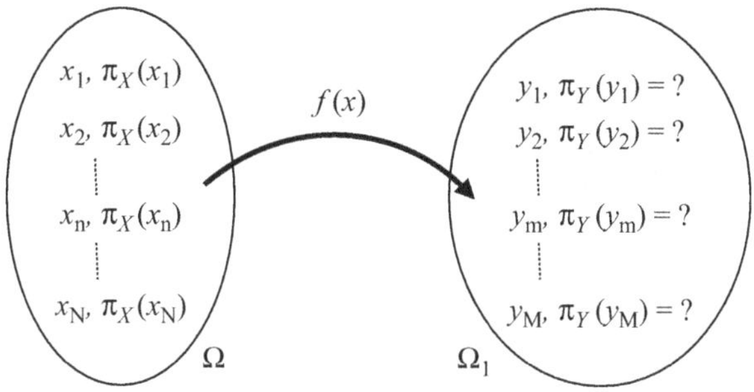

Fig. 2.7 Possibilistic extension principle

Example Let us consider an initial set of alternatives $\Omega = \{-3, -2, -1, 0, 1, 2, 3\}$ on which the possibility distribution π_X is given as follows:

$$\pi_X(-3) = 0.3; \pi_X(-2) = 0.8; \pi_X(-1) = 0.9; \pi_X(0) = 1; \pi_X(1) = 0.7; \pi_X(2) = 0.5; \text{and } \pi_X(3) = 0.4.$$

And, consider the functional relation f: $\Omega \rightarrow \Omega_1 = \{0, 1, 4, 9\}$ given by the equation $y = f(x) = x^2$. Using Zadeh's extension principle, the induced possibility distribution π_Y of a possibilistic variable Y defined over $\Omega_1$is thus given by:

$$\begin{aligned}
&\pi_Y(0) = \pi_X(0) = 1;\\
&\pi_Y(1) = max\,\{\pi_X(-1) = 0.9, \pi_X(1) = 0.7\} = 0.9;\\
&\pi_Y(4) = max\,\{\pi_X(-2) = 0.8, \pi_X(2) = 0.5\} = 0.8; \text{and}\\
&\pi_Y(9) = max\,\{\pi_X(-3) = 0.3, \pi_X(3) = 0.4\} = 0.4.
\end{aligned}$$

2.2.5 *Specificity Concept and Minimal Specificity Principle (MSP)*

The concept of specificity was originally introduced by Yager [10] (in the framework of fuzzy set theory) to measure the degree to which a fuzzy set contains one and only one singleton. This measure was then extended to indicate the degree to which a possibility distribution allows one and only one alternative as possible and to measure the amount of information contained in a possibility distribution. Therefore, a natural use of this concept is to "compare" different possibility distributions defined on the same set of alternatives.

Given two possibility distributions $\pi_1(.)$ and $\pi_2(.)$, both defined on the same set of alternatives Ω, $\pi_1(.)$ is said to be *more specific* than $\pi_2(.)$, [11], if and only if (see Fig. 2.8):

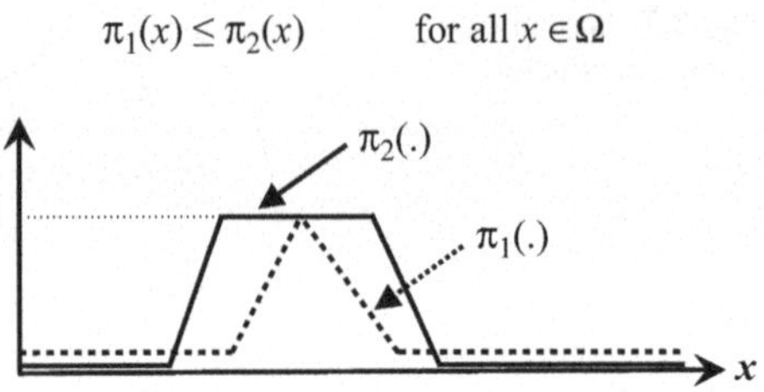

Fig. 2.8 Specificity concept

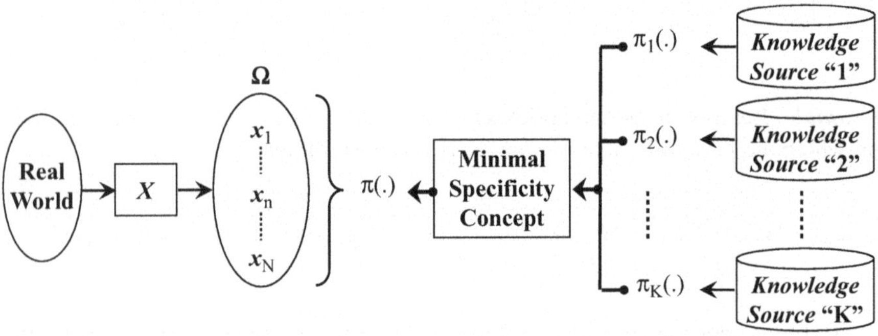

Fig. 2.9 Minimal specificity principle

$$\pi_1(x) \leq \pi_2(x) \quad \text{for all}\, x \in \Omega$$

Clearly, the more specific $\pi(.)$, the more informative it is ($\pi_2(.)$ is said less informative than $\pi_1(.)$). In particular, in the case of total ignorance, i.e., $\forall x{\in}\Omega$, $\pi_{TI}(x) = 1$, $\pi_{TI}(.)$ contains no information at all (since all alternatives are fully possible). The possibility distribution $\pi_{TI}(.)$ is also said to be *vacuous*.

The concept of specificity is not only used for the comparison of different possibility distributions. It is also used in order to merge several possibilistic sources of knowledge. This use is generally referred to as the *minimal specificity principle* (MSP) (Fig. 2.9). In fact, let X be possibilistic variable on the set of alternatives Ω for which different knowledge sources (based on "partial" possibilistic pieces of evidence) are available: π_1, $\pi_2 \ldots \pi_K$. We are interested in a global possibilistic modeling (i.e., obtaining a global possibility distribution π on Ω) by inferring the partial possibilistic evidences π_1, $\pi_2 \ldots \pi_K$ under the two following conditions:

I. π is at least as *precise* as each partial possibilistic evidences π_k, $k = 1, 2, \ldots, K$.
II. π does not include possibilistic knowledge additional to different partial possibilistic evidences π_k, $k = 1, 2, \ldots, K$.

We consider the concept of specificity as a possibilistic information precision criterion. Therefore, the first condition can be expressed as that the distribution π is at least as specific as each π_k, $k = 1, 2, \ldots, K$: $\pi(x) \leq \pi_k(x)$, for all $x{\in}\Omega$ and $k = 1, 2, \ldots, K$. This leads to consider $\pi(x) \leq min\{\pi_k(x)\}$, for all $x{\in}\Omega$ and $k = 1, 2, \ldots, K$, which means that the "more" we know about X (i.e., the more specific our knowledge about

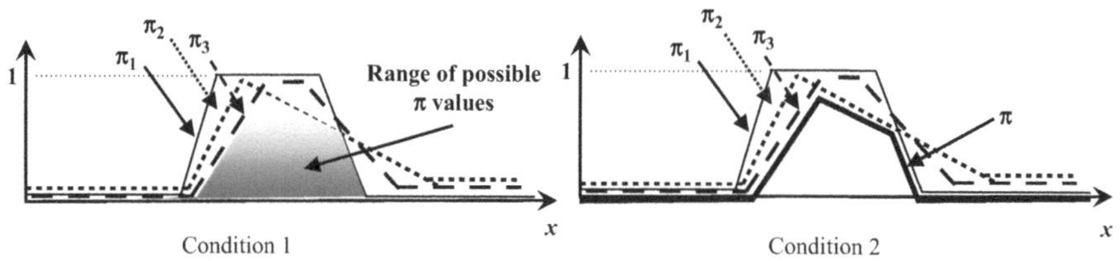

Fig. 2.10 Minimal specificity principle: uncertain range of possible values for X

X is), the smaller the uncertain range of possible values for X becomes (Fig. 2.10). Thus, when inferring π_k, $k = 1, 2, \ldots, K$, it is reasonable that π should be at least as specific as each available piece of information.

The second condition concerns the selection of the most specific possibility distribution, π, that we are allowed to infer from π_k, $k = 1, 2, \ldots, K$.

From the first condition, $min\{\pi_k(x)\}$ (for all $x \in \Omega$ and $k = 1, 2, \ldots, K$) combines all the available pieces of information. Recall also that a possibility distribution is said to be *inferable* from π_k, $k = 1, 2, \ldots, K$ if and only if it is at most as specific as any combination of these distributions. This leads to select π such that $\pi(x) \geq min\{\pi_k(x)\}$, $\forall x \in \Omega$ and $k = 1, 2, \ldots, K$. Nevertheless, selecting $\pi(x) \geq min\{\pi_k(x)\}$ means additional knowledge. This leads to the minimal specificity principle.

When the available pieces of information are represented by a set of possibilistic constraints, the possibility distribution that accounts for it is the least specific one that satisfies the constraints:

$$\pi(x) = min\{\pi_k(x)\}, \quad \text{for all } x \in \Omega \text{ and } k = 1, \ 2, \ \ldots, \ K$$

π is said to be obtained by reasonable inference from π_k, $k = 1, 2, \ldots, K$.

2.3 Possibility and Necessity Measures

A possibility distribution evaluates to what extent each alternative from Ω can be the true occurring value of the uncertain information. More generally, it is possible to determine to what extent it is possible that the occurring (but unknown) alternative belongs to any event A defined over Ω (i.e., A is a subset of Ω). In possibility theory, the knowledge about events is represented by means of two set measures of uncertainty: the *possibility measure* and the *necessity measure*.

The possibility measure is a set function indicating how likely it is for an event to occur (i.e., the true alternative belongs to this event). The necessity measure, on the other hand, is another set function indicating how certain is an event to occur. These set functions are especially simple to work with because they can be completely characterized by point functions. Moreover, because the occurrence of an event cannot be certain unless it is possible, these two measures are closely related to each other. In order to "build" our intuition, let us consider that we are concerned by crisp

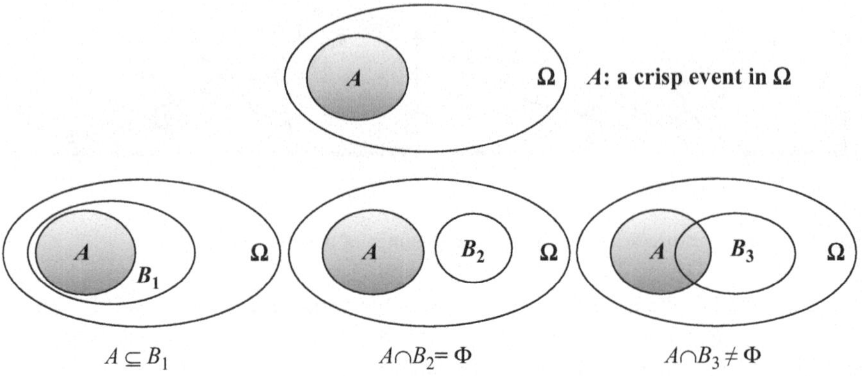

Fig. 2.11 Intuitive concept of a possibility measure

events or evidences (i.e., crisp subsets from the set of alternatives Ω). Let A denote our *prior* knowledge encoded by the possibility distribution π_A, i.e., we are certain that the true alternative belongs to the crisp evidence A. Consider the three situations (Fig. 2.11), where we have to infer our *prior* knowledge into other crisp events: B_1, B_2, and B_3 all contained in Ω.

The uncertainty associated with the knowledge that the true alternative belongs to A makes the clear determination of whether another event is true or false not always attainable. Using Fig. 2.11, we fully understand and accept that B_1 is valid (i.e., fully possible and fully certain) and B_2 is not true (i.e., fully impossible) and can't guarantee the truth of occurrence of B_3 (we can only say that B_3 is possible but not certain). Therefore, the attainment of a clear answer to questions concerning the occurrence "truth" of other events in our knowledge is not always attainable. In the case where the considered events are ambiguous, i.e., can be encoded by epistemic possibility distributions, two measures have been introduced by L. Zadeh [12] to help. These are the measures of possibility and necessity. These measures of possibility and necessity can be seen as respectively providing upper and lower (optimistic and pessimistic) bounds, on the answer to the question of occurrence of B given our *prior* knowledge encoded by π_A.

2.3.1 Possibility Measure

Let Ω denote an exhaustive set of mutually exclusive alternatives. A *possibility measure*, Π, is a mapping from the power set of Ω, i.e., $\mathcal{P}(\Omega)$, to the real unit interval [0,1]:

$$\begin{array}{rccc} \Pi : & \mathcal{P}(\Omega) & \rightarrow & [0,1] \\ & A \subseteq \Omega & \rightarrow & \Pi(A) \end{array}$$

satisfying the following requirements:

1. $\Pi(\Phi) = 0$ (where Φ denotes the empty set).
2. $\Pi(\cup_{j \in J} A_j) = max_{j \in J} \Pi(A_j)$, for any family $\{A_j, j \in J\}$ of elements from $\mathcal{P}(\Omega)$ (i.e., each A_j is a subset of Ω).

In the case on a finite set of alternatives $\Omega = \{x_1, x_2, x_3, \ldots, x_N\}$, the second condition of the possibility measure definition can be reduced to a simpler form:

$$\Pi(A \cup B) = max\,(\Pi(A), \Pi(B)), \quad \forall A, B \in \mathcal{P}(\Omega)$$

This means that the occurrence of one of the two events A or B, taken indifferently, receives the same possibility measure as the occurrence of the most possible event.

For any event $A \in \mathcal{P}(\Omega)$, $\Pi(A)$ is called the *possibility measure* of A. $\Pi(A)$ can be interpreted as representing the extent to which it is possible that the true alternative belongs to A.

An event $A \in \mathcal{P}(\Omega)$ is said to be a *fully possible* event (resp. *fully impossible*) if its possibility measure $\Pi(A)$ is equal to 1 (resp. null).

A possibility measure Π is called *normal* (or *consistent with the available knowledge*) if $\Pi(\Omega) = 1$.

Here follow some interesting properties of possibility measures:

1. Possibility measures are less restricting than probability measures in the sense that the possibility degree of an event A (i.e., subset of Ω) is not necessarily determined by the knowledge of the possibility measure of its complement A^c.
2. In the case of a normal possibility measure, if we consider any subset A of Ω and its complement A^c, then at least one of them is fully possible:

$$\forall A \in \mathcal{P}(\Omega) \text{ then } max\,(\Pi(A), \Pi(A^c)) = 1$$

(This can be simply deduced from $\Pi(A \cup A^c) = \Pi(\Omega) = max(\Pi(A), \Pi(A^c)) = 1$).

3. The possibility measure Π is *monotonous* with respect to the inclusion of subsets of Ω:

$$\text{if}: \quad B \subseteq A \,\text{then}\, \Pi(B) \leq \Pi(A)$$

4. The possibility measure of the intersection of two subsets A and B of Ω is not determined from the possibility measure of the possibility measures of these two subsets. The only information we obtain from the definition of the possibility measure is the following (Fig. 2.12):

$$\Pi(A \cap B) \leq min\,(\Pi(A), \Pi(B)), \quad \forall A, B \in \mathcal{P}(\Omega)$$

5. Two subsets A and B of Ω can be individually possible (i.e., $\Pi(A) \neq 0$ and $\Pi(B) \neq 0$), but their joint, or simultaneous occurrence, is fully impossible (i.e., $\Pi(A \cap B) = 0$).

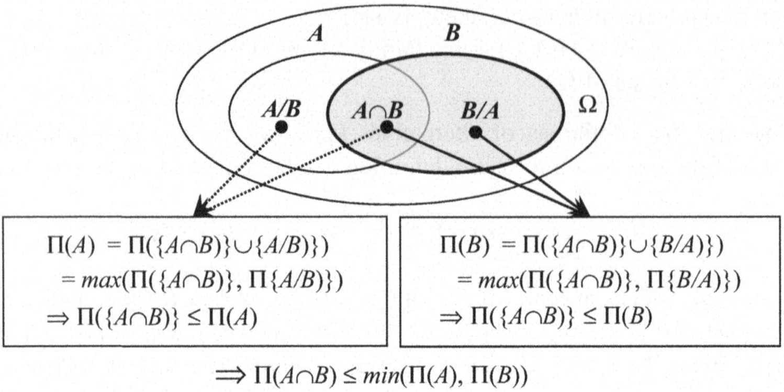

Fig. 2.12 Possibility measure of events intersection

Example (disease identification) Let $\Omega = \{d_1, d_2, d_3, d_4, d_5\}$ denote an exhaustive and mutually exclusive set of diseases and that our state of knowledge concerning an examined patient's disease is as follows:

- It is fully possible to be in presence of disease d_1 or d_2.
- It is relatively possible to be in presence of disease d_1, d_3, or d_5.
- Disease d_5 is somehow possible.
- Disease d_4 is considered as impossible.

This state of knowledge can be represented by the following potential possibility measure:

$$\Pi(\{d_1, d_2\}) = 1, \Pi(\{d_1, d_3, d_5\}) = 0.8, \Pi(\{d_5\}) = 0.5 \text{ and } \Pi(\{d_4\}) = 0.$$

From this measure, we can deduce the following knowledge:

- $\Pi(\{d_1, d_2, d_4\}) = \Pi(\{d_1, d_2\} \cup \{d_4\}) = max(\Pi(\{d_1, d_2\}), \Pi(\{d_4\})) = max(1, 0) = 1$ (i.e., it is absolutely possible that the patient's disease is one of $\{d_1, d_2, d_4\}$).
- $\Pi(\{d_4, d_5\}) = \Pi(\{d_4\} \cup \{d_5\}) = max(\Pi(\{d_4\}), \Pi(\{d_5\})) = max(0, 0.5) = 0.5$ (i.e., it is somehow possible that the patient's disease is one of $\{d_4, d_5\}$).
- $\Pi(\{d_1\}) = \Pi(\{d_1, d_2\} \cap \{d_1, d_3, d_5\}) \leq min(\Pi(\{d_1, d_2\}), \Pi(\{d_1, d_3, d_5\})) = min(1, 0.8) = 0.8$ (i.e., it is relatively possible that the patient's disease is one of $\{d_4, d_5\}$).

For any possibility measure, Π, there exists a corresponding punctual possibility distribution:

$$\begin{aligned} \pi : \Omega &\rightarrow [0, 1] \\ x &\rightarrow \pi(x) = \Pi(\{x\}) \end{aligned}$$

This possibility distribution can be used to retrieve the possibility measure of any subset $A \in \mathcal{P}(\Omega)$ by considering the maximum possibility value of its constituting alternatives: $\Pi(A) = max_{x \in A}[\pi(x)]$.

Remark: Let us consider the medical diagnosis example where $\Omega = \{d_1, d_2, d_3, d_4, d_5\}$ denotes the set of five diseases and the medical state of knowledge is represented with the following possibility measure:

$$\Pi(\{d_1, d_2\}) = 1, \Pi(\{d_1, d_3, d_5\}) = 0.8, \Pi(\{d_5\}) = 0.5 \text{ and } \Pi(\{d_4\}) = 0.$$

The following interesting question arises: what is the possibility measure of each individual disease, i.e., $\Pi(\{d_n\}) = \pi(d_n)$, $n = 1, \ldots, 5$? If it is straightforward that $\Pi(\{d_4\}) = 0$ and $\Pi(\{d_5\}) = 0.5$, it is less intuitive for $\Pi(\{d_1\})$, $\Pi(\{d_2\})$, and $\Pi(\{d_3\})$. Given the basic properties, we can deduce that $\Pi(\{d_1, d_3\}) = 0.8$ and $\Pi(\{d_1, d_2\}) = 1$. At this level, we have two "families" of solutions:

$$\Pi(\{d_1\}) = 0.8, \Pi(\{d_2\}) = 1 \text{ and } \Pi(\{d_3\}) \leq 0.8;$$
$$\Pi(\{d_3\}) = 0.8, \Pi(\{d_2\}) = 1 \text{ and } \Pi(\{d_1\}) \leq 0.8.$$

This means that several "possibility distributions" can lead to the same possibility measure! The reason behind this "indetermination" of a unique possibility distribution is simply that the medical knowledge expressed by the expert in terms of the possibility measure does not cover all the $32 = 2^5$ events (i.e., all subsets of Ω). In other words, the possibility measure, assumed to encode the medical state of knowledge, is not fully defined.

Conversely, the association between possibility measures and possibility distributions is straightforward when a possibility distribution is, first, defined, and then, the corresponding possibility measure is "constructed." In fact, when a possibility distribution is defined over Ω (i.e., $\pi(x)$ is known for all alternatives $x \in \Omega$), then it is straightforward to compute $\Pi(A)$ (for all $A \in \mathcal{P}(\Omega)$) as (Fig. 2.13):

$$\Pi(A) = max_{x \in A} \pi(x), \quad \forall A \in \mathcal{P}(\Omega)$$

In other words, the possibility measure of a subset $A \in \mathcal{P}(\Omega)$ is the possibility degree of its best alternatives.

It is important to notice the similarity in the construction of the possibility measure with the probability measure. In fact, if $f_X(x)$ denotes the probability density function of a random variable X defined over the set of alternatives Ω, then the probability measure of the subset $A \in \mathcal{P}(\Omega)$ is defined by:

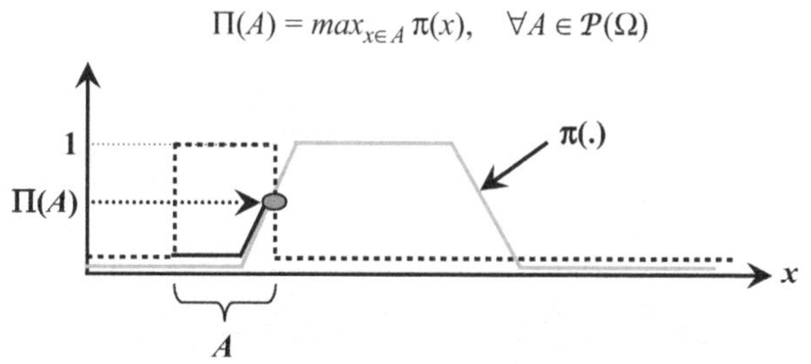

Fig. 2.13 Possibility measure of a crisp set information

$$\Pr\{A\} = \int_{x\in A} f_X(x)\mathbf{1I}_A(x)dx$$

where $\mathbf{1I}_A(x)$ denotes the characteristic function of the subset A. Notice that the product operation is replaced by the *minimum* operation between the possibility distribution $\pi(x)$ and the characteristic function $\mathbf{1I}_A(x)$, and the integral operation is replaced by the *maximum* operation.

2.3.2 Necessity Measure

The possibility measure provides information concerning the fact that an event can occur, but it is not sufficient to describe the uncertainty about this event and to obtain a conclusion, concerning its occurrence, from the available knowledge. For instance, if $\Pi(A) = 1$, the event A is fully possible, but we can have, at the same time, $\Pi(A^C) = 1$, which proves that we have an absolute uncertainty about the occurrence of A. A solution to this uncertainty characterization problem is to complete the "knowledge" related to the occurrence of A by means of a dual set function called the *necessity measure*.

A *necessity measure*, denoted by $\mathcal{N}$, is defined as a mapping from the power set of Ω, i.e., $\mathcal{P}(\Omega)$, to the real unit interval [0,1]:

$$\begin{aligned} \mathcal{N}: \quad \mathcal{P}(\Omega) &\rightarrow [0,1] \\ A \subseteq \Omega &\rightarrow \mathcal{N}(A) \end{aligned}$$

and satisfying the following requirements:

1. $\mathcal{N}(\Phi) = 0$ and $\mathcal{N}(\Omega) = 1$;
2. $\mathcal{N}\left(\cap_{j\in J}A_j\right) = inf_{j\in J}\mathcal{N}(A_j)$, for any family $\{A_j, j \in J\}$ of elements from $\mathcal{P}(\Omega)$ where *inf* denotes the "infimum" operator.

In the case on a finite set of alternatives $\Omega = \{x_1, x_2, x_{3,}..., x_N\}$, the second condition can be reduced into a simpler form:

$$\mathcal{N}(A \cap B) = min\,(\mathcal{N}(A), \mathcal{N}(B)), \quad \forall A, B \in \mathcal{P}(\Omega)$$

For any $A\in\mathcal{P}(\Omega)$, $\mathcal{N}(A)$ is called the necessity measure of the event A. $\mathcal{N}(A)$ can be interpreted as the extent to which it is certain that the true alternative belongs to A. If $\mathcal{N}(A) = 0$, we have no certainty about the occurrence of the event A; if $\mathcal{N}(A) = 1$, we are absolutely certain that A occurs. Here follow some interesting properties of the necessity measure:

- The necessity measure $\mathcal{N}$ is a *monotonous* function with respect to the inclusion of subsets of Ω:

$$\text{If } B \subseteq A \text{ then } \mathcal{N}(B) \leq \mathcal{N}(A)$$

- The necessity measure of the union of two subsets A and B of Ω is not determined from the necessity measure of these subsets. The only information we obtain from the definition of the necessity measure is the following:

$$\mathcal{N}(A \cup B) \geq max\,(\mathcal{N}(A), \mathcal{N}(B)), \quad \forall A, B \in \mathcal{P}(\Omega)$$

- In the case of a normal measure (i.e., $\Pi(\Omega) = 1$), if we consider any subset A of Ω and its complement A^c, then at least one of them is fully uncertain, i.e.,

$$min\,(\mathcal{N}(A), \mathcal{N}(A^c)) = 0, \quad \forall A \in \mathcal{P}(\Omega)$$

The duality between possibility and necessity measures can be pointed out by the following theorem:

Theorem For a given set of alternatives Ω and a possibility measure Π defined on Ω, the particular set measure defined by $\forall A \in \mathcal{P}(\Omega)$, $\mathcal{N}(A) = 1 - \Pi(A^C)$ is a necessity measure on Ω (where A^c denotes the complement of A in Ω) that is dual to Π.

Proof We are certain that A occurs (i.e., $\mathcal{N}(A) = 1$) if and only if A^C is impossible (i.e., $\Pi(A^C) = 0$), and thus, $\Pi(A) = 1$. Consider two sets A and $B \in \mathcal{P}(\Omega)$ and let us show that:

$$\mathcal{N}(A \cup B) = min\,(\mathcal{N}(A), \mathcal{N}(B))$$

$$\begin{aligned}
\mathcal{N}(A \cap B) &= 1 - \Pi\left((A \cap B)^C\right) = 1 - \Pi\left(A^C \cup B^C\right) \\
&= 1 - max\left\{\Pi\left(A^C\right), \Pi\left(B^C\right)\right\} \\
&= 1 - max\,\{1 - \mathcal{N}(A), 1 - \mathcal{N}(B)\} \\
&= 1 - [1 - min\,\{\mathcal{N}(A), \mathcal{N}(B)\}] \\
&= min\,\{\mathcal{N}(A), \mathcal{N}(B)\}
\end{aligned}$$

Note that if a possibility measure Π is associated with a possibility distribution π, we can define the dual necessity measure by:

$$\mathcal{N}(A) = 1 - \Pi\left(A^C\right) = 1 - max_{x \in A^C}\pi(x) = inf_{x \in A^C}[1 - \pi(x)]$$

where A^C is the complement set of A. This simply means that the same possibility distribution is sufficient to precisely define both possibility and necessity measures.

Let us note that when dealing with a normal possibility distribution, the necessity of a subset A cannot be positive as long as its possibility measure is smaller than 1, i.e., if $\Pi(A) < 1$, then $\mathcal{N}(A) = 0$. This can be very easily understood, since

$\Pi(A) < 1$ implies that $\Pi(A^C) = 1$ (recall that $\Pi(\Omega) = \Pi(A \cup A^C) = 1$), thus $\mathcal{N}(A) = 1 - \Pi(A^C) = 0$. This is very intuitive: a set can only be compulsory when it is fully positive (it is important to notice that this property is not fulfilled if subnormal possibility distributions are considered).

2.3.3 Duality Relevant Properties of Possibility and Necessity Measures

Let:

- Ω denote an exhaustive set of mutually exclusive alternatives.
- π a possibility distribution defined on Ω and encoding an ambiguous piece of information for which Π and $\mathcal{N}$ denote the possibility and the necessity measures.
- A is a subset (crisp or fuzzy) encoding an observed evidence.

Knowing that the *prior* possibilistic information (encoded by π) is assumed as normalized, then we can establish the following properties:

1. $\Pi(A) = \mathcal{N}(A) = 1$ means that the occurrence of the evidence A is *certain*.
2. $\Pi(A) = 1$ and $\mathcal{N}(A) > 0$ means that the evidence A is *normally true*.
3. $\Pi(A) = 1$ and $\mathcal{N}(A) = 0$ means that the evidence A implies a state of *total ignorance* (since both A and A^C are fully possible).
4. $\Pi(A) < 1$ and $\mathcal{N}(A) = 0$ means that the evidence A is *normally false*.
5. $\Pi(A) = 0$ and $\mathcal{N}(A) = 0$ means that the evidence A is *certainly false*.

The normalization of the possibility distribution π ensures that:

$$\mathcal{N}(A) \leq \Pi(A), \quad \forall A \in \mathcal{P}(\Omega)$$

which expresses an intuitive fact that an evidence A should be possible to some extent before being certain. When A is a crisp evidence (i.e., $A \subseteq \Omega$), the above inequality strengthens into $\mathcal{N}(A) > 0 \Rightarrow \Pi(A) = 1$. Here follows some relevant properties linking possibility and necessity measures, $\forall A,B \subseteq \Omega$ (normalized possibility distributions are considered):

1. $\Pi(A) \geq \mathcal{N}(A)$;
2. $max\{\Pi(A), 1 - \mathcal{N}(A)\} = 1$;
3. $\Pi(A) + \Pi(A^C) \geq 1$, and $\mathcal{N}(A) + \mathcal{N}(A^C) \leq 1$;
4. $max\{\Pi(A), \Pi(A^C)\} = 1$, and $min\{\mathcal{N}(A), \mathcal{N}(A^C)\} = 0$;
5. $\Pi(A \cup B) = max\{\Pi(A), \Pi(B)\}$, and $\mathcal{N}(A \cap B) = min\{\mathcal{N}(A), \mathcal{N}(B)\}$;
6. $\Pi(A \cap B) \leq min\{\Pi(A), \Pi(B)\}$, and $\mathcal{N}(A \cup B) \geq max\{\mathcal{N}(A), \mathcal{N}(B)\}$;
7. $\Pi(A) + \mathcal{N}(A^C) = 1$, and $\Pi(A^C) + \mathcal{N}(A) = 1$;
8. $\Pi(A) < 1 \Rightarrow \mathcal{N}(A) = 0$, and $\mathcal{N}(A) > 0 \Rightarrow \Pi(A) = 1$.

The case of subnormal possibility distributions will be discussed later.

2.3.4 Relative Possibility and Necessity Measures of Ambiguous Events

Possibility and necessity measures have been defined for crisp events, i.e., crisp subsets from Ω. Nevertheless, ambiguous events represented as fuzzy sets on Ω are frequently observed. For instance, if we are given the possibilistic information Paul is *Young* (associated with the possibility distribution of Paul's age $\pi_{Young}(.)$), we may be interested by knowing the possibility that Paul's age is *around* 18 years (ambiguous event represented by the membership function $\mu_{A\,=\,Around\;18}(.)$), or we may look forward updating, or inferring, the *prior* possibilistic knowledge encoded by $\pi_{Young}(.)$ by integrating the additional available ambiguous piece of information that Paul's age is *around* 18 years. Let Ω be an exhaustive set of mutually exclusive alternatives, $\pi(.)$ a normal possibility distribution defined on Ω and encoding our possibilistic state of knowledge, and let A denote an ambiguous event (i.e., A is a fuzzy set defined on Ω, with $\mu_A(x)$ as its membership function) (Fig. 2.14).

For each elementary alternative $x \in \Omega$, $\pi(x)$ is the grade of possibility that "x" is the unique occurring alternative. By contrast, $\mu_A(x)$ is the degree of membership or "satisfaction" of the alternative "x" of the ambiguous information represented by the fuzzy event A. Let's recall the basic definitions:

- *Supp*(A), resp. *Supp*(π), denoting the support subset, i.e., crisp subset of alternatives for which $\mu_A(x) > 0$, resp. $\pi(x) > 0$
- *Core*(A), resp. *Core*(π), denoting the core subset, i.e., crisp subset of alternatives for which $\mu_A(x) = 1$, resp. $\pi(x) = 1$

For a given alternative "x," the conjunctive fusion of these two values (i.e., *min* ($\mu_A(x)$, $\pi(x)$)) characterizes in a natural way the *degree of agreement* (or *compatibility*) between the occurrence possibility and the membership degree of the alternative "x." This reasoning can be extended to all the alternatives from Ω leading to the definition of the *compatibility distribution* between the *prior* knowledge encoded by the possibility distribution π and the ambiguous information encoded by the fuzzy event A:

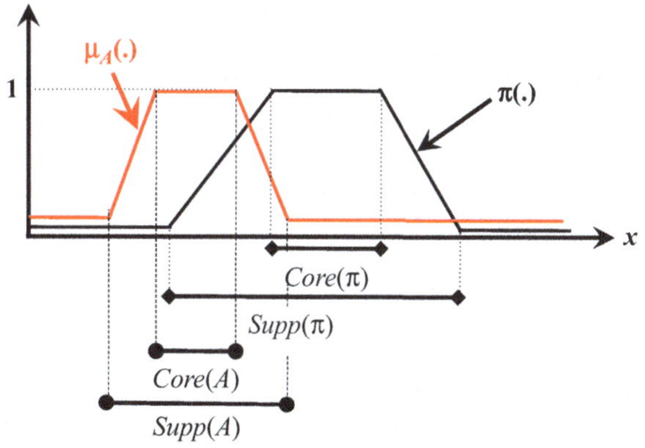

Fig. 2.14 *Prior* possibilistic knowledge, π, and additional ambiguous event A

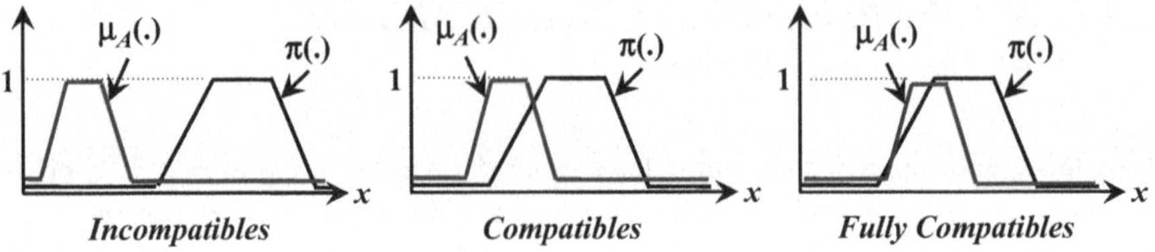

Fig. 2.15 Different compatibility forms

$$\begin{aligned} A \cap \pi : \Omega &\rightarrow [0,1] \\ x &\rightarrow min\,(\mu_A(x), \pi(x)) \end{aligned}$$

The *prior* knowledge (encoded by the possibility distribution π) and the ambiguous information (encoded by the fuzzy event A) are called (Fig. 2.15):

- *Compatibles*: if $Supp(A) \cap Supp(\pi) \neq \Phi$ (i.e., if at least one alternative "x_0" has simultaneously a non-null occurrence possibility degree, $\pi(x_0) \neq 0$, and a non-null membership degree to the fuzzy set A, $\mu_A(x_0) \neq 0$)
- *Fully Compatibles*: if $Core(A) \cap Core(\pi) \neq \Phi$ (i.e., if at least one alternative "x_0" has simultaneously a unity occurrence possibility degree, $\pi(x_0) = 1$, and a unit membership degree to the fuzzy set A, $\mu_A(x_0) = 1$)
- *Incompatibles*: if $Supp(A) \cap Supp(\pi) = \Phi$ (i.e., none of the alternatives "$x \in \Omega$" has simultaneously a non-null occurrence possibility degree, $\pi(x_0) \neq 0$, and a non-null membership degree to the fuzzy set A, $\mu_A(x_0) \neq 0$)

Recall that possibility and necessity measures are defined for crisp events, i.e., subsets from Ω, and allow characterizing the occurrence of any crisp event through two indicators: the possibility and the necessity measures. Therefore, in order to characterize the occurrence of an ambiguous event A in the framework of *prior* knowledge encoded by the possibility distribution π, L. Zadeh [13] has extended the definition of possibility and necessity measures. This extension is realized through two steps:

1. Possibilistic knowledge adjustment: a compatibility distribution $A \cap \pi$ is considered, instead of π, as the "adjustment" applied to the *prior* possibilistic knowledge π by the additional available ambiguous event A (i.e., $A \cap \pi$ is considered as representing the "new *prior*" possibilistic knowledge encoding the matching between π and the ambiguous event A) (Fig. 2.16b).
2. The *possibility measure of the fuzzy set A relative to π* is defined as the maximal degree of compatibility (Fig. 2.16c):

$$\Pi(A/\pi) = max_{x \in \Omega} A \cap \pi(x) = max_{x \in \Omega}[min\,(\mu_A(x), \pi(x))]$$

$\Pi(A/\pi)$, also called the *possibility degree of matching* [14], measures the extent to which the fuzzy event A is "compatible" or "suitable" with the *prior* possibility knowledge π. In other words, $\Pi(A/\pi)$ estimates the extent to which π and A refer to the same alternative. Notice that this approach of computing the possibility measure

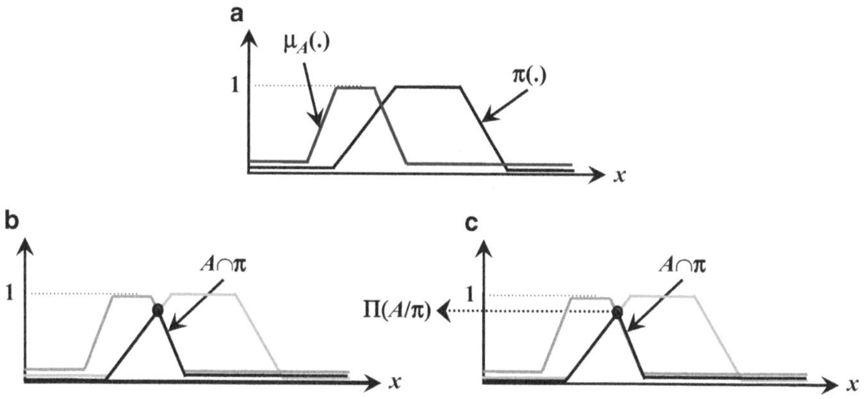

Fig. 2.16 *Prior* possibilistic adjustment and possibility measure of ambiguous event A. (**a**) Possibility distribution (π) and the ambiguous evidence A, (**b**) Degree of agreement of π and A, (**c**) Possibility degree of matching

of the ambiguous event A is in fact an extension of the crisp event possibility measure computation. In fact, if A is a crisp event (i.e., $\mu_A(x) = \mathbf{1I}_A(.) = 1$ if $x \in \Omega$; and $\mu_A(x) = \mathbf{1I}_A(.) = 0$ if $x \notin \Omega$), then $\Pi(A/\pi)$ is reduced to the "originally defined" possibility measure $\Pi(A)$. The dual necessity measure of the fuzzy set A relative to π is, also, defined as follows:

$$\begin{aligned}\mathcal{N}(A/\pi) &= 1 - \Pi\left(A^C/\pi\right) = 1 - \ max_{x\in\Omega}\, A^C \cap \pi(x)\\ &= Inf_{x\in\Omega}\left[max\left(\mu_A(x), 1 - \mu(x)\right)\right]\end{aligned}$$

$\mathcal{N}(A/\pi)$, called the *necessity degree of matching* [14], estimates the extent to which it is necessary that the alternative to which the *prior* possibility knowledge π refers is among the ones compatible with the fuzzy event A.

Example Assume that the *prior* available information related to the height of Peter is that it is between 1.70 m and 1.80 m. This *prior* information can be encoded as an imprecise type possibilistic distribution (Fig. 2.17a). Given the additional ambiguous information that Peter is *Tall* (*Tall* is encoded as a fuzzy event having the membership function given in Fig. 2.17a), we may look forward to compute the degrees of possibility/necessity of matching of the additional information *Tall* with the *prior* possibilistic knowledge.

As already shown, to compute the possibility degree of matching, the first step consists of computing the compatibility distribution between the *prior* knowledge encoded by the possibility distribution π and the ambiguous information encoded by the fuzzy event A: $A\cap\pi(x) = min(\mu_A(x), \pi(x))$, for all alternatives $x \in \Omega$.

Finally, the possibility degree of matching is given as the maximum value (Fig. 2.17b) of the obtained compatibility distribution. Concerning the necessity degree of matching, the complementary possibility distribution, i.e., $1 - \pi$, is first computed.

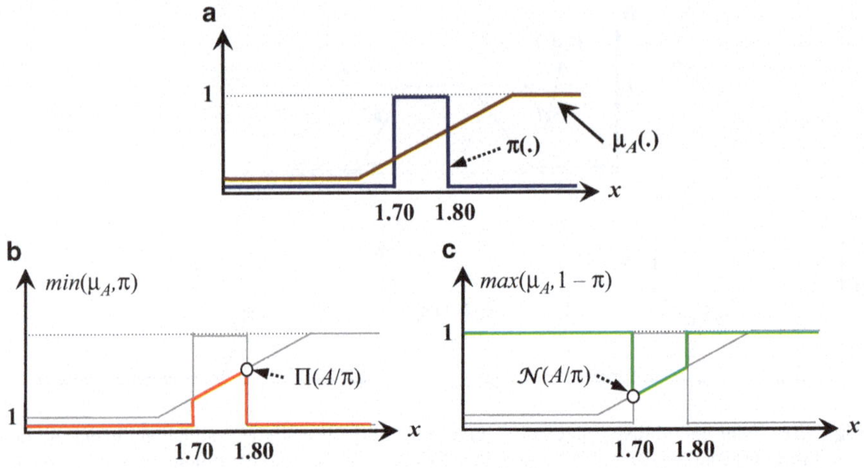

Fig. 2.17 Computing the possibility/necessity degrees of matching of a possibilistic knowledge, π, and an additional ambiguous event A. (**a**) Possibilistic knowledge and additional fuzzy event A, (**b**) Computing the possibility degree of matching, (**c**) Computing the necessity degree of matching

The second step consists of computing $max(\mu_A(x),1-\pi(x))$, for each alternative $x \in \Omega$. Finally, the necessity degree of matching will be the smallest value: $\mathcal{N}(A/\pi) = Inf_{x\in\Omega}[max(\mu_A(x), 1-\pi(x))]$ (Fig. 2.17c).

2.3.5 *Important Properties of Possibility/Necessity Degrees of Matching*

Let Ω be a finite set of alternatives, π a normal possibility distribution, and A, A_1, A_2, and B normal fuzzy sets (having as membership functions: $\mu_A(.)$, $\mu_{A1}(.)$, $\mu_{A2}(.)$ and $\mu_B(.)$) encoding the corresponding ambiguous information:

1. $\Pi(A/\pi) = 0$ if and only if $Supp(A) \cap Supp(\pi) = \Phi$ (Fig. 2.18a), where Φ denotes the empty set; $\Pi(A/\pi) = 1$ if and only if $Core(A) \cap Core(\pi) \neq \Phi$ (Fig. 2.18b).
2. $\mathcal{N}(A/\pi) = 0$ if and only if $Supp(A^C) \cap Core(\pi) \neq \Phi$ (i.e., at least one fully possible alternative has null membership to the fuzzy event A) (Fig. 2.18c).
   - $\mathcal{N}(A/\pi) = 1$ if and only if $Supp(\pi) \subseteq Core(A)$ (i.e., all possible alternatives have full membership to the fuzzy event A) (Fig. 2.18d).
   - $\mathcal{N}(A/\pi) > 0$ if and only if $Core(\pi) \subset Supp(A)$, notice the strict inclusion.
3. An important consequence of the definition of the possibility degree of fuzzy sets matching is the F-additivity property expressed by:

$$\Pi(A_1 \cup A_2/\pi) = max[\Pi(A_1/\pi), \Pi(A_2/\pi)]$$

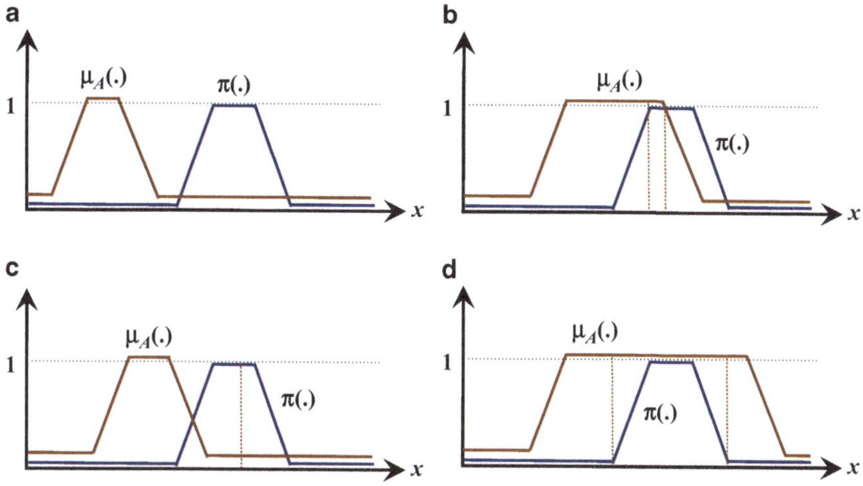

Fig. 2.18 Cases of possibility/necessity degrees of matching of a possibilistic knowledge, $\boldsymbol{\pi}$, and an additional ambiguous event A. (**a**) $\Pi(A/\pi) = 0$ ($Supp(A) \cap Supp(\pi) = \Phi$), (**b**) $\Pi(A/\pi) = 1$ ($Core(A) \cap Core(\pi) \neq \Phi$), (**c**) $\mathcal{N}(A/\pi) = 0$ ($Supp(A^C) \cap Core(\pi) \neq \Phi$), (**d**) $\mathcal{N}(A/\pi) = 1$ ($Supp(\pi) \subseteq Core(A)$)

where A_1 and A_2 are arbitrary fuzzy sets defined over Ω. The demonstration of this property is straightforward:

$$\begin{aligned}\Pi(A_1 \cup A_2/\pi) &= max_{x\in\Omega}\left[min\left\{\mu_{A_1}\cup\mu_{A_2}(x), \pi(x)\right\}\right]\\ &= max_{x\in\Omega}\left[min\left\{max\left(\mu_{A_1}(x), \mu_{A_2}(x)\right), \pi(x)\right\}\right]\\ &= max_{x\in\Omega}\left[max\left\{min\left(\mu_{A_1}(x), \pi(x)\right), min\left(\mu_{A_2}(x), \pi(x)\right)\right\}\right]\\ &= max\left[max_{x\in\Omega}\left\{min\left(\mu_{A_1}(x), \pi(x)\right)\right\}, max_{x\in\Omega}\left\{min\left(\mu_{A_2}(x), \pi(x)\right)\right\}\right]\\ &= max\left[\Pi(A_1/\pi), \Pi(A_2/\pi)\right]\end{aligned}$$

Moreover, we can clearly show that:

- $\Pi(A_1 \cap A_2/\pi) \leq min\ [\Pi(A_1/\pi), \Pi(A_2/\pi)]$;
- $\mathcal{N}(A_1 \cap A_2/\pi) = min\ [\mathcal{N}(A_1/\pi), \mathcal{N}(A_2/\pi)]$;
- $\mathcal{N}(A_1 \cup A_2/\pi) \geq max\ [\mathcal{N}(A_1/\pi), \mathcal{N}(A_2/\pi)]$;

4. $\mathcal{N}(A/\pi) \leq \Pi(A/\pi)$ is always satisfied when considering a normal possibility distribution π and a normal fuzzy set A.
5. The following relation $max\,[\Pi(A/\pi), 1 - \mathcal{N}(A/\pi)] = 1$ is no longer satisfied when A is a fuzzy set. In fact, we may have, for a given fuzzy set A, $\Pi(A/\pi) < 1$ and $\Pi(A^C/\pi) < 1$.
6. If $\mu_A(.) = \pi(.)$, then $\Pi(A/\pi) = 1$ and $\mathcal{N}(A/\pi) \geq 1/2$.
7. $\mathcal{N}(A/\pi) = 1$ if and only if $\mu_A(.)$ is the membership function of a crisp subset $A \subseteq \Omega$ which stands for $\mu_A(.)$ and $\pi(.)$.

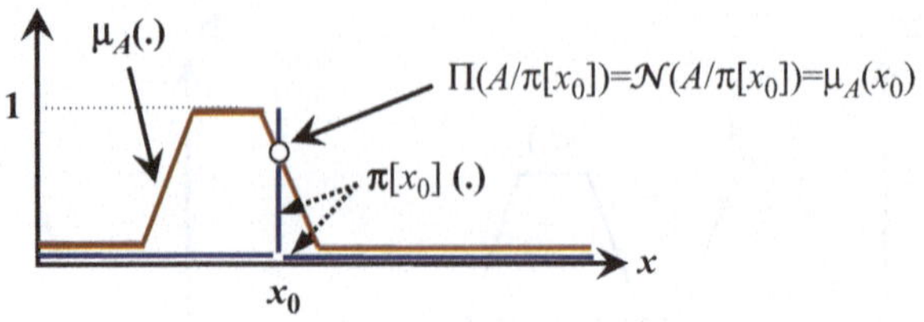

Fig. 2.19 Possibility/necessity degrees of matching of a precise possibilistic knowledge, $\pi_{[x_0]}$, and an additional ambiguous event *A*

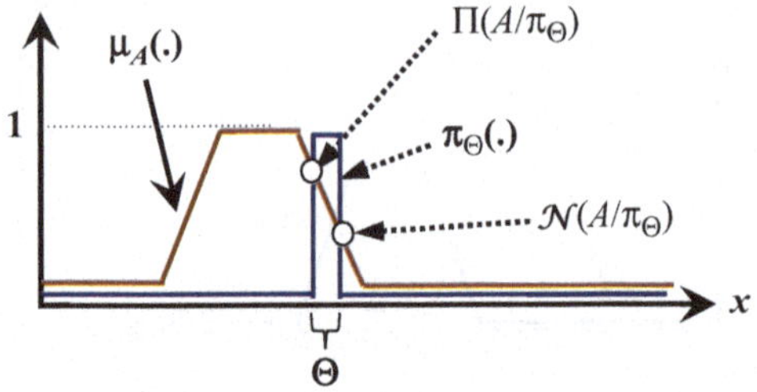

Fig. 2.20 Degrees of possibility/necessity of matching of a precise possibilistic knowledge, π_Θ, and an additional ambiguous event *A*

Example 1 If $\pi_{[x_0]}$ denotes a precise-type possibility distribution (i.e., $\pi_{[x_0]}(x) = 1$ if $x = x_0$, and $\pi_{[x_0]}(x) = 0$ for all $x \neq x_0$) (Fig. 2.19), then the two matching possibility and necessity degrees are equal: $\Pi\left(A/\pi_{[x_0]}\right) = \mathcal{N}\left(A/\pi_{[x_0]}\right) = \mu_A(x_0)$

Example 2 If π_Θ denotes an imprecise-type possibility distribution (i.e., $\pi_\Theta(x) = 1$ if $x \in \Theta$, and $\pi_\Theta(x) = 0$ for all $x \notin \Theta$) (Fig. 2.20), then $\mathcal{N}(A/\pi_\Theta) \leq \Pi(A/\pi_\Theta)$. Moreover, the more precise Θ is, the closer to zero $\Pi(A/\pi_\Theta) - \mathcal{N}(A/\pi_\Theta)$ is. However, the converse is false (i.e., if $\Pi(A/\pi) = \mathcal{N}(A/\pi)$, then this does not necessarily lead to say that Θ is precise). For instance, if π refers to a crisp event and the additional knowledge is a crisp event identical to π, i.e., $A = \pi$, then we can easily show that $\Pi(A/\pi) = \mathcal{N}(A/\pi)$.

2.4 Subnormal Possibility Distributions

Let π denote a possibility distribution defined as a mapping from the set of alternatives Ω to the unit interval [0, 1]:

$$\begin{aligned} \pi : \Omega &\rightarrow [0, 1] \\ x &\rightarrow \pi(x) \end{aligned}$$

As previously mentioned, π reflects the subjective knowledge related to the possibility of occurrence of different alternatives as being the unique and true occurring one. Indeed, possibility distributions can arguably be advocated as a more natural representation of human uncertain knowledge than probability distributions. π is viewed as describing possible values that could be assigned to some

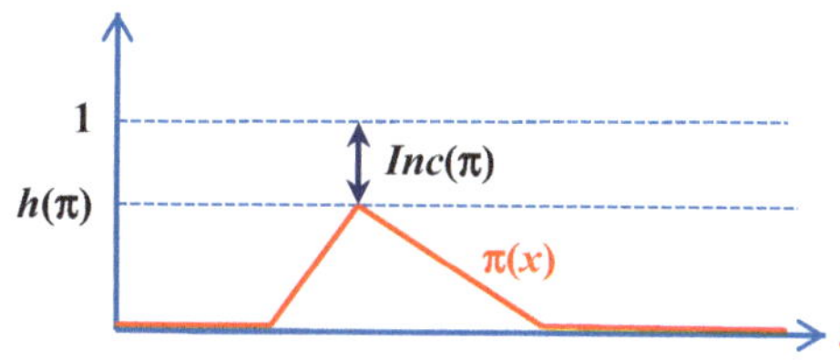

Fig. 2.21 Subnormal possibility distribution

unknown variable X, therefore $\pi(x)$ represents to what extent it is possible that X is the true alternative: $\pi(x) = Poss\{X = x_{True}\}$. Two subjective aspects are encoded within the possibility distribution π. The first aspect concerns the fact that partial possibility degrees of occurrence (i.e., belonging to the open interval [0,1]) are affected to different alternatives. This aspect has a great importance in terms of ordering, or ranking, these partial possibility degrees attributed to different alternatives. The second subjective aspect encoded within the possibility distribution π is related to the height of π, $h(\pi)$, defined as (see Fig. 2.21): $h(\pi) = max_{x \in \Omega}\pi(x)$. This value is frequently called the *degree of consistency* of π with the available knowledge. In fact, a possibility distribution π is said to be *normal*, or *consistent with the available knowledge*, if and only if:

$$h(\pi) = max_{x\in\Omega}\pi(x) = 1$$

that is, the supremum of $\pi(x)$ over Ω is unity. In other words, this means that the "normality" (or the consistency) character of a possibility distribution describes the fact that "at least" one of the alternatives contained in Ω is totally possible. Otherwise, the possibility distribution is said to be *subnormal* or *inconsistent with the available knowledge*.

The value $Inc(\pi) = 1 - max_{x \in \Omega}\pi(x) = 1 - h(\pi)$ is called the *degree of inconsistency* of the of the knowledge encoded by the possibility distribution π. If $Inc(\pi) > 0$ (i.e., $h(\pi) = max_{x \subset \Omega}\pi(x) < 1$)), then the possibility distribution π is called *subnormal*. This case is interpreted by saying that the available knowledge on the occurrence of different alternatives, expressed by π, is not *consistent*. In other words, there is no available alternative which is in complete agreement with the unknown true alternative. If the considered possibility distribution, π, is normalized, then $h(\pi) = 1$ and $Inc(\pi) = 0$.

In possibility theory literature, a possibility distribution is generally considered as being normalized; otherwise, one of the existing normalizing methods is applied transforming, thus, the subnormal possibility distribution into a normalized one.

It is worthwhile to notice that imposing the use of normalized possibility distributions or normalizing subnormal ones leads to the fact that the important informative inconsistency aspect will "vanish" from the possibility distribution encapsulating our state of knowledge concerning the occurrence of different alternatives. In fact, in real-world problems, subnormal possibility distributions are frequently encountered. Here follow some situations [15] where subnormal distributions are encountered:

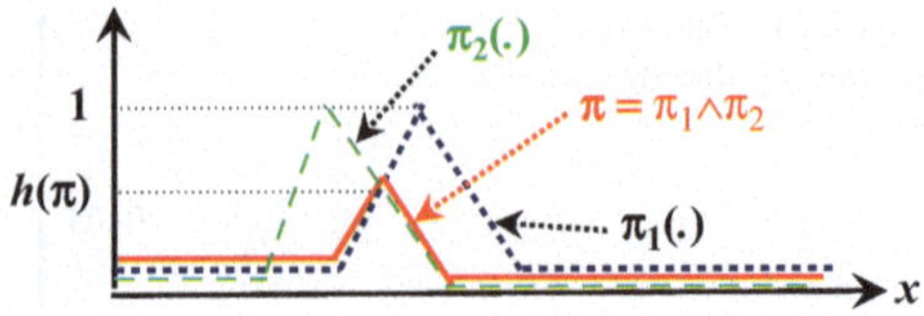

Fig. 2.22 Subnormal distribution generated by the conjunctive fusion of two possibility distributions

1. The lack of exhaustiveness of the set of alternatives. In this case, the universe of alternatives Ω does not contain all potential alternatives (for instance, all thematic classes present in an analyzed remotely sensed image of a given region or all potential targets classes in an automatic target recognition problem). Therefore, it is very natural that the available epistemic knowledge could not be consistent while affecting possibility degrees to different alternative.
2. The available knowledge concerning the potential occurrence of different alternatives is partially informative, and partial aspect is generally expressed by a human expert in terms of inconsistency (i.e., through a subnormal possibility distribution).
3. Subnormality is the reflection of some conflict between epistemic information sources. It usually arises from the combination of epistemic information from different sources when there is some conflict between the observations of the sources. For instance, if we consider merging two sources of knowledge (expressed as two possibility distributions π_1 and π_2 on the same set of alternatives Ω) (Fig. 2.22) using the *min* conjunctive fusion operator, then the resulting possibility distribution is "generally" subnormal. Namely, if $1 - max_{x \in \Omega}[\pi_1(x) \wedge \pi_2(x)] \neq 1$, then the conjunction fusion issued distribution (i.e., $\pi_1 \wedge \pi_2$) is a subnormal possibility distribution. This situation is generally "described" as a conflict between π_1 and π_2.

2.4.1 *Possibility Distributions Normalization Methods*

Normalization is defined as the process that turns a subnormal possibility distribution π into a normal one π'. Several normalization methods have been proposed in the literature. Let us consider a possibility distribution π assumed to be subnormal: $h(\pi) = max_{x \in \Omega}[\pi(x)] < 1$. In order to "transform" the subnormal distribution π into a normalized one, three major normalization methods are used: ordinal, numerical, and inconsistency shift methods. These methods are detailed in this section as well as their effects on the possibility and necessity measures.

2.4.1.1 Ordinal Normalization

The *ordinal normalization* (also called *minimal* or *min-based* normalization) method is the most elementary approach. It consists in considering the alternatives with the greatest degree of possibility and "shifting" their possibility degree to unity:

$$\pi'(x) = \begin{cases} 1 & \text{if } \pi(x) = h(\pi) \\ \pi(x) & \text{otherwise} \end{cases}$$

Using this method, the "new" possibility and necessity measures resulting from the normalized possibility distribution π' are given as ($\forall A \subseteq \Omega$):

$$\Pi'(A) = \begin{cases} 1 & \text{if } \Pi(A) = h(\pi) \\ \Pi(A) & \text{otherwise} \end{cases}$$

and

$$\mathcal{N}'(A) = 1 - \Pi'\left(A^C\right) = \begin{cases} 0 & \text{if } \Pi\left(A^C\right) = h(\pi) \\ 1 - \Pi(A) & \text{if } \Pi\left(A^C\right) < h(\pi) \end{cases}$$

2.4.1.2 Numerical (or Ratio) Normalization

This *numerical* (also called *ratio*) *normalization* method is based on dividing $\pi(x)$ by its greatest value:

$$\pi'(x) = \frac{\pi(x)}{h(\pi)} = \frac{\pi(x)}{max_{x\in\Omega}\pi(x)}$$

Using this method, the "new" possibility and necessity measures resulting from the normalized possibility distribution π' are given as ($\forall A \subseteq \Omega$):

$$\Pi'(A) = max_{x\subset A}\left[\pi'(x)\right] = max_{x\in A}\left[\frac{\pi(x)}{h(\pi)}\right] = \frac{1}{h(\pi)}max_{x\in A}\left[\pi(x)\right]$$

$$\Rightarrow \qquad \Pi'(A) = \frac{1}{h(\pi)}\Pi(A)$$

$$\text{and;} \quad \mathcal{N}'(A) = 1 - \Pi'\left(A^C\right) = 1 - \frac{1}{h(\pi)}\Pi\left(A^C\right) = 1 - \frac{1}{h(\pi)}[1 - \mathcal{N}(A)]$$

$$\Rightarrow \qquad \mathcal{N}'(A) = \frac{1}{h(\pi)}\mathcal{N}(A) - \frac{1 - h(\pi)}{h(\pi)}$$

2.4.1.3 Inconsistency Shifting Normalization

In this method, also called *maximal normalization* method, the normalization process is conducted [16] by adding the *lake of normalization*, or lake of evidence, (i.e., the degree of inconsistency) to all the elements in the set of alternatives:

$$\pi'(x) = \pi(x) + Inc(\pi) = \pi(x) + 1 - max_{x\in\Omega}\,\pi(x), \quad \forall x \in \Omega$$

Using this method, the "new" possibility and necessity measures resulting from the normalized possibility distribution π' are given as ($\forall A \subseteq \Omega$):

$$\begin{aligned} &\Pi'(A) = max_{x\in A}[\pi'(x)] = max_{x\in A}[\pi(x) + Inc(\pi)] = max_{x\in A}[\pi(x)] + Inc(\pi) \\ \Rightarrow \quad &\Pi'(A) = \Pi(A) + Inc(\pi) = \Pi(A) + 1 - h(\pi) \end{aligned}$$

and

$$\begin{aligned} &\mathcal{N}'(A) = 1 - \Pi'(A^C) = 1 - [\Pi(A^C) + 1 - h(\pi)] = 1 - \Pi(A^C) - [1 - h(\pi)] \\ \Rightarrow \quad &\mathcal{N}'(A) = \mathcal{N}(A) - [1 - h(\pi)] \end{aligned}$$

This normalization method (also called Yager's normalization method) presents an interesting property that we call *informational preserving*. In fact, let $\Omega = \{x_1, x_2, \ldots, x_N\}$ denote a finite set of alternatives and $\pi : \Omega \rightarrow [0, 1]$ an ordered possibility distribution on Ω (i.e., different values $\pi(x_\nu)$ are arranged in a decreasing order such that $1 = \pi(x_1) \geq \pi(x_2) \geq \pi(x_3) \geq \ldots \geq \pi(x_N)$). The total amount of information supplied by the distribution π is expressed through the U-uncertainty measure of π [19]:

$$\begin{aligned} U(\pi) &= \sum_{n=2}^{N} (\pi(x_n) - \pi(x_{n+1}))\ \log_2(n) + [1 - \pi(x_1)]\ \log_2(N) \\ &= \sum_{n=2}^{N} \left[\{1 - [\pi(x_1) - \pi(x_n)]\}\ \log_2 \frac{n}{n-1}\right] \end{aligned}$$

and as $1 - [\pi'(x_1) - \pi'(x_n)] = 1 - [\pi(x_1) - \pi(x_n)]$ ($\pi'(.)$ is obtained by a simple shift of $\pi(.)$), thus $U'(\pi) = U(\pi)$, meaning that both π' and π are equally informative.

These different possibility distributions normalization methods are illustrated in Fig. 2.23.

2.4.2 Dubois's Alternative Necessity Measure

Let us recall the "fundamental" property, when considering normal distributions, that makes the link between the possibility and necessity measures $\forall A \subseteq \Omega$, $\mathcal{N}(A) \leq \Pi(A)$ which is considered as a "*satisfactory inequality*" (since an event "must" be possible before being certain, i.e., occurring necessarily). Nevertheless, it has been noticed for a long time [Pra-80] that this inequality no longer holds when π is a subnormal possibility distribution. In order to tackle this (major) inconvenient, Yager [17] proposed to "substitute" the necessity measure by another "certainty" measure defined by:

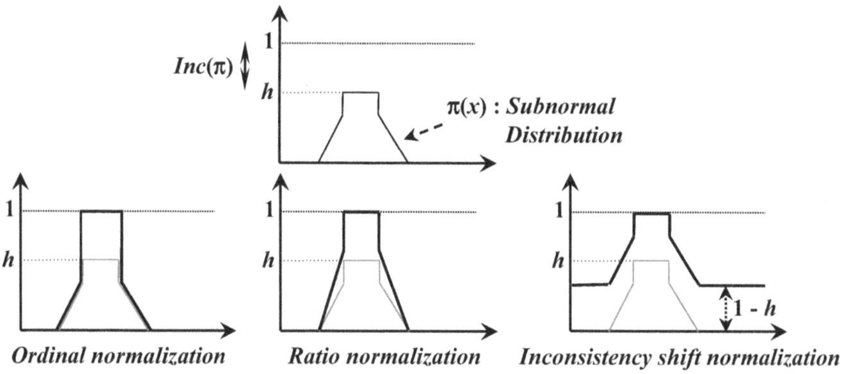

Fig. 2.23 Normalization methods

$$\begin{aligned} \text{Cert}: \quad \mathcal{P}(\Omega) &\rightarrow [0,1] \\ A &\rightarrow \text{Cert}(A) = \min(\Pi(A), \mathcal{N}(A)) \end{aligned}$$

Using this certainty measure, the "satisfactory inequality" remains satisfied even if the considered possibility distribution is subnormal:

$$\forall A \in \Omega, \text{Cert}(A) \leq \Pi(A)$$

This new certainty measure suffers from two major drawbacks [18]:

(i) Cert(.) has no physical interpretation.
(ii) The basic necessity property is no longer satisfied:
$\forall A, B \subseteq \Omega$, $\text{Cert}(A \cap B) \neq \min\{\text{Cert}(A), \text{Cert}(B)\}$ (whereas, it is always the case of the necessity measure $\mathcal{N}(.)$).

As an alternative to Yager's proposition, Dubois has suggested a more interesting definition of the necessity measure:

$$\begin{aligned} \mathcal{N}_D: \quad \mathcal{P}(\Omega) &\rightarrow [0,1] \\ A &\rightarrow \mathcal{N}_D(A) = \Pi(\Omega) - \Pi(A^C) \end{aligned}$$

where $\Pi(\Omega) = h(\pi) = \max_{x \in \Omega} \pi(x)$.

In this definition, the greatest degree of possibility attached to an alternative of Ω plays the role of a "reference point" with respect to which the degree of uncertainty is estimated:

- $\mathcal{N}_D(.)$ reduces to the basic necessity measure $\mathcal{N}(.)$ when the considered possibility distribution π is normal.
- $\mathcal{N}_D(.)$ preserves the main following properties ($\forall A, B \subseteq \Omega$):

(i) $\Pi(A \cup B) = \max\{\Pi(A), \Pi(B)\}$;
(ii) $\mathcal{N}_D(A \cap B) = \min\{\mathcal{N}_D(A), \mathcal{N}_D(B)\}$;
(iii) $\mathcal{N}_D(A) \leq \Pi(A)$.

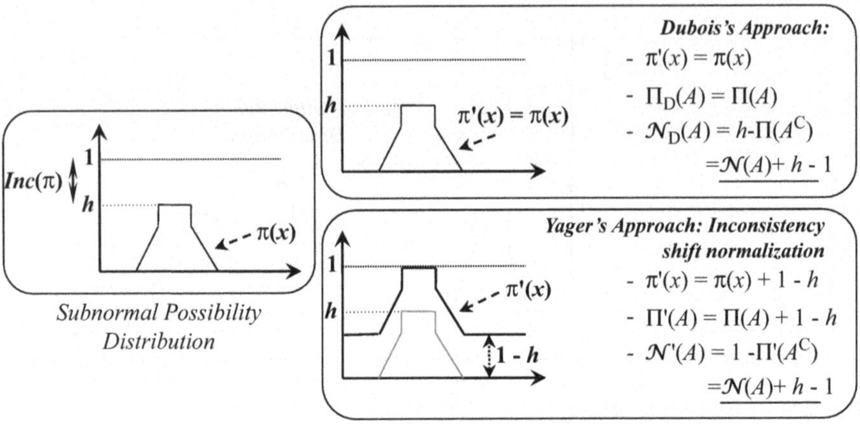

Fig. 2.24 Yager's and Dubois's approaches concerning subnormal distributions

It is important to notice that the new necessity measure proposed by Dubois (i.e., $\mathcal{N}_D(A) = \Pi(\Omega) - \Pi(A^C)$) and the necessity measure obtained by the inconsistency shift normalization, proposed by Yager) are strictly the same. Nevertheless, Dubois's approach does not modify the original possibility distribution (for normalization issues), and this leads to the fact that Dubois's possibility measure remains unchanged. This is not the case in Yager's approach, where the subnormal possibility distribution has been normalized, and thus the new possibility measure is shifted by the same amount: degree of inconsistency (Fig. 2.24).

2.4.3 Normal Versus Subnormal Distributions Properties

In this section, a comparative presentation of different properties related the three cases: normal possibility distributions, subnormal possibility distributions, and subnormal possibility distributions where Dubois's alternative necessity measure is applied. In fact, possibility theory was introduced as a derived solution for the fuzzy sets theory to take into account both the uncertainty (related to the true alternative "discovery") and the ambiguity (related to the available epistemic knowledge). This was achieved through the definition of the possibility distribution concept as well as the two set possibility, $\Pi(.)$, and necessity, $\mathcal{N}(.)$, measures.

The possibility distribution is assumed to capture and to model the subjective side of the available knowledge, and the possibility and necessity measures are assumed to capture the imprecision related to the occurrence of any event A (i.e., subset $A \subseteq \Omega$) by means of an interval $[\mathcal{N}(A), \Pi(A)]$. This interval characterization of the occurrence of an event A "imposes" the condition: $\mathcal{N}(A) \leq \Pi(A)$ which expresses an "intuitive" fact that an event must be possible to some extent before it can begin to be certain.

When the basic possibility distribution is considered as normal, then this condition is always fulfilled. This explains the major reason for which most studies in

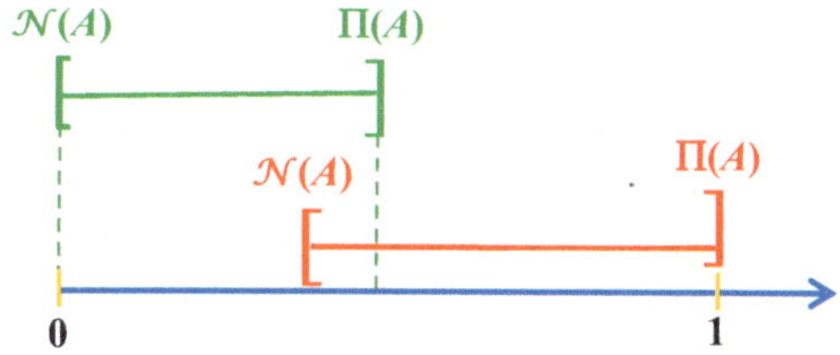

Fig. 2.25 Necessity-possibility measures interval configurations for normalized possibility distributions

Table 2.1 Main characteristics of the normal, subnormal and subnormal (using Dubois's alternative necessity measure) possibility distributions

Normal possibility distribution	Subnormal possibility distribution	Dubois's alternative necessity measure
$\pi : \Omega \to [0, 1]$ $x \to \pi(x)$ $\exists x_0 \in \Omega, \pi(x_0) = 1$ $\Pi(A) = max_{x \in A}[\pi(x)]$ $\mathcal{N}(A) = 1 - \Pi(A^c)$	$\pi' : \Omega \to [0, 1]$ $x \to \pi'(x)$ $h(\pi') = max_{x \in \Omega}[\pi'(x)] < 1$ $1 - h(\pi')$: degree of inconsistency $\Pi'(A) = max_{x \in A}[\pi'(x)]$ $\mathcal{N}'(A) = 1 - \Pi'(A^c)$	$\pi'' : \Omega \to [0, 1]$ $x \to \pi''(x)$ $h(\pi'') = max_{x \in \Omega}[\pi''(x)] < 1$ $1 - h(\pi'')$: degree of inconsistency $\Pi_D(A) = max_{x \in A}[\pi''(x)]$ $\mathcal{N}_D(A) = h(\pi'') - \Pi(A^c)$
$\Pi(\Omega) = 1$ $\mathcal{N}(\Omega) = 0$	$\Pi'(\Omega) = h(\pi')$ $\mathcal{N}'(\Omega) = 0$	$\Pi_D(\Omega) = h(\pi'')$ $\mathcal{N}_D(\Omega) = h(\pi'')$
$max(\Pi(A), \Pi(A^C)) = 1$	$max(\Pi'(A), \Pi'(A^C)) = h(\pi')$	$max(\Pi_D(A), \Pi_D(A^C)) = h(\pi'')$
$min\left(\mathcal{N}(A), \mathcal{N}(A^C)\right) = 0$	$\mathcal{N}'(A) > 0$ and $\mathcal{N}'(A^C) > 0$ $min\left(\mathcal{N}'(A), \mathcal{N}'(A^C)\right) = 1 - h(\pi')$	$min\left(\mathcal{N}_D(A), \mathcal{N}_D(A^C)\right) = 0$
$\mathcal{N}(A) \le \Pi(A)$: guaranteed	$\mathcal{N}'(A) \le \Pi'(A)$: not guaranteed	$\mathcal{N}_D(A) \le \Pi_D(A)$: guaranteed

possibility theory "mainly" consider normal possibility distributions. Nevertheless, the "cost" of this requirement comes from other relations linking both possibility and necessity measures which are:

$$\text{If } \mathcal{N}(A) > 0 \Rightarrow \Pi(A) = 1 \text{ and, If } \Pi(A) < 1 \Rightarrow \mathcal{N}(A) = 0$$

meaning that at least one of upper/lower interval bounds is either zero or one, which strongly limits the "interest" of this occurrence modeling (Fig. 2.25).

The first case (i.e., $[\mathcal{N}(A) = 0, \Pi(A) < 1]$) is if all fully possible alternative (i.e., alternatives x for which $\pi(x) = 1$) belong to A^c (i.e., the complement subset of A). In fact, in this case $\mathcal{N}(A) = 1 - \Pi(A^c) = 0$. In the second case (i.e., at least one alternative belonging to A is fully possible), then:

- $\Pi(A) = 1$, and $\mathcal{N}(A) = 1 - \Pi(A^c) = 0$ (if another fully possible alternative belongs to A^c). This will lead to $[\mathcal{N}(A) = 0, \Pi(A) = 1]$.
- $\Pi(A) = 1$, and $\mathcal{N}(A) = 1 - \Pi(A^c) > 0$ (if A^c does not contain any fully possible alternative). This will lead to $[\mathcal{N}(A) > 0, \Pi(A) = 1]$.

Table 2.1 provides a brief synthesis of the main characteristics of the three potential situations: normal possibility distribution, subnormal possibility distribution, and subnormal possibility distribution using Dubois's alternative necessity measure ($\forall A, B \subseteq \Omega$).

As it can be witnessed, the use of Dubois's alternative necessity measure guarantees the validity of the main property limiting the necessity by the possibility of the occurrence of any event (i.e., $\mathcal{N}_D(A) \leq \Pi_D(A), \forall A,B \subseteq \Omega$) while maintaining the important informative facet related to the inherent possibility distribution inconsistency.

References

1. D. Dubois, H. Prade, The legacy of 50 years of fuzzy sets: A discussion. Fuzzy Sets Syst. **281**, 21–31 (2015)
2. T. Denoeux, 40 years of Dempster-Shafer theory. Int. J. Approx. Reason. **79**, 1–6 (2016)
3. L. Zadeh, Fuzzy sets as the basis for a theory of possibility. Fuzzy Sets Syst. **1**, 3–28 (1978)
4. D. Dubois, H. Prade, Possibility theory: Qualitative and quantitative aspects, in *Quantified Representation of Uncertainty and Imprecision*, (Springer, Dordrecht, 1998), pp. 169–226
5. H. Prade, Reasoning with fuzzy default values, in Proceedings of the 5th International Symposium on Multiple-Valued Logic (1985), pp. 191–197
6. D. Dubois, H. Prade, *Possibility Theory: An Approach to Computerized Processing of Uncertainty* (Plenum Press, New York, 1988)
7. R.R. Yager, Approximate reasoning as a basis for rule-based expert systems. IEEE Trans. Syst. Man Cybern. **4**, 636–643 (1984)
8. L.A. Zadeh, Fuzzy sets. Inf. Control. **8**, 338–353 (1965)
9. L.A. Zadeh, Possibility theory and soft data analysis, in *Fuzzy Sets, Fuzzy Logic, and Fuzzy Systems: Selected Papers*, ed. by L. A. Zadeh, (World Scientific, Singapore, 1996), pp. 481–541
10. R.R. Yager, Measuring tranquility and anxiety in decision making: An application of fuzzy sets. Int. J. Gen. Syst. **8**(3), 139–146 (1982)
11. R.R. Yager, On the specificity of a possibility distribution. Fuzzy Sets Syst. **50**, 279–292 (1992)
12. L.A. Zadeh, Fuzzy sets and information granularity. Adv. Fuzzy Set Theory Appl. **11**, 3–18 (1979)
13. L.A. Zadeh, Fuzzy sets as a basis for a theory of possibility. Fuzzy Sets Syst. **1**, 3–28 (1978)
14. D. Dubois, H. Prade, Tolerant fuzzy pattern matching: an introduction, in *Fuzziness in Database Management Systems*, (Springer, Heidelberg, 1995), pp. 42–58
15. M. Oussalah, On the normalization of subnormal possibility distributions: New investigations. Int. J. Gen. Syst. **31**, 277–301 (2002)
16. R.R. Yager, On the relationship of methods of aggregating evidence in expert systems. Cybern. Syst. **16**, 1–21 (1985)
17. R.R. Yager, A modification of the certainty measure to handle subnormal distributions. Fuzzy Sets Syst. **20**, 317–324 (1986)
18. D. Dubois, H. Prade, An alternative approach to the handling of subnormal possibility distributions:—A critical comment on a proposal by Yager. Fuzzy Sets Syst. **24**, 123–126 (1987)
19. M. Higashi, G.J. Klir, Measures of uncertainty and information based on possibility distributions, Int. J. Gen. Syst. **9**(1), 43–58 (1983)

Chapter 3
Joint Possibility Distributions and Conditioning

3.1 Introduction

Let X be a variable taking values in a finite, exhaustive and exclusive set of alternatives $\Omega = \{x_1, x_2, \ldots, x_N\}$. The exhaustivity (resp. exclusivity) of Ω means that the alternative value taken by the variable X belongs to Ω (resp. two alternatives cannot simultaneously occur). Uncertainty related to the variable X means that the unique true alternative, x_{n_0}, taken by X is unknown. If the available knowledge about the true value is based on statistical data and given as a probability measure Pr (where $\Pr\{X = x_n\}$ refers to the probability that the true value is x_n), then the variable X is called a random variable. In the case where the available knowledge is "weaker" than probabilities (subjective knowledge, ambiguous...), and expressed as a possibility distribution π_X (which acts as an elastic constraint on the alternatives that can be assigned to X) then, the variable X is called a possibilistic variable defined on the set of alternatives Ω. In this case, $\pi_X(x_n) = Poss\{X = x_n\}, \forall\, x_n \in \Omega$ represents to what extent it is possible that x_n is the true alternative.

In real-world applications, we generally have to simultaneously deal with several possibilistic variables. Possibility distributions fusion topic considers the situation where several possibilistic distributions are defined over the same set of alternatives Ω. In order to obtain a "single" possibilistic variable, jointly encoding the considered possibilistic variables, the expert knowledge injected into the fusion process concerns the reliability of different distributions. This knowledge is used for the selection of the possibilistic fusion operator to be used. In other words, possibility distributions fusion topic doesn't study the joint relationship between different possibility distributions and does not target to construct a single multidimensional possibilistic variable. In fact, two situations can be encountered. The first situation concerns the case where several "separate" possibilistic variables are considered (Case-1), and the second situation is related to multidimensional possibilistic variables (Case-2).

B. Solaiman, É. Bossé, *Possibility Theory for the Design of Information Fusion Systems*, Information Fusion and Data Science,
https://doi.org/10.1007/978-3-030-32853-5_3

Case-1 Let us assume that we are dealing with two possibilistic variables X, resp. Y, (related to the same "physical" phenomenon) and associated with the possibility distribution π_X, resp. π_Y, defined on the set of alternatives Ω_1, resp. Ω_2:

$$\begin{aligned} \pi_X : \quad & \Omega_1 \to [0,1] \quad \text{and,} \quad & \pi_Y : \quad & \Omega_2 \to [0,1] \\ & x \to \pi_X(x) & & y \to \pi_Y(y) \end{aligned}$$

For instance, in a pattern recognition application (resp. a medical diagnosis application), the variable X may refer to an observed feature f_1 (resp. symptom s_1) of an object (resp. patient) from a given class of objects (resp. from a set of patients having a given disease), and Y may refer to another observed feature f_2 (resp. symptom s_2) of the same object (resp. patient).

$\pi_X(x)$, resp. $\pi_Y(y)$, represents the degree with which the feature f_1, resp. f_2, appears for the observed object (resp. symptom s_1, symptom s_2, appears for the observed patient).

Case-2 Let us assume that the bidimensional possibilistic variable Z is defined over the product set of alternatives $\Omega = \Omega_1 \times \Omega_2 = \{(x, y), x \in \Omega_1, y \in \Omega_2\}$:

$$\begin{aligned} \pi_Z : \quad & \Omega_1 \times \Omega_2 \to [0,1] \\ & (x, y) \to \pi_Z(x, y) \end{aligned}$$

For instance, in a medical diagnosis, resp. pattern recognition application, Ω_1 represents the set of symptoms (resp. the set of potentially observed features) and Ω_2 represents the set of diseases (resp. classes) and $\pi_Z(x, y)$ represents the degree with which symptom x appears in disease y (resp. the feature x appears in class y). Notice that in this case, π_Z is defined over the Cartesian set product $\Omega_1 \times \Omega_2$ and not on the individual sets of alternatives Ω_1 and Ω_2.

In the first case, the question of the evaluation of a joint possibilistic variable $X \times Y$ (defined on the Cartesian product $\Omega_1 \times \Omega_2$) is raised where the objective is to compute the joint possibility degree of (x, y) using the basic possibilistic degrees $\pi_X(x)$ and $\pi_Y(y)$. This process is called *joint possibility distribution evaluation*.

In the second case, the inverse question of evaluating individual possibilistic degrees $\pi_X(x)$ and $\pi_Y(y)$ using the joint possibility distribution $\pi_Z(x, y)$ is raised. This process is called *marginal possibility distributions evaluation*.

In both cases, the mutual relationship between both possibilistic variables is studied in terms of conditioning, i.e., the occurrence possibility degree of one possibilistic variable given the knowledge that a precise alternative (or an event composed of several alternatives) of the second possibilistic variable is assumed as occurred. This chapter is devoted to detail and to study these different questions.

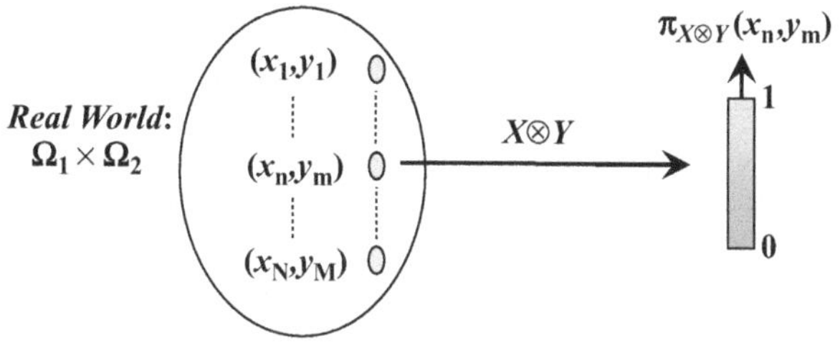

Fig. 3.1 Joint possibilistic information defined on $\Omega_1 \times \Omega_2$

3.2 Joint and Marginal Possibility Distributions

Let $\Omega_1 = \{x_1, x_2 \ldots, x_N\}$ (resp. $\Omega_2 = \{y_1, y_2 \ldots y_M\}$) denote a finite exhaustive and exclusive set of alternatives, and $\Omega = \Omega_1 \times \Omega_2$ is the Cartesian product alternative set of Ω_1 and Ω_2 (i.e., $\Omega = \Omega_1 \times \Omega_2 = \{(x_n, y_m), x_n \in \Omega_1, y_m \in \Omega_2\}$). Now, let us consider the two dimensional possibilistic variable $X \otimes Y$ defined on $\Omega_1 \times \Omega_2$ where the available possibilistic knowledge is encoded by the possibility distribution:

$$\begin{aligned} \pi_{X\otimes Y} : \Omega = \Omega_1 \times \Omega_2 \quad &\rightarrow \quad [0, 1] \\ (x, y) \quad &\rightarrow \quad \pi_{X\otimes Y}(x, y) \end{aligned}$$

The value $\pi_{X\otimes Y}(x, y)$ indicates to what extent it is possible for the couple of alternatives (x, y) with $x \in \Omega_1$ and $y \in \Omega_2$ to occur simultaneously (Fig. 3.1).

The distribution $\pi_{X\otimes Y}(.,.)$ is called the *joint possibility distribution.*

The *prior* knowledge of the joint possibility distribution $\pi_{X\otimes Y}(x, y)$ allows to "generate" two "partial" possibilistic information elements X on Ω_1 and Y on Ω_2. To see this, note that:

$$\begin{aligned} \pi_X(x) &= Poss[X = x] \\ &= Poss[(X = x) \cap (Y = y_1 \text{ or } Y = y_2 \text{ or} \ldots \text{or } Y = y_M)] \\ &= Poss[(X = x) \cap (\text{all}\, y \in \Omega_2)] \\ &= \pi_{X\otimes Y}((X = x), (\text{all}\, y \in \Omega_2)) \\ &= \max_{y\in\Omega_2} \pi_{X\otimes Y}(x, y) \end{aligned}$$

and similarly,

$$\begin{aligned} \pi_Y(y) &= \pi_{X\otimes Y}((\text{all}\, x \in \Omega_1), (Y = y)) \\ &= \max_{x\in\Omega_1} \pi_{X\otimes Y}(x, y) \end{aligned}$$

This leads to define the *marginal possibilistic variable* by the two following possibility distributions:

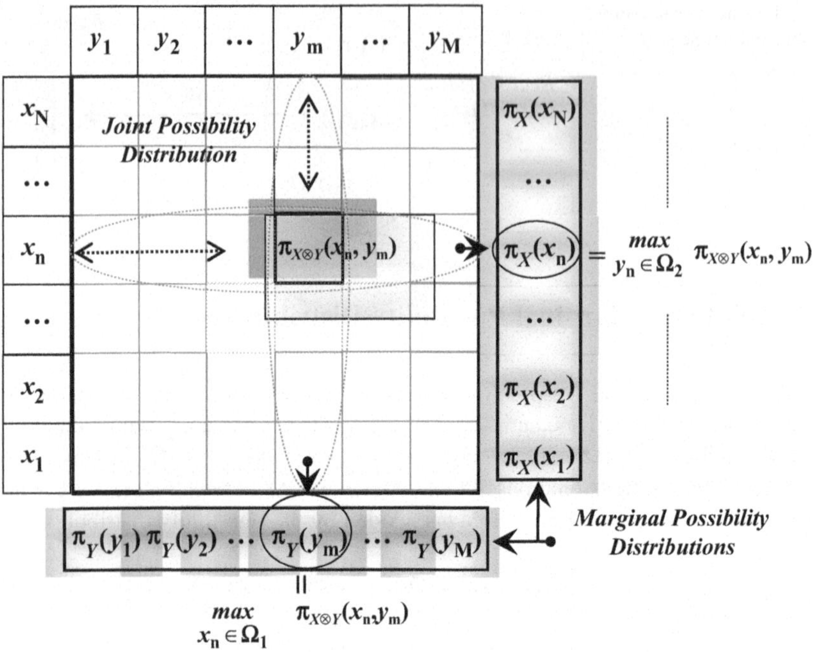

Fig. 3.2 Joint and marginal possibility distributions

$$\pi_X(x) = \max_{y \in \Omega_2} \pi_{X \otimes Y}(x, y) = [\pi_{X \otimes Y}(x, y) \downarrow \Omega_1], \quad \text{for any}\, x \in \Omega_1$$

$$\pi_Y(y) = \max_{x \in \Omega_1} \pi_{X \otimes Y}(x, y) = [\pi_{X \otimes Y}(x, y) \downarrow \Omega_2], \quad \text{for any}\, y \in \Omega_2$$

where $[\pi_{X \otimes Y}(x, y) \downarrow \Omega_1]$ (resp. $[\pi_{X \otimes Y}(x, y) \downarrow \Omega_2]$) denotes the projection of $\pi_{X \otimes Y}(x, y)$ on Ω_1 (resp. Ω_2). These marginal possibility distributions are sometimes written as:

$$\pi_X(x) = \pi_{X \otimes Y}(x, \Omega_2)$$
$$\pi_Y(y) = \pi_{X \otimes Y}(\Omega_1, y)$$

meaning that $\pi_X(x)$ (resp. $\pi_Y(y)$) is the possibility degree for the "partial" possibilistic variable X (resp. Y) defined on Ω_1 (resp. Ω_2) to observe the alternative $X = x$ (resp. $Y = y$) whatever the alternative taken by the "partial" possibilistic information Y (resp. X).

The joint and marginal possibility distributions concepts are illustrated in Fig. 3.2.

It is important to notice that the normal property of the joint possibility distribution (i.e., $\exists (x_0, y_0) \in \Omega_1 \times \Omega_2$ for which $\pi_{X \otimes Y}(x_0, y_0) = 1$) is preserved through the

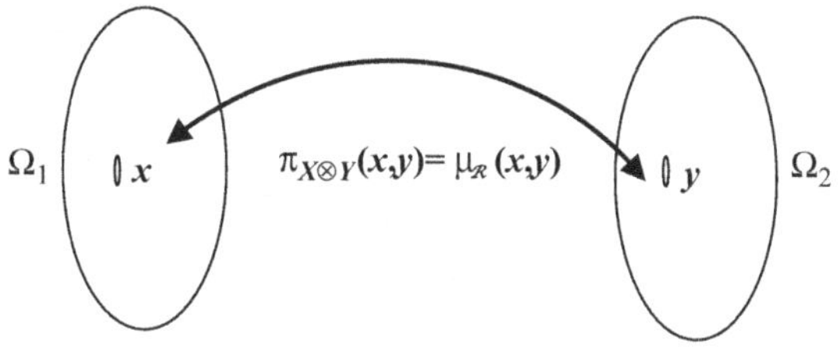

Fig. 3.3 Joint possibility distribution as a fuzzy relation

projections leading to the marginal possibility distributions (i.e., if $\pi_{X\otimes Y}$ is a normal distribution, then both π_X and π_Y are also normal possibility distributions).

Notice that the joint possibility distribution can be regarded as a fuzzy relation $\mathscr{R}$ over the Cartesian product alternatives set $\Omega_1 \times \Omega_2$ where $\pi_{X\otimes Y}(x, y) = \mu_{\mathscr{R}}(x, y)$ as illustrated in Fig. 3.3.

This vision means that the marginal possibility distributions can be considered as the "fuzzy" projection of the fuzzy relation $\mathscr{R}$ over the two alternative sets Ω_1 and Ω_2.

Marginal possibility distributions $\pi_X(x)$ and $\pi_Y(y)$ describe possibilistic information regarding the individual variables X and Y. Similarly, the joint possibility distribution $\pi_{X\otimes Y}(x, y)$ describes information regarding a "relation" between the two variables X and Y.

Assuming given the marginal possibility distributions $\pi_X(x)$ and $\pi_Y(y)$ is clearly not sufficient to determine a "unique" joint possibility distribution $\pi_{X\otimes Y}(x, y)$ that may have generated these marginal distributions. However, it is straightforward to show that the joint possibility distribution is upper bounded by the minimum value of the two generated marginal possibility distributions:

Theorem $\pi_{X\otimes Y}(x, y) \leq \; min\; (\pi_X(x), \pi_Y(y))$.

Proof For each $x \in \Omega_1 : \quad Poss[(X = x) \cap (\text{all}\, y \in \Omega_2)] = \pi_X(x)$,
and for each $y \in \Omega_2 : \quad Poss[(\text{all}\, x \in \Omega_1) \cap (Y = y)] = \pi_Y(y)$;
Therefore, for each pair $(x, y) \in \Omega_1 \times \Omega_2$,

$$\begin{aligned}
\pi_{X\otimes Y}(x,y) = Poss[(X=x)\cap(Y=y)] &= Poss[(\{x\}\times\Omega_2)\cap(\Omega_1\times\{y\})] \\
&\leq min\;(Poss[\{x\}\times\Omega_2], Poss[\Omega_1\times\{y\}]) \\
&= min\;(Poss[\{x\}], Poss[\{y\}]) \\
&= min\;\{\pi_X(x), \pi_Y(y)\}
\end{aligned}$$

Hence, $\pi_{X\otimes Y}(x, y) \leq \; min\; (\pi_X(x), \pi_Y(y))$.

Example 1 Let $\Omega_1 = \{x_1, x_2, x_3\}$ and $\Omega_2 = \{y_1, y_2, y_3, y_4\}$ and the joint possibility distribution $\pi_{X\otimes Y}(.,.)$ defined on the Cartesian product alternatives set $\Omega_1 \times \Omega_2$ be as described in Fig. 3.4a.

The marginal possibility distribution π_{X_1} (resp. π_{Y_1}) obtained by the projection of $\pi_{X\otimes Y}(.,.)$ on Ω_1 (res. Ω_2) are given by:

a

	y_1	y_2	y_3	y_4	π_{X_1}
x_3	0.2	0.2	0.8	0	**0.8**
x_2	1	0.4	0	0.1	**1**
x_1	0.7	0.5	0.1	0	**0.7**
π_{Y_1}	**1**	**0.5**	**0.8**	**0.1**	

b

	y_1	y_2	y_3	y_4	π_{X_2}
x_3	0.6	0.4	0.8	0.1	**0.8**
x_2	1	0.2	0.7	0.1	**1**
x_1	0.7	0.5	0.5	0	**0.7**
π_{Y_2}	**1**	**0.5**	**0.8**	**0.1**	

Fig. 3.4 Joint possibility distribution (s) defined on $\{x_1, x_2, x_3\} \times \{y_1, y_2, y_3, y_4\}$

$$\pi_{X_1}(x_1) = max\,(0.7, 0.5, 0.1, 0) = 0.7; \quad \pi_{Y_1}(y_1) = max(0.2, 1, 0.7) = 1;$$
$$\pi_{X_1}(x_2) = max\,(1, 0.4, 0, 0.1) = 1; \quad \pi_{Y_1}(y_2) = max\,(0.2, 0.4, 0.5) = 0.5;$$
$$\pi_{X_1}(x_3) = max\,(0.2, 0.2, 0.8, 0) = 0.8; \quad \pi_{Y_1}(y_3) = max\,(0.8, 0, 0.1) = 0.8;$$
$$\pi_{Y_1}(y_4) = max\,(0, 0.1, 0) = 0.1.$$

Therefore, $(\pi_{X_1}) = [\pi_{X_1}(x_1), \pi_{X_1}(x_2), \pi_{X_1}(x_3)] = [0.7, 1, 0.8]$ and $(\pi_{Y_1}) = [\pi_{Y_1}(y_1), \pi_{Y_1}(y_2), \pi_{Y_1}(y_3), \pi_{Y_1}(y_4)] = [1, 0.5, 0.8, 0.1]$.

Notice that the joint possibility distribution defined in Fig. 3.4b, different from the one defined in Fig. 3.4a, generates the same two marginal possibility distributions, i.e., $(\pi_{X_2}) = (\pi_{X_1})$ and $(\pi_{Y2}) = (\pi_{Y1})$.

3.3 Cylindrical Extension of Non-interactive Possibilistic Variables

Let us consider the situation where two possibilistic variables (X, π_1) and (Y, π_2) are separately defined on two finite and disjoint alternative sets $\Omega_1 = \{x_1, x_2 .. x_N\}$ and $\Omega_2 = \{y_1, y_2 .. y_M\}$:

$$\begin{array}{cccccc} \pi_1 : \Omega_1 & \rightarrow & [0, 1] & \pi_2 : \Omega_2 & \rightarrow & [0, 1] \\ x & \rightarrow & \pi_1(x) & y & \rightarrow & \pi_2(y) \end{array}$$

and that these variables do not interact, i.e., no joint relation linking both of them. The question raised at this level is:

How to "construct" a joint possibility distribution, $\pi_{X \times Y}(.,.)$, $\forall\,(x\,y) \in \Omega_1 \times \Omega_2$ in such a way that π_1 and π_2 are the marginal distributions (i.e., $\pi_X(x) = \max_{y \in \Omega_2} \pi_{X \times Y}(x, y) = \pi_1(x)$, and $\pi_X(x) = \max_{x \in \Omega_1} \pi_{X \times Y}(x, y) = \pi_2(y) \forall x \in \Omega_1$ and $\forall y \in \Omega_2$)?

As previously shown, see Example 1, Fig. 3.4, several joint possibility distributions may generate the same marginal distributions.

The basic idea of the *cylindrical extension* (i.e., a "kind" of joint possibility distribution) is to "infer" the two elementary possibilistic variables X and Y into a "common" possibilistic variable on the Cartesian product set of alternatives $\Omega = \Omega_1 \times \Omega_2$:

$$\begin{aligned} \pi_{1\times 2} : \Omega_1 \times \Omega_2 &\rightarrow [0,1] \\ (x,y) &\rightarrow \pi_{1\times 2}(x,y) = T(\pi_1(x), \pi_2(y)) \end{aligned}$$

where T denotes a t-norm operator and $\pi_{1\times 2}(x, y)$, called a *T-product possibility distribution*, represents the inferred degree of possibility of $(x, y) \in \Omega = \Omega_1 \times \Omega_2$.

In order to be as specific as the basic available possibilistic distributions π_1 and π_2, $\pi_{1\times 2}(.,.)$ should respect the two following constraints:

$$\begin{aligned} & \pi_{1\times 2}(x,y) \leq \pi_1(x) \quad \text{for all}\, x \in \Omega_1 \\ \text{and} \qquad & \pi_{1\times 2}(x,y) \leq \pi_2(y) \quad \text{for all}\, y \in \Omega_2 \end{aligned}$$

In other words, the "constructed" joint possibility distribution should be upper bounded by the minimum value of the two possibility distributions, i.e.,

$$\pi_{1\times 2}(x,y) \leq min\ (\pi_1(x), \pi_2(y))$$

Therefore, using the minimal specificity principle, the only possibility distribution that can be obtained by a reasonable inference from π_1 and π_2 is given by:

$$\pi_{1\times 2}(x,y) = min\ \{\pi_1(x), \pi_2(y)\} \quad \text{for all}\ (x,y) \in \Omega = \Omega_1 \times \Omega_2$$

$\pi_{1\times 2}(.,.)$ is called the *cylindrical extension* of π_1 and π_2.

In this case, the two possibilistic variables (X, π_1) and (Y, π_2) are said to be *separable*, and the joint possibility distribution $\pi_{1\times 2}$ expresses the minimal constraint (i.e., for each $x \in \Omega_1$ and $y \in \Omega_2$ the values $\pi_{1\times 2}(x, y)$ are the largest acceptable values and this for all joint possibility distributions having π_1 and π_2 as marginal distributions).

Notice that the cylindrical extension $\pi_{1\times 2}(.,.)$ is obtained by inference from π_1 and π_2, whereas, the joint possibility distribution, $\pi_{X\otimes Y}(.,.)$ is "directly" defined on the Cartesian product set $\Omega = \Omega_1 \times \Omega_2$. This leads to the fact that the joint possibility distribution $\pi_{X\otimes Y}(.,.)$ "includes" information about the relationship between X and Y.

Therefore, $\pi_{X\otimes Y}(.,.)$ is somehow "more informative" than the cylindrical extension $\pi_{1\times 2}(.,.)$.

It is important to notice that the cylindrical extension $\pi_{1\times 2}(.,.)$ does not account for the relational links between the two possibilistic information X and Y. Using the minimal specificity principle, we can write: $\pi_{X\otimes Y}(x, y) \leq \pi_{1\times 2}(x, y) = min\ \{\pi_1(x), \pi_2(y)\}$, for all $(x, y) \in \Omega = \Omega_1 \times \Omega_2$. Particularly, it may happen that for some (x, y) is impossible because of a link between x and y, whereas $\pi_{1\times 2}(x, y) = min\ \{\pi_1(x), \pi_2(y)\}$ is different from zero. In other words, if there is an unknown link between x and y, then the possibility distribution $\pi_{1\times 2}(x, y) = min\ \{\pi_1(x), \pi_2(y)\}$ provides upper bounds on the degrees of possibility and gives conclusions which are always correct but might be too little informative.

Remarks

- *T-independence*: In the framework of possibility theory, let (X, π_1) and (Y, π_2) denote two possibilistic variables, respectively, defined on the sets of alternatives Ω_1 and Ω_2, respectively; and (Z, π_Z) is an arbitrarily possibilistic variable defined on the Cartesian product set of alternatives $\Omega_1 \times \Omega_2$. (Z, π_Z) is called a T-product possibilistic variable if and only if, there exist a t-norm (T) for which:

$$\pi_Z(x, y) = T[\pi_1(x), \pi_2(y)] \quad \forall (x, y) \in \Omega_1 \times \Omega_2$$

  Moreover, in this case, the two possibilistic variable (X, π_1) and (Y, π_2) are called *T-independent*. Therefore, in the case of non-interactive possibilistic variables, the joint distribution constructed as the cylindrical extension of π_1 and $\pi_2 : \pi_{1\times 2}(x, y) = min\ \{\pi_1(x), \pi_2(y)\}, \quad \forall\ (x, y) \in \Omega_1 \times \Omega_2$ is, simply, the T-product of π_1 and π_2 where the minimum operator is the considered t-norm.

- *Possibilistic non-interactivity* vs. *stochastic independence*: Possibilistic non-interactivity plays a similar role as stochastic independence in probability theory with the difference that stochastic independence doesn't lead to joint probability distribution, $f_{X,\ Y}(x, y)$ upper bounding by the product of the marginal probability distributions (i.e., $f_{X,\ Y}(x, y) < f_X(x).\ f_Y(y)$ is not verified), while $\pi_{X\times Y}(x, y) = min\ (\pi_1(x), \pi_2(y))$, for all $(x, y) \in \Omega_1 \times \Omega_2$, when the two possibilistic variables (X, π_1) and (Y, π_2) are non-interactive.

  The reason for this difference is that the stochastic independence reflects the correlation absence between random variables, whereas possibilistic non-interactivity reflects the lack of knowledge concerning the relationship between possibilistic variables.

Example 2 Let $\Omega_1 = \{x_1, x_2, x_3\}$ and $\Omega_2 = \{y_1, y_2, y_3, y_4\}$ and consider the following non-interactive possibilistic variables X and Y associated with the possibility distributions:

$$\pi_1 = [\pi_1(x_1), \pi_1(x_2), \pi_1(x_3)] = [0.7, 1, 0.8]; \text{and}$$

$$\pi_2 = [\pi_2(y_1), \pi_2(y_2), \pi_2(y_3), \pi_2(y_4)] = [1, 0.5, 0.8, 0.1].$$

Using $\pi_{X\times Y}(x, y) = min\ (\pi_1(x), \pi_2(y))$, the resulting joint possibility distribution is given in Fig. 3.5.

Notice that this joint possibility distribution is different from the one given in Example 1.

Fig. 3.5 Joint possibility distribution, $\pi_{X \times Y}(x, y)$, of non-interactive possibilistic variables X and Y

$\pi_1(x)$ =[	$\pi_2(y)$ = [	1	0.5	0.8	0.1]
		y_1	y_2	y_3	y_4
0.7	x_3	0.7	0.5	0.7	0.1
1	x_2	1	0.5	0.8	0.1
0.8	x_1	0.8	0.5	0.8	0.1
]					

Now, if we consider the obtained joint distribution, the "deduced" marginal possibility distributions will be:

$$\begin{aligned} \pi_X(x_1) &= max\,(0.7, 0.5, 0.7, 0.1) = 0.7; & \pi_Y(y_1) &= 1; \\ \pi_X(x_2) &= max\,(1, 0.5, 0.8, 0.1) = 1; & \pi_Y(y_2) &= 0.5; \\ \pi_X(x_3) &= max\,(0.8, 0.5, 0.8, 0.1) = 0.8; & \pi_Y(y_3) &= 0.8; \\ & & \pi_Y(y_4) &= 0.1. \end{aligned}$$

Leading to (in this example): $\pi_X = \pi_1$ and $\pi_Y = \pi_2$.

3.3.1 Projections of a Cylindrical Extension

Using the cylindrical extension $\pi_{1\times 2}(x, y)$, the two following projection possibility distributions can be defined:

$$\pi_{1\downarrow}(x) = \max_{y \in \Omega_2} \pi_{1\times 2}(x, y), \quad \text{for any } x \in \Omega_1$$

$$\pi_{2\downarrow}(y) = \max_{x \in \Omega_1} \pi_{1\times 2}(x, y), \quad \text{for any } y \in \Omega_2$$

Notice that $\pi_{1\downarrow}(.) \neq \pi_1(.)$ and $\pi_{2\downarrow}(.) \neq \pi_2(.)$.

Example 3 Let $\Omega_1 = \{x_1, x_2, x_3\}$ and $\Omega_2 = \{y_1, y_2, y_3, y_4\}$ and the two possibility distributions π_1 and π_2 defined on the alternative sets Ω_1 and Ω_2 by:

$$\pi_1(x_1) = 0.2, \pi_1(x_2) = 0.8, \pi_1(x_3) = 1.0;$$
$$\pi_2(y_1) = 0.4, \pi_2(y_2) = 0.9, \pi_2(y_3) = 0.7, \pi_2(y_4) = 1.0;$$

The cylindrical extension $\pi_{1\times 2}(., .)$ is thus obtained as:
$\pi_{1\times 2}(x_n, y_m) = min\ \{\pi_1(x_n), \pi_2(y_m)\}$ for all $(x_n, y_m) \in \Omega = \Omega_1 \times \Omega_2$, Fig. 3.6.

The projection possibility distributions obtained from the cylindrical extension $\pi_{1\times 2}(., .)$ are given by:

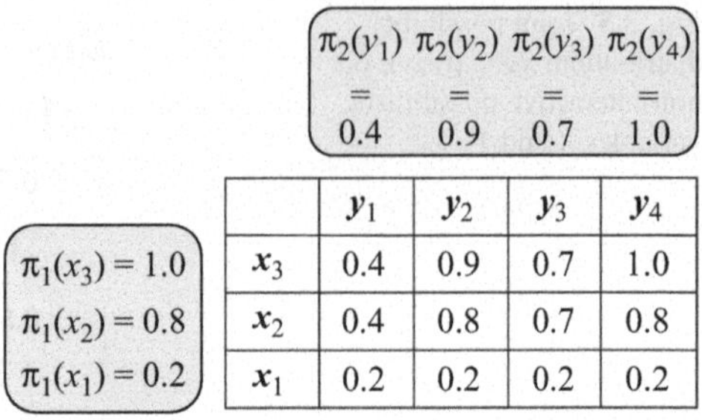

Fig. 3.6 Cylindrical extension possibility distribution

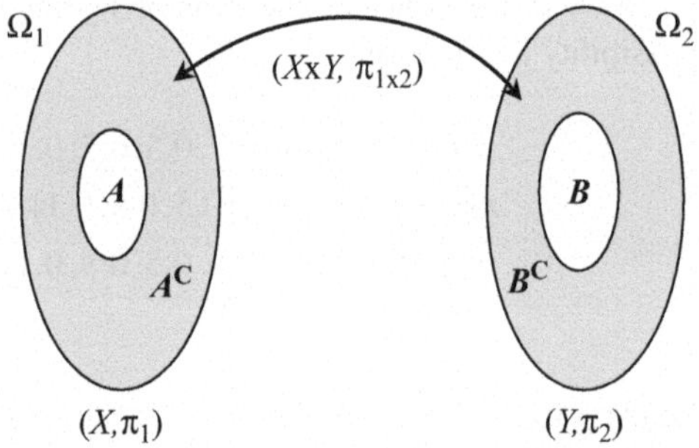

Fig. 3.7 Joint possibility and necessity measures

$$\pi_{1\downarrow}(x_1) = \max_{y_m \in \Omega_2} \pi_{1\times 2}(x_1, y_m) = 1.0,$$

$$\pi_{1\downarrow}(x_2) = \max_{y_m \in \Omega_2} \pi_{1\times 2}(x_2, y_m) = 0.8, \text{ and}$$

$$\pi_{1\downarrow}(x_3) = \max_{y_m \in \Omega_2} \pi_{1\times 2}(x_3, y_m) = 0.2.$$

Similarly:

$$\pi_{2\downarrow}(y_1) = 0.4, \pi_{2\downarrow}(y_2) = 0.9, \pi_{2\downarrow}(y_3) = 0.7 \text{ and } \pi_{2\downarrow}(y_4) = 1.0.$$

3.3.2 *Joint Possibility and Necessity Measures*

Let (X, π_1) and (Y, π_2) denote two non-interactive possibilistic variables defined on two finite alternative sets Ω_1 and Ω_2, and let $A \subseteq \Omega_1$ (resp. $B \subseteq \Omega_2$) denote a crisp subset of the X-set of alternatives: Ω_1, (resp. Y-set of alternatives: Ω_2), Fig. 3.7.

Π_1 and $\mathcal{N}_1$, resp. Π_2 and $\mathcal{N}_2$, denote the possibility and necessity measures associated with (X, π_1), resp. (Y, π_2). Finally, $\pi_{1\times 2}$ denotes the joint possibility distribution, Fig. 3.7, generated by the two non-interactive (X, π_1) and (Y, π_2), i.e.,

$$\pi_{1\times 2}(x, y) = min\,(\pi_1(x), \pi_2(y)), \quad \forall (x, y) \in \Omega_1 \times \Omega_2$$

The Cartesian product and the Cartesian coproduct crisp subsets of A and B are defined as follows:

$A \times B = \{(x, y)/x \in A \text{ and } y \in B\}$: The Cartesian product;
$A + B = \left(A^{C} \times B^{C}\right)^{C} = \Omega_1 \times \Omega_2/\left\{(x, y)/x \in A^{C} \text{ and } y \in B^{C}\right\}$: The Cartesian coproduct

The possibility and necessity measures associated with the joint possibility distribution, $\pi_{1\times 2}(.,.)$, are given by:

$$\Pi_{1\times 2}(A \times B) = \max_{(x, y)\in A\times B} \pi_{1\times 2}(x, y)$$

$$\mathcal{N}_{1\times 2}(A \times B) = 1 - \max_{(x, y)\in (A\times B)^{C}} \pi_{1\times 2}(x, y)$$

and

$$\Pi_{1\times 2}(A + B) = \max_{(x, y)\in A+B} \pi_{1\times 2}(x, y)$$

$$\mathcal{N}_{1\times 2}(A + B) = 1 - \max_{(x, y)\in (A+B)^{C}} \pi_{1\times 2}(x, y) = 1 - \max_{(x, y)\in A^{C}\times B^{C}} \pi_{1\times 2}(x, y)$$

Two *marginal possibility and necessity measures* can thus be induced:

$\Pi_X(A) = \Pi_1(A) = \Pi_{1\times 2}(A \times \Omega_2)$ $\quad$ $\Pi_Y(B) = \Pi_2(B) = \Pi_{1\times 2}(\Omega_1 \times B)$
$\mathcal{N}_X(A) = \mathcal{N}_1(A) = 1 - \Pi_{1\times 2}\left(A^{C} \times \Omega_2\right)$ $\quad$ $\mathcal{N}_Y(B) = \mathcal{N}_2(B) = 1 - \Pi_{1\times 2}\left(\Omega_1 \times B^{C}\right)$

Moreover, the following decomposability properties can be verified straightforward:

$$\Pi_{1\times 2}(A \times B) = min\,(\Pi_1(A), \Pi_2(B))$$

$$\mathcal{N}_{1\times 2}(A \times B) = min\,(\mathcal{N}_1(A), \mathcal{N}_2(B))$$

$$\Pi_{1\times 2}(A + B) = max\,(\Pi_1(A), \Pi_2(B))$$

$$\mathcal{N}_{1\times 2}(A + B) = max\,(\mathcal{N}_1(A), \mathcal{N}_2(B))$$

Example 4 Let $\Omega_1 = \{d_1, d_2, d_3, d_4\}$ and $\Omega_2 = \{s_1, s_2, s_3, s_4, s_5, s_6\}$ denote exhaustive sets of mutually exclusives products, and the joint possibility distribution $\pi_{1\times 2}(.,.)$ defined on the Cartesian product universe of discourse $\Omega_1 \times \Omega_2$ and representing the possibility that a customer may buy simultaneously (Table 3.1).

The marginal possibility distribution π_X (resp. π_Y) obtained by the projection of $\pi_{1\times 2}(.,.)$ on Ω_1 (resp. Ω_2) are given by:

Table 3.1 Joint possibility distribution $\pi_{1\times 2}(.,.)$

	s_1	s_2	s_3	s_4	s_5	s_6
d_1	0.1	0.2	0.7	1.0	0.0	0.4
d_2	0.1	0.0	0.5	0.8	0.7	0
d_3	0.5	0.4	0.4	0.2	0.3	0.3
d_4	0.4	0.0	0.8	0.6	0.1	0.6

$$\begin{aligned}
&\pi_X(d_1) = 1.0; \quad \pi_Y(s_1) = 0.5;\\
&\pi_X(d_2) = 0.8; \quad \pi_Y(s_2) = 0.4;\\
&\pi_X(d_3) = 0.5; \quad \pi_Y(s_3) = 0.8;\\
&\pi_X(d_4) = 0.8; \quad \pi_Y(s_4) = 1.0;\\
&\qquad\qquad\qquad \pi_Y(s_5) = 0.7;\\
&\qquad\qquad\qquad \pi_Y(s_6) = 0.6.
\end{aligned}$$

The possibility measure that a customer buys one of the products from $\{s_4, s_5\}$ equals to one, $\Pi_Y(\{s_4, s_5\}) = 1.0$, and, similarly:

$$\Pi_X(\{d_3, d_4\}) = 0.8, \Pi_X(\{d_1, d_2, d_3\}) = 1.0, \Pi_Y(\{s_1, s_2, s_6\}) = 0.6\ldots$$

In the case where A and B are two fuzzy sets defined on Ω_1 and Ω_2 with membership functions, respectively, given by $\mu_A(x)$ and $\mu_B(y)$, the Cartesian product of the fuzzy sets A and B, $A \times B$, is defined as:

$$\begin{aligned}
A \times B : \Omega_1 \times \Omega_2 &\rightarrow [0, 1]\\
(x, y) &\rightarrow \mu_{A\times B}(x, y) = \min\{\mu_A(x), \mu_B(y)\};
\end{aligned}$$

The Cartesian coproduct of the fuzzy sets A and B, $A + B$, is defined as:

$$\begin{aligned}
A + B : \Omega_1 \times \Omega_2 &\rightarrow [0, 1]\\
(x, y) &\rightarrow \mu_{A+B}(x, y) = 1 - \min\{\mu_A{}^C(x), \mu_B{}^C(y)\};
\end{aligned}$$

Similarly to the case of crisp sets of events, the decomposability properties (see Fig. 3.8) are defined as follows:

1. $\Pi_{1\times 2}(A \times B) = \min\{\Pi_1(A), \Pi_2(B)\}$
2. $\mathcal{N}_{1\times 2}(A \times B) = \min\{\mathcal{N}_1(A), \mathcal{N}_2(B)\}$
3. $\Pi_{1\times 2}(A + B) = \max\{\Pi_1(A), \Pi_2(B)\}$
4. $\mathcal{N}_{1\times 2}(A + B) = \max\{\mathcal{N}_1(A), \mathcal{N}_2(B)\}$

where Π_1 and $\mathcal{N}_1$ (resp. Π_2 and $\mathcal{N}_2$) are the possibility and necessity measures of matching associated with the possibility distribution π_1 on Ω_1 (resp. π_2 on Ω_2).

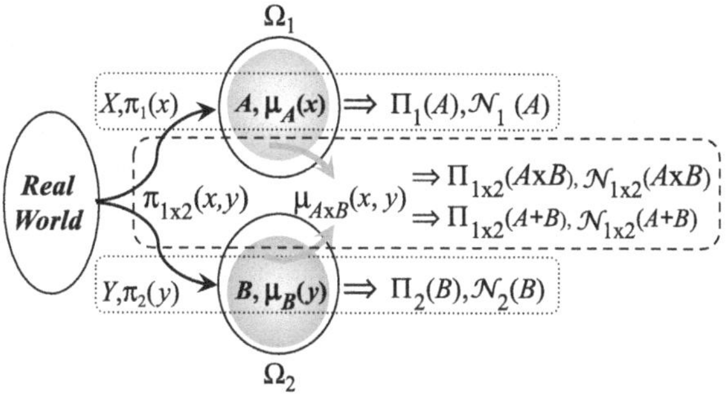

Fig. 3.8 Decomposability properties of the cylindrical possibilistic extension concept

3.4 Conditioning Under the Knowledge of the Joint Possibility Distribution

Information conditioning is a fundamental issue in all theories and approaches dealing with information processing. The aim of conditioning is to study the mutual influence and impact between alternatives or events (i.e., subsets of alternatives) from different alternative sets. Let us consider the following context:

- $\Omega_1 = \{x_1, x_2 \ldots x_N\}$ and $\Omega_2 = \{y_1, y_2 \ldots y_M\}$ denote two finite sets of alternatives.
- $\Omega = \Omega_1 \times \Omega_2$ denotes the Cartesian product of Ω_1 and Ω_2.
- $\pi_{X \otimes Y}(.,.)$ denotes a joint possibility distribution encoding the joint possibilistic information $X \otimes Y$ defined on $\Omega_1 \times \Omega_2$.
- $\pi_{X \otimes Y}(.,.)$ is assumed a priori known.

Within this context and the assumption that the two possibilistic variables (X, π_X) and (Y, π_Y) at the origin of $\pi_{X \otimes Y}$ do interact, we may ask the following question:

Under the assumption of knowing that an alternative $y \in B (B \subseteq \Omega_2)$ has occurred, can we measure the degree to which it is possible that a given alternative $x \in \Omega_1$ appears?

In other terms, our target is to somehow evaluate the interaction between the two possibilistic variables through the estimation of the possibility $\pi_{X/B}(x/B)$ of occurrence of an alternative $x \in \Omega_1$ given that $y \in B$ has occurred.

The *conditional possibility distribution* is thus defined as a mapping:

$$\begin{array}{rcl} \pi_{X/B}(./B) : \Omega_1 & \rightarrow & [0, 1] \\ x & \rightarrow & \pi_{X/B}(x/B) \end{array}$$

and the question is then how to compute this conditional possibility distribution when the only available information is the joint possibility distribution $\pi_{X \otimes Y}(.,.)$?

In general, the joint possibility distribution $\pi_{X \otimes Y}(.,.)$ can be expressed via the following equation:

$$\pi_{X\otimes Y}(x,y) = min\left\{\pi_Y(y), \pi_{X/\{y\}}(x/y)\right\}$$

Before detailing the different approaches proposed for the definition of conditioning rules, two important possibilistic concepts are, hereafter, introduced: *possibilistic independence* and *possibilistic non-interaction*.

Let $\pi_X(x)$ and $\pi_Y(y)$ denote the marginal possibility distributions and consider two crisp sets: $A \subseteq \Omega_1$ and $B \subseteq \Omega_2$.

Possibilistic independence:

- The possibilistic variable (X, π_X) is said to be *independent* of the possibilistic variable (Y, π_Y) if and only if: $\pi_{X/\{y\}}(x/y) = \pi_X(x)$ for all $x \in \Omega_1$, and all $y \in \Omega_2$.
- Similarly, the possibilistic variable (Y, π_Y) is said to be *independent* of the possibilistic variable (X, π_X) if and only if: $\pi_{Y/\{x\}}(y/x) = \pi_Y(y)$ for all $x \in \Omega_1$, and all $y \in \Omega_2$.
- The two possibilistic variables (X, π_X) and (Y, π_Y) are said to be *independent* if and only if: $\pi_{X/\{y\}}(x/y) = \pi_X(x)$ and $\pi_{Y/\{x\}}(y/x) = \pi_Y(y)$ for all $x \in \Omega_1$, and all $y \in \Omega_2$.
- Similarly, the two crisp sets A and B are said to be *independent* if and only if:

$$\pi_{X/B}(x/y) = \pi_X(x) \text{ and } \pi_{Y/A}(y/x) = \pi_Y(x) \quad \text{for all } x \in A, \text{and all } y \in B.$$

- If the two possibilistic variables (X, π_X) and (Y, π_Y) are independent, the joint possibility distributions $\pi_{X\otimes Y}$ is said to *avoid sure loss under epistemic independence* and is called *independent joint possibility distributions*.

Possibilistic non-interaction:

Two crisp sets A and B ($A \subseteq \Omega_1$ and $B \subseteq \Omega_2$) are said to have *non-interaction* if and only if:

$$\pi_{X\otimes Y}(x,y) = min\{\pi_X(x),\ \pi_Y(y)\} \quad \text{for all } x \in A, \text{ and all } y \in B$$

Non-interaction aspect between possibilistic variables has been mentioned (and used in order to generate a joint possibility distribution) without giving a specific technical meaning. Nevertheless, this technical meaning can be precisely defined as follows:

Two possibilistic variables (X, π_X) and (Y, π_Y) are said to be *non-interactive* if and only if:

$$\pi_{X\otimes Y}(x,y) = min\{\pi_X(x),\ \pi_Y(y)\} \quad \text{for all } x \in \Omega_1, \text{ and all } y \in \Omega_2$$

In possibility theory, the two notions of independence and non-interaction do not necessarily coincide. All depends on the way the conditional possibility distribution is defined. Nevertheless, assume the two independent possibilistic variables (X, π_X) and (Y, π_Y): $\pi_{X/\{y\}}(x/y) = \pi_X(x)$ and $\pi_{Y/\{x\}}(y/x) = \pi_Y(y)$ for all $x \in \Omega_1$, and all $y \in \Omega_2$, then, replacing this property in the definition of the conditional possibility distribution leads to:

$$\begin{aligned}\pi_{X\otimes Y}(x,y) &= min\left\{\pi_Y(y),\ \pi_{X/\{y\}}(x/y)\right\}\\ &= min\left\{\pi_X(x),\ \pi_{Y/\{x\}}(y/x)\right\}\\ &= min\left\{\pi_X(x),\ \pi_Y(y)\right\}\end{aligned}$$

Hence, possibilistic independence implies possibilistic non-interaction. The converse, however, is not necessarily true. Moreover, and even though the conditioning issue has been studied as a counterpart of conditional probability distributions, there is no unique meaningful definition of conditioning, unlike in probability theory. Several conditioning rules have been proposed and are presented in the following subsections.

3.4.1 Zadeh's Conditioning Rule

In his paper [1], L. Zadeh has proposed the first conditioning rule that simply consists in equating conditional degrees of possibility with the joint possibility distribution:

$$\begin{aligned}\pi_{ZA}(./y_m):\quad &\Omega_1 &\rightarrow\quad &[0,1]\\ &x_n &\rightarrow\quad &\pi_{ZA}(x_n/y_m) = \pi_{X\otimes Y}(x_n, y_m)\end{aligned}$$

for all $(x_n, y_m) \in \Omega_1 \times \Omega_2$, Fig. 3.9.

For instance, using the example given in Fig. 3.4a, we obtain:

$$\begin{aligned}\pi_{ZA}(./y_1) &= [\pi_{ZA}(x_1/y_1), \pi_{ZA}(x_2/y_1), \pi_{ZA}(x_3/y_1)] = [0.7, 1.0, 0.2]\\ \pi_{ZA}(./y_2) &= [\pi_{ZA}(x_1/y_2), \pi_{ZA}(x_2/y_2), \pi_{ZA}(x_3/y_2)] = [0.5, 0.4, 0.2]\\ \pi_{ZA}(./y_3) &= [\pi_{ZA}(x_1/y_3), \pi_{ZA}(x_2/y_3), \pi_{ZA}(x_3/y_3)] = [0.1, 0.0, 0.8]\\ \pi_{ZA}(./y_4) &= [\pi_{ZA}(x_1/y_4), \pi_{ZA}(x_2/y_4), \pi_{ZA}(x_3/y_4)] = [0.0, 0.1, 0.0]\end{aligned}$$

As can be noticed, a major "disadvantage" of Zadeh's conditioning rule is that, even when the joint possibility distribution is normal, it may produce subnormal conditional possibility distributions. This is, for instance, the case of $\pi_{ZA}(./y_2)$, $\pi_{ZA}(./y_3)$ and $\pi_{ZA}(./y_4)$.

It is straightforward to show that:

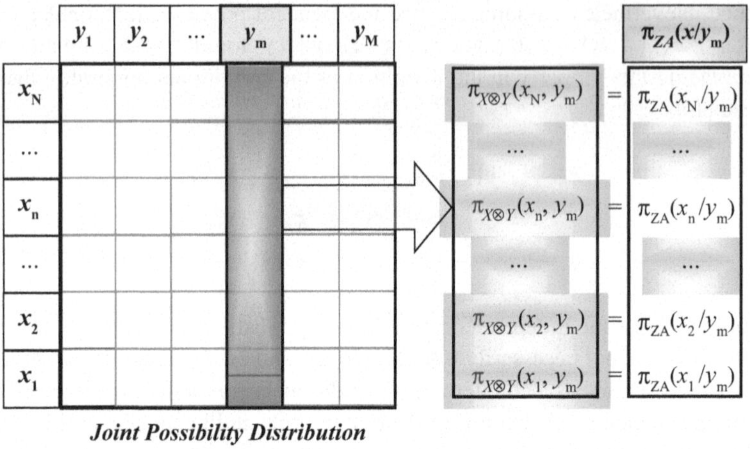

Fig. 3.9 Zadeh's conditioning rule

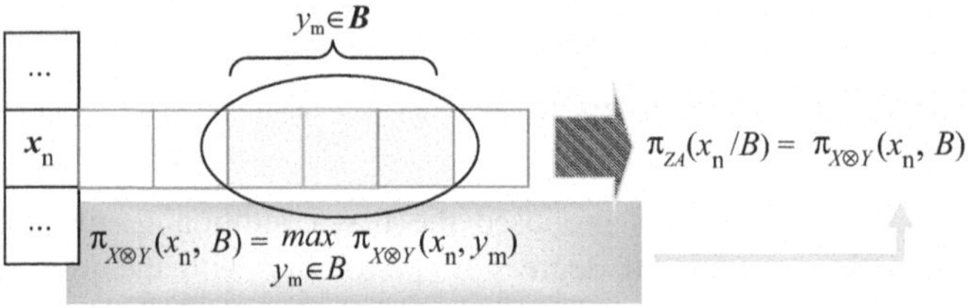

Fig. 3.10 Zadeh's conditioning rule applied on a subset of alternatives

$$\pi_X(x) = \max_{y \in \Omega_2} [\pi_{ZA}(x/y)]$$

and
$$\pi_Y(y) = \max_{x \in \Omega_1} [\pi_{ZA}(y/x)]$$

Notice that this conditional definition rule assumes that the "conditioning event" is a single alternative $y_m \in \Omega_2$. It can be simply extended to the case where the conditioning event is a subset of alternatives $B \subseteq \Omega_2$:

$$\pi_{ZA}(x_n/B) = \pi_{X\otimes Y}(x_n, B) = \max_{y \in B} \pi_{X\otimes Y}(x_n, y_m)$$

for all $x_n \in \Omega_1$, Fig. 3.10.

Using the example given in Fig. 3.4a, we can show that:

$$\begin{aligned}\pi_{ZA}(./\{y_2, y_3\}) &= [\pi_{ZA}(x_1/\{y_2, y_3\}), \pi_{ZA}(x_2/\{y_2, y_3\}), \pi_{ZA}(x_3/\{y_2, y_3\})] \\ &= [max\{0.2, 0.8\}, \; max\{0.4, 0.0\}, \; max\{0.5, 0.1\}] = [0.8, 0.4, 0.5]\end{aligned}$$

In the case where the possibilistic variables (X, π_X) and (Y, π_Y) are independent, we have:

$$\pi_{ZA}(x_n/y_m) = \pi_{X\otimes Y}(x_n, y_m) = \pi_X(x_n) \quad \text{for all } (x_n, y_m) \in \Omega_1 \times \Omega_2$$

3.4.2 *Hisdal's Conditioning Rule*

Similarly to the definition of Bayes' rule, E. Hisdal [2] has proposed to define the conditional possibility distribution as the solution of the equation:

$$\pi_{X\otimes Y}(x, y) = min\{\pi_X(y), \pi_{HI}(x/y)\}$$

for all $(x, y) \in \Omega_1 \times \Omega_2$. This equation is the natural translation of the conditional probability definition formula (i.e., $\Pr\{A \cap B\} = \Pr\{B\}. \Pr\{A/B\}$). Notice that the minimum operator plays the same role as the product operator in probability theory.

It is also important to notice that if we assume the independence of the two possibilistic variables (X, π_X) and (Y, π_Y), then Hisdal's equation is reduced to:

$$\pi_{X\otimes Y}(x, y) = min\{\pi_Y(y), \pi_X(x)\}$$

which means the equivalence between Hisdal's and Zadeh's definitions of conditioning rules. Unfortunately, Hisdal's definition does not have a unique solution (allowing to retrieve $\pi_{HI}(x/y)$). In fact, the solutions to the equation proposed by Hisdal are given by:

$$\pi_{HI}(x/y) = \begin{cases} \pi_{X\otimes Y}(x, y) & \text{if } \pi_{X\otimes Y}(x, y) < \pi_Y(y) \\ [\pi_{X\otimes Y}(x, y), 1] & \text{if } \pi_{X\otimes Y}(x, y) = \pi_Y(y) \end{cases}$$

The right hand side of $\pi_{HI}(x/y)$ indicates that a conditional possibility distribution can take any value between $\pi_{X\otimes Y}(x, y)$ and 1 when $\pi_Y(y) = \pi_{X\otimes Y}(x, y)$, see Fig. 3.11.

Therefore, conditional possibilities are "interval-valued" rather than "point-valued." Notice that Zadeh's conditioning rule (i.e., $\pi_{ZA}(x_n/y_m) = \pi_{X\otimes Y}(x_n, y_m)$) is one of the potential solutions of Hisdal's conditioning rule. The extension of this rule to the case of conditioning by the occurrence of a subset of alternatives $B \subseteq \Omega_2$ (i.e., $\pi_{HI}(x/B)$) can be conducted similarly to Zadeh's rule. This leads to:

$$\pi_{HI}(x/B) = \begin{cases} \pi_{X\otimes Y}(x, B) & \text{if } \pi_{X\otimes Y}(x, B) < \Pi_Y(B) \\ [\pi_{X\otimes Y}(x, B), 1] & \text{if } \pi_{X\otimes Y}(x, B) = \Pi_Y(B) \end{cases}$$

where, $\Pi_Y(B) = \max_{y\in B} \pi_{X\otimes Y}(x_n, y_m)$ and $\Pi_{X\otimes Y}(x, B) = \max_{y\in B} \pi_{X\otimes Y}(x_n, y_m)$.

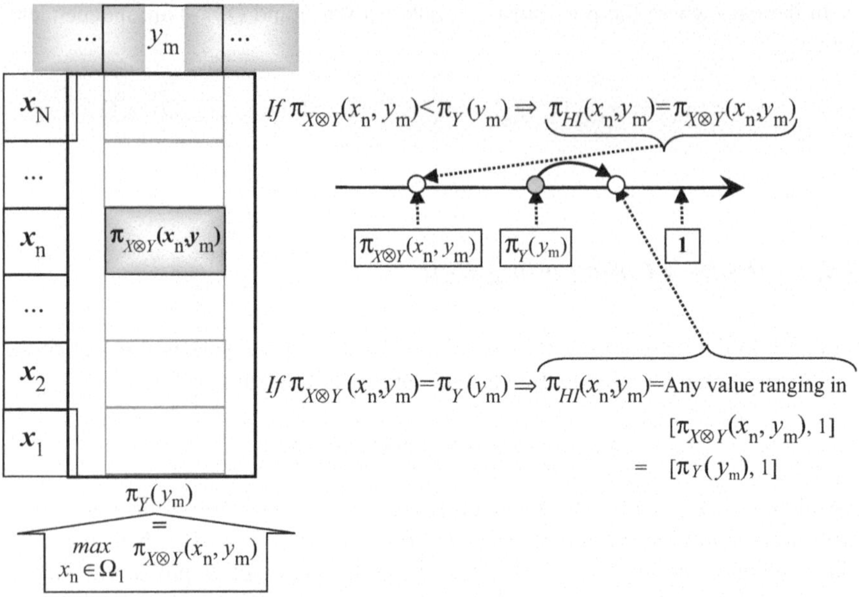

Fig. 3.11 Hisdal's conditioning rule

Table 3.2 Joint possibility distribution $\pi_{X\otimes Y}(.,.)$

	s_1	s_2	s_3	s_4	s_5	s_6
d_1	0.1	0.2	0.7	1.0	0.0	0.4
d_2	0.1	0.0	0.5	0.8	0.7	0
d_3	0.5	0.4	0.4	0.2	0.3	0.3
d_4	0.4	0.0	0.8	0.6	0.1	0.6

Example Let us reconsider the example previously given in Fig. 3.7:

- $\Omega_1 = \{d_1, d_2, d_3, d_4\}$ and $\Omega_2 = \{s_1, s_2, s_3, s_4, s_5, s_6\}$ (denoting exhaustive sets of mutually exclusives products).
- The joint possibility distribution $\pi_{X\otimes Y}(.,.)$ defined on the Cartesian product universe of discourse $\Omega_1 \times \Omega_2$ and representing the possibility that a customer may buy simultaneously (Table 3.2).

The two marginal possibility distributions π_X and π_Y obtained by the projection of $\pi_{X\otimes Y}(.,.)$ on Ω_1 and Ω_2 are given by $\pi_X(.) = [1.0, 0.8, 0.5, 0.8]$ and $\pi_Y(.) = [0.5, 0.4, 0.8, 1.0, 0.7, 0.6]$. Let us compute, according to Hisdal, the conditional possibility distribution: $\pi_{HI}(./s_3)$. The first step consists on considering the joint possibility distribution: $\pi_{X\otimes Y}(., s_3) = [0.7, 0.5, 0.4, 0.8]$ and the marginal possibility degree $\pi_Y(s_3) = 0.8$, Fig. 3.12.

Hisdal's conditional possibility distribution is thus defined as "any" possibility distribution of the form: $\pi_{HI}(./s_3) = [0.7, 0.5, 0.4]$, any value ranging in [0.8, 1]. For instance, the following possibility distributions constitute some of the solutions to Hisdal's definition:

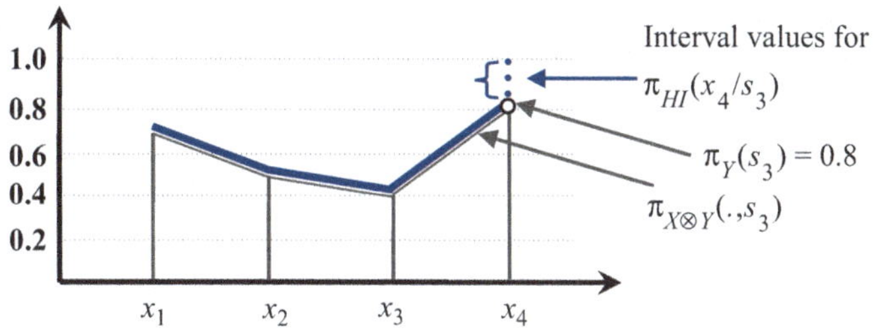

Fig. 3.12 Example of Hisdal's conditioning rule computation

- $[0.7, 0.5, 0.4, 0.8]$;
- $[0.7, 0.5, 0.4, 0.9]$;
- $[0.7, 0.5, 0.4, 1]\ldots$

Notice that the first solution represents the solution proposed by L. Zadeh:

$$\pi_{HI}(./s_3) = [0.7, 0.5, 0.4, 0.8] = \pi_{ZA}(./s_3).$$

Remark

If the conditional possibility distribution $\pi_{HI}(x/y)$ and the marginal possibility distribution $\pi_Y(y)$ are given, then substituting $\pi_{X\otimes Y}(x, y) = min\ \{\pi_Y(y), \pi_{HI}(x/y)\}$ in the definition of the marginal possibility distribution leads to the determination of $\pi_X(x)$:

$$\begin{aligned}\pi_X(x) &= \underset{y\in\Omega_2}{max}\, \pi_{X\otimes Y}(x, y)\\ &= \underset{y\in\Omega_2}{max}\left[min\ \{\pi_Y(y), \pi_{HI}(x/y)\}\right]\end{aligned}$$

Here follows some important comments on Hisdal's conditioning rule:

- Hisdal's conditioning rule exhibits a substantial difficulty since that the resulting conditional possibility distribution is not defined uniquely.
- An arbitrary solution to the equation proposed by Hisdal is not necessarily a normal possibility distribution.
- The conditional possibility distribution proposed by Zadeh's conditioning rule (i.e., $\pi_{ZA}(x/y) = \pi_{X\otimes Y}(x, y)$) is a particular solution of Hisdal's equation. It corresponds to the least specific (or, the smallest) possibility distribution of Hisdal's solutions. In other words, Hisdal's conditioning rule constitutes an "extension" of Zadeh's one.
- By Hisdal's formulation, independence between two crisp sets implies non-interaction and not the converse [2].

In order to obtain normal possibility distributions verifying Hisdal's equation, several approaches are proposed in the literature:

- Dubois and Prade [3] proposed to apply ordinal normalization to Zadeh's conditional possibility distribution (i.e., by shifting the "most possible" values to one). This will lead to a normal possibility distribution which is, also, a solution of Hisdal's equation:

$$\pi_{DP}(x/y) = \begin{cases} \pi_{X\otimes Y}(x,y) & \text{if} \quad \pi_{X\otimes Y}(x,y) < \pi_Y(y) \\ 1 & \text{if} \quad \pi_{X\otimes Y}(x,y) = \pi_Y(y) \end{cases}$$

- A. Ramer [4] has suggested to "arbitrarily" pick only one of the values of $x_0 \in \Omega_1$ for which $\pi_{X\otimes Y}(x_0, y_m) = \pi_Y(y_m)$, normalize this selected value (i.e., putting $\pi_{X\otimes Y}(x_0/y_m) = 1$), leading thus to a normal possibility distribution. This normalization method has the disadvantage of requiring an arbitrary choice whenever there is more than one x that maximizes $\pi_{X\otimes Y}(x, y_m)$.

3.4.3 *Dempster's Conditioning Rule*

Since possibility measures are upper probabilities induced by multivalued mappings [5], the rule proposed by A. Dempster [6] for conditioning upper probabilities can be applied to the problem of defining conditional possibility distributions.

In a similar way to Bayesian conditioning, Dempster considers the conditional possibility distribution as the solution of the equation $\pi_{X\otimes Y}(x, y) = \pi_y(y).\ \pi_{DE}(x/y)$. This yields, for $x \in \Omega_1$, $y \in \Omega_2$:

$$\pi_{DE}(x/y) = \begin{cases} \dfrac{\pi_{X\otimes Y}(x,y)}{\pi_Y(y)} & \text{if} \quad \pi_Y(y) > 0 \\ 0 & \text{if} \quad \pi_Y(y) = 0 \end{cases}$$

Recalling that

$$\pi_Y(y) = \max_{x\in\Omega_1} \pi_{X\otimes Y}(x,y),$$

we can say that Dempster's conditional possibility distribution is obtained by applying a numerical normalization to Zadeh's conditional possibility distribution (i.e., division of all joint possibility values by their height). The special case where all joint possibility values are nulls (since their maximum value $\pi_Y(y)$ is null) is processed by transforming the conditional possibility distribution to the total ignorance possibility distribution where the least committal value (i.e., 1) is attributed to all conditional possibilities (i.e., $\pi_{DE}(x/y) = 1$, for all $x \in \Omega_1$). The extension of this rule to the case of conditioning by the occurrence of a subset of alternatives $B \subseteq \Omega_2$ is also given by:

$$\pi_{DE}(x/B) = \begin{cases} \dfrac{\pi_{X\otimes Y}(x,B)}{\Pi_Y(B)} & \text{if} \quad \Pi_Y(B) > 0 \\ 0 & \text{if} \quad \Pi_Y(B) = 0 \end{cases}$$

3.4.4 Nguyen's Conditioning Rule

Nguyen [7] has proposed a different approach than the one proposed by Hisdal in order to define the conditional possibility distribution. He defined the "normalized conditional possibility distribution" such that, by analogy with what is found in probability theory, the possibilistic concept of non-interaction (similar to the concept of statistical independence). Let Ω_1 and Ω_2 denote two alternative sets, and $\pi_{X\otimes Y}$ the joint possibility distribution defined on the Cartesian product $\Omega = \Omega_1 \times \Omega_2 = \{(x_n, y_m), x_n \in \Omega_1, y_m \in \Omega_2\}$. We assume that $\pi_{X\otimes Y}$ is a normal distribution. Therefore, the marginal possibility distributions:

$$\pi_X(x) = \max_{y\in\Omega_2} \pi_{X\otimes Y}(x,y)$$
$$\pi_Y(y) = \max_{x\in\Omega_1} \pi_{X\otimes Y}(x,y)$$

are also normal possibility distributions.

Nguyen's conditional possibility distribution, $\pi_{NG}(x/y)$, is assumed to be expressed as:

$$\pi_{NG}(x/y) = \pi_{X\otimes Y}(x,y).\alpha(\pi_X(x), \pi_Y(y))$$

where the function $\alpha(.,.)$ is called the normalization function. This expression is "inspired" from the conditional probability definition as $\Pr\{A/B\} = \Pr\{A \cap B\}/\Pr\{B\}$.

The function $\alpha(.,.)$ is determined from the two following requirements:

$$\alpha : [0,1]\times]0,1] \quad \rightarrow \quad R^+$$

1. $\pi_{NG}(x/y) \in [0,1], \forall x \in \Omega_1, \quad \forall y \in \Omega_2$.
2. $min(\pi_X(x), \pi_Y(y)).\ \alpha(\pi_X(x), \pi_Y(y)) = \pi_X(x), \quad \forall x \in \Omega_1$.

The second requirement for the definition of $\pi_{NG}(x/y)$ corresponds to the case where $\pi_X(x)$ and $\pi_Y(y)$ are separable, i.e., the joint distribution $\pi_{X\otimes Y}(x,y) = min\,(\pi_X(x), \pi_Y(y))$ expresses the minimal constraint between $\pi_X(x)$ and $\pi_Y(y)$. In this case, $\pi_{NG}(x/y)$ should equal the projection $\pi_X(x)$. In other words, the concept of possibilistic non-interactivity plays the same role as statistical independence for probability theory.

Nguyen [7] has proved that, in the case where $\pi_Y(y) \neq 0$, the only solution for these two constraints is given by:

$$\pi_{NG}(x/y) = \begin{cases} \pi_{X\otimes Y}(x,y) & \text{if } \pi_X(x) \leq \pi_Y(y) \\ \pi_{X\otimes Y}(x,y).\dfrac{\pi_X(x)}{\pi_Y(y)} & \text{if } \pi_X(x) > \pi_Y(y) \end{cases}$$
$$= \frac{\pi_{X\otimes Y}(x,y)}{\pi_Y(y)}.\,max\,(\pi_X(x), \pi_Y(y))$$
$$= \pi_{X\otimes Y}(x,y).\,max\left(1, \frac{\pi_X(x)}{\pi_Y(y)}\right)$$

(Nguyen has left the question as to what should be done if $\pi_Y(y) = 0$ unanswered).

Similarly, we have:

$$\pi_{NG}(y/x) = \begin{cases} \pi_{X\otimes Y}(x,y) & \text{if } \pi_Y(y) \leq \pi_X(x) \\ \pi_{X\otimes Y}(x,y).\dfrac{\pi_Y(y)}{\pi_X(x)} & \text{if } \pi_Y(y) > \pi_X(x) \end{cases}$$
$$= \frac{\pi_{X\otimes Y}(x,y)}{\pi_X(x)}.\,max\,(\pi_X(x), \pi_Y(y))$$
$$= \pi_{X\otimes Y}(x,y).\,max\left(1, \frac{\pi_Y(y)}{\pi_X(x)}\right)$$

Important Properties:

1. $\pi_{NG}(y/x)$ is a normal possibility distribution.

Proof given that $\pi_{X\otimes Y}$ is a normal distribution, assume that for some $(x_0, y_0) \in \Omega_1 \times \Omega_2$, we have $\pi_{X\otimes Y}(x_0, y_0) = 1$. Therefore, the two marginal possibility distributions will verify:

$$\pi_X(x_0) = \max_{y\in\Omega_2} \pi_{X\otimes Y}(x_0, y) = 1 \text{ and}$$
$$\pi_Y(y_0) = \max_{x\in\Omega_1} \pi_{X\otimes Y}(x, y_0) = 1.$$

As a consequence, we have $\pi_{NG}(x_0/y_0) = 1$ leading to the fact that $\pi_{NG}(x/y)$ is also a normal possibility distribution.

2. Nguyen conditional possibility distributions $\pi_{NG}(x/y)$ and $\pi_{NG}(y/x)$ satisfy the following properties:

$$\pi_{X\otimes Y}(x,y) \leq \pi_{NG}(x/y) \leq \pi_X(x)$$
$$\pi_{X\otimes Y}(x,y) \leq \pi_{NG}(y/x) \leq \pi_Y(y)$$

Proof given that $\pi_{NG}(x/y) = \pi_{X\otimes Y}(x,y).\,max\left(1, \frac{\pi_X(x)}{\pi_Y(y)}\right)$ and $max\left(1, \frac{\pi_X(x)}{\pi_Y(y)}\right) \geq 1$, leads straightforward to: $\pi_{X\otimes Y}(x, y) \leq \pi_{NG}(x/y)$ (a similar reasoning gives: $\pi_{X\otimes Y}(x, y) \leq \pi_{NG}(y/x)$). Let us prove the second part of the given inequalities.

$$\text{As}\quad \pi_X(x) = \max_{y\in\Omega_2} \pi_{X\otimes Y}(x,y), \text{ and}$$
$$\pi_Y(y) = \max_{x\in\Omega_1} \pi_{X\otimes Y}(x,y)$$

This implies that $\pi_X(x) \geq \pi_{X\otimes Y}(x, y)$ and $\pi_Y(y) \geq \pi_{X\otimes Y}(x, y)$, therefore:

- if $\pi_X(x) \leq \pi_Y(y)$, we have: $\pi_{NG}(x/y) = \pi_{X\otimes Y}(x, y) \leq \pi_X(x)$, and
- if $\pi_X(x) > \pi_Y(y)$, we have:

$$\begin{aligned}
\pi_{NG}(x/y) &= \pi_{X\otimes Y}(x,y).\frac{\pi_X(x)}{\pi_Y(y)} \\
&= \frac{\pi_{X\otimes Y}(x,y)}{\pi_Y(y)}.\pi_X(x)\,\pi_X(x), \text{ and} \\
&\leq \pi_X(x) \quad (\text{since } \pi_Y(y) \geq \pi_{X\otimes Y}(x,y)).
\end{aligned}$$

This means that in all cases, we have: $\pi_{NG}(x/y) \leq \pi_X(x)$ (a similar reasoning will lead to $\pi_{NG}(y/x) \leq \pi_Y(y)$).

3. In the case of Nguyen conditional possibility distributions, the following relation:

$$\pi_{NG}(x/y).\pi_Y(y) = \pi_{NG}(y/x).\pi_X(x)$$

is satisfied (where "." denotes the mathematical product).

Proof given that:

- $\pi_{NG}(x/y) = \pi_{X\otimes Y}(x,y).\,max\left(1, \frac{\pi_X(x)}{\pi_Y(y)}\right)$ and
- $\pi_{NG}(y/x) = \pi_{X\otimes Y}(x,y).\,max\left(1, \frac{\pi_Y(y)}{\pi_X(x)}\right)$,

$$\begin{aligned}
\Rightarrow \quad \pi_{NG}(y/x) &= \pi_{NG}(x/y).\frac{1}{max\left(1, \frac{\pi_X(x)}{\pi_Y(y)}\right)}.max\left(1, \frac{\pi_Y(y)}{\pi_X(x)}\right) \\
&= \pi_{NG}(x/y)\frac{\pi_Y(y)}{\pi_X(x)}
\end{aligned}$$
$$\Rightarrow \pi_{NG}(x/y).\pi_Y(y) = \pi_{NG}(y/x).\pi_X(x)$$

4. In the case of Nguyen conditional possibility distributions, the following properties hold:

$$\pi_X(x) = \max_{y \in \Omega_2} \pi_{NG}(x/y)$$

$$\pi_y(y) = \max_{x \in \Omega_1} \pi_{NG}(y/x)$$

Proof given the property: $\pi_{X\otimes Y}(x, y) \leq \pi_{NG}(x/y) \leq \pi_X(x)$, we can write:

$$\max_{y \in \Omega_2} \pi_{X\otimes Y}(x, y) \leq \max_{y \in \Omega_2} \pi_{NG}(x/y) \leq \pi_X(x) = \max_{x \in \Omega_1} \pi_X(x)$$

but as $\pi_X(x) = \max_{y \in \Omega_2} \pi_{X\otimes Y}(x, y)$, we obtain: $\pi_X(x) \leq \max_{y \in \Omega_2} \pi_{NG}(x/y) \leq \pi_X(x)$, thus,

$$\pi_X(x) = \max_{y \in \Omega_2} \pi_{NG}(x/y)$$

Similarly, we have $\pi_Y(y) = \max_{x \in \Omega_1} \pi_{NG}(y/x)$.

3.4.5 *Causal Link Conditioning*

One of the major visions of the conditioning concept is the deductive vision where two possibilistic variables are linked through a causal link (i.e., implication) relation.

Consider two possibilistic variables (X, π_1) and (Y, π_2) defined on two finite and disjoint sets of alternatives: $\Omega_1 = \{x_1, x_2 \ldots x_N\}$ and $\Omega_2 = \{y_1, y_2 \ldots y_M\}$.

The basic idea of the causal link conditioning vision can be formulated as follows: let (X, π_1) and (Y, π_2) denote two possibilistic variables defined, respectively, on Ω_1 and Ω_2 (where the possibility distributions π_1 and π_2 are assumed to be known) and that there exists a causal link from X to Y expressed as an implication fuzzy rule of the form, Fig. 3.13:

$$\text{if } (X \,\text{is}\, A) \text{ then } (Y \,\text{is}\, B)$$

where A and/or B are fuzzy sets such that their membership functions are numerically identical to their possibility degrees:

$$\forall x \in \Omega_1, \forall y \in \Omega_2 : \mu_A(x) = \pi_1(x) \text{ and } \mu_B(y) = \pi_2(y)$$

Under this causal link vision, the conditional possibilistic distribution:

Fig. 3.13 Causal link conditioning

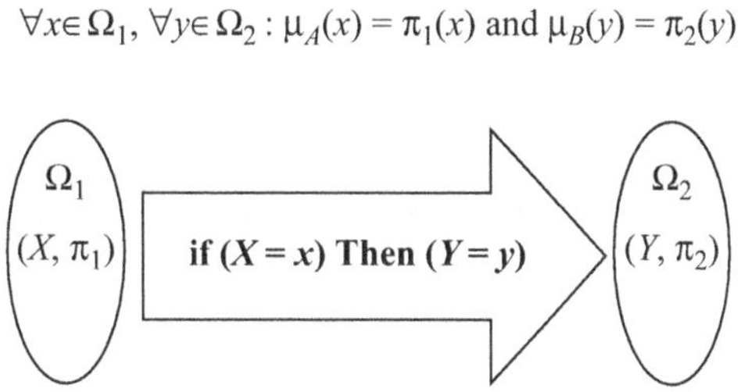

$$\begin{aligned} \pi_{Y/X} : \Omega_2 &\rightarrow [0, 1] \\ y &\rightarrow \pi_{Y/X}(y/x) \end{aligned}$$

and this for all $x \in \Omega_1$, is considered as the restriction of the possible values of Y for a given value of X: $\pi_{Y/X} = \mu_{A \rightarrow B}$ (i.e., $\pi_{Y/X}$ is the membership function of $A \rightarrow B$ where $\rightarrow$ denotes a fuzzy implication operator). Several ways may be used to define implication operator types. For instance, following [8]:

$$\pi_{Y/X}(y/x) = T^*(\pi_1(x), \pi_2(y))$$

where $T^*(\pi_1(x), \pi_2(y)) = Sup\{a \in [0, 1], T(\pi_1(x), a) \leq \pi_2(y)\}$, and $T(\pi_1(x), a)$ is a t-norm.

The obtained conditional possibility distribution depends on the considered t-norm:

- *Godel-Browver conditioning*: $T(u, v) = min\ (u, v)$

$$\forall x \in \Omega_1, y \in \Omega_2 : \pi_{Y/X}(y/x) = \begin{cases} 1 & \text{If } \pi_1(x) \leq \pi_2(y) \\ \pi_2(y) & \text{If } \pi_1(x) > \pi_2(y) \end{cases}$$

- *Goguen-Gaines conditioning*: $T(u, v) = u.\ v$

$$\forall x \in \Omega_1, y \in \Omega_2 : \pi_{Y/X}(y/x) = \begin{cases} 1 & \text{If } \pi_1(x) = 0 \\ min\left(1, \dfrac{\pi_2(y)}{\pi_1(x)}\right) & \text{If } \pi_1(x) \neq 0 \end{cases}$$

- *Lukasiewiez conditioning*: $T(u, v) = max\ (0, u + v - 1)$

$$\forall x \in \Omega_1, y \in \Omega_2 : \pi_{Y/X}(y/x) = min\ (1, 1 - \pi_1(x) + \pi_2(y))$$

- *Additive generator conditioning*: $T(u,v) = f^{(-1)}(f(u) + f(v))$

Where $f: [0,1] \to [0,\infty[$ is a continuous, strictly decreasing function such that $f(1) = 0, f(0) < 1$, with,

$$f^{(-1)}(u) = \begin{cases} 1 & \text{If } u \leq 0; \\ f^{(-1)}(u) & \text{If } 0 \leq u \leq f(0) \\ 0 & \text{If } u > f(0) \end{cases}$$

In this case:

$$\forall x \in \Omega_1, y \in \Omega_2: \quad \pi_{Y/X}(y/x) = f^{(-1)}[f(\pi_2(y)) - f(\pi_1(x))]$$

Moreover, B. Bouchon [8] has shown the following important property:

Property If $\pi_{Y/X}(y/x) = T^*(\pi_1(x), \pi_2(y))$, then the joint possibility distribution of the possibilistic variable (X, Y) is given by: $\pi_{(X,Y)}(x,y) = \min(\pi_1(x), \pi_2(y))$.

Proof Recall that $T^*(\pi_1(x), \pi_2(y)) = Sup\{a \in [0,1], T(\pi_1(x), a) \leq \pi_2(y)\}$ where $T(.,.)$ is a *t*-norm. Given the well-known property of *t*-norms:

$T(u,v) \leq \min(u,v)$, then, $\forall a \in [0,1]$ we have $T(\pi_1(x), a) \leq \pi_2(y)$, the case where $\pi_1(x) \leq \pi_2(y)$ gives $T(\pi_1(x), 1) \leq \pi_2(y)$ and then $\pi_{Y/X}(y/x) = 1$ gives $T(\pi_{Y/X}(y/x), \pi_1(x)) = \pi_1(x)$.

3.5 Conditioning and Belief Revision

Conditional possibility distributions have an important place in all information processing approaches: a pathology decision conditioned by the observation of a given symptom, a pattern recognition conditioned by a given observed feature, etc. Belief revision is the process that consists in modifying a set of initial beliefs in order to integrate a new piece of information. In the framework of possibility theory, the term "belief" refers to the knowledge encoded as possibility distributions.

3.5.1 Crisp Event-Based Possibilistic Revision

Let us consider a normal possibilistic variable (X, π_X) defined on the set of alternatives Ω, and let B be a crisp event, $B \subseteq \Omega$, representing a new *sure* piece of information, Fig. 3.14, which is assumed not contradictive with our initial knowledge (i.e., $\Pi_X(B) > 0$, or equivalently, $Supp(B) \cap Supp(\pi_X) \neq \Phi$).

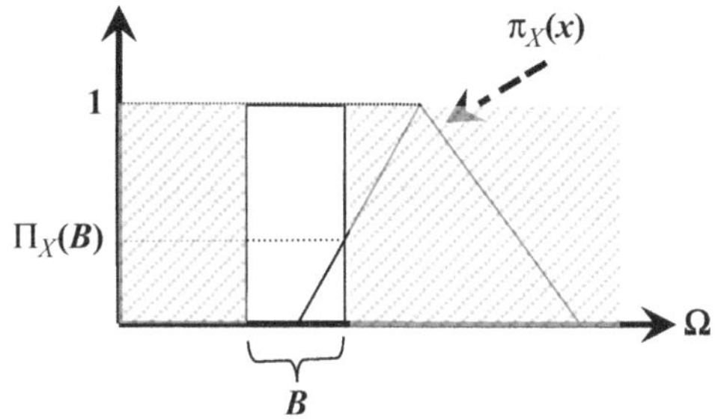

Fig. 3.14 Conditioning by a sure and not contradictive piece of information B

The question arising naturally is how should we "modify" the initial possibility distribution π_X consequently to this new, sure and not contradictive piece of information $B \in \Omega$?

Some "*natural*" properties for defining the conditional possibility distribution $\pi'(.) = \pi_X(./B)$, also called the revised possibility distributions, should be verified:

A1. $\pi'(.)$ should be normalized (i.e., $max_{x \in \Omega}\pi'(x) = 1$).

A2. $\forall x \notin B : \pi'(x) = 0$. This property is assumed to confirm the fact that B is a sure piece of information. Thus, impossible alternatives should remain impossible after conditioning.

A3. $\forall x_1, x_2 \in B : \pi_X(x_1) > \pi_X(x_2)$ implies $\pi'(x_1) > \pi'(x_2)$. This means that the conditional possibility distribution should not alter the relative order between original degrees of possibility.

A4. If $\mathcal{N}_X(B) > 0$ then: $\forall x \in B : \pi'(x) = \pi_X(x).\mathcal{N}(B) > 0$ implies that $\Pi_X(B) = 1$ (since π_X is considered as a normal possibility distribution). In other words, this property means that B contains some fully possible alternatives, thus, the possibilistic knowledge "revision" due to conditioning should not alter the original degrees of possibility.

Unfortunately, these properties do not guarantee to obtain a unique conditional possibility distribution and may lead to subnormal conditional possibility distributions (if all fully possible alternatives are outside B, due to A2).

The first step for the possibilistic belief revision process consists on encoding the piece of information B as an imprecise possibilistic variable (Y, π_Y) where $\pi_Y(x) = 1$ if $x \in B$ and $\pi_Y(x) = 0$ if $x \notin B$ (π_Y is defined on the same set of alternatives Ω).

Considering the two possibilistic variables (X, π_X) and (Y, π_Y) as non-interactive, the second step consists on using the cylindrical extension concept in order to generate the joint possibility distribution $\pi_{X\times Y}$:

$$\begin{aligned} \pi_{X\times Y}(x,x) &= min\,(\pi_X(x), \pi_Y(x)) &= \pi_X(x) \quad &\text{if}\, x \in B \\ & &= 0 \quad &\text{if}\, x \notin B \end{aligned}$$

At this level, the application of different conditioning rules can be conducted in order to obtain the revised possibility distribution: $\pi'(x)$, Fig. 3.14.

- *Zadeh's conditioning rule:*

  The application of Zadeh's conditioning rule, i.e.,

$$\begin{aligned} \pi'(x) = \pi_{ZA}(x/B) &= \pi_{X \times Y}(x, x) \\ &= \pi_X(x) \qquad \text{if}\, x \in B \\ & \quad\; 0 \qquad\qquad \text{if}\, x \notin B \end{aligned}$$

  Notice that, unless $\Pi_X(B) = 1$, the resulting revised possibility distribution π' is subnormal.

- *Hisdal's conditioning rule*:

  As previously mentioned, the application of Hisdal's conditioning rule doesn't lead to a unique revised possibility distribution. The "set" of potential revised possibility distributions is given by:

$$\pi'(x) = \pi_{HI}(x/B) = \begin{cases} \pi_X(x) & \text{if} \quad \pi_X(x) < \Pi_Y(B) \\ [\pi_X(x), 1] & \text{if} \quad \pi_X(x) = \Pi_Y(B) \end{cases}$$

  Similarly to Zadeh's conditioning rule, unless $\Pi_X(B) = 1$, the resulting Hisdal's revised possibility distribution is subnormal.

  A normal solution, called the *min-based conditioning*, was proposed by [9]. It consists on choosing the least specific Hisdal's revised possibility distribution (i.e., assigning the possibility degree 1 only to the alternatives of highest possibility degrees):

$$\pi'(x) = \pi_D(x/B) = \begin{cases} 1 & \text{if} \quad x \in B, \pi_X(x) = \Pi_Y(B) \\ \pi_X(x) & \text{if} \quad x \in B, \pi_X(x) < \Pi_Y(B) \\ 0 & \text{if} \quad x \notin B \end{cases}$$

  Notice that this solution is identical to the ordinal-normalized Zadeh's revised possibility distribution.

- *Dempster's Conditioning Rule:*

  The application of Dempster's conditioning rule, i.e.,

$$\pi'(x) = \pi_{DE}(x/B) = \begin{cases} \dfrac{\pi_X(x)}{\Pi_Y(B)} & \text{if} \quad x \in B \\ 0 & \text{if} \quad x \notin B \end{cases}$$

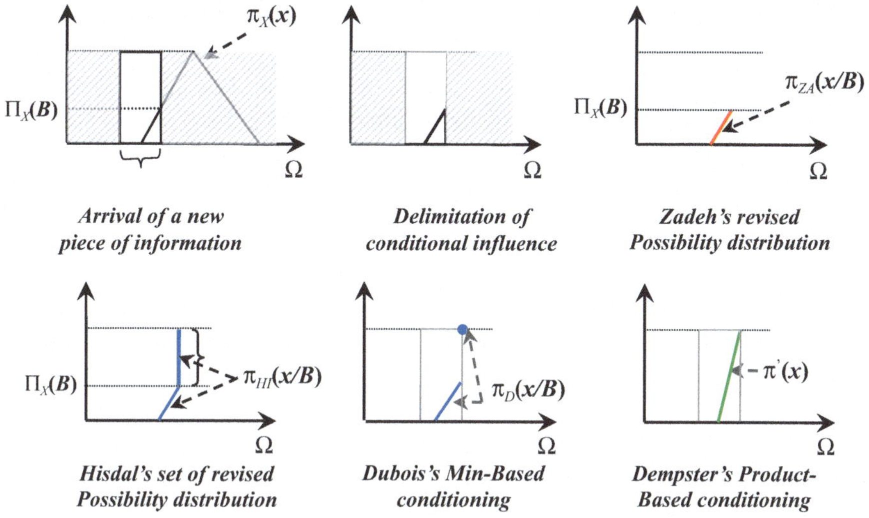

Fig. 3.15 Conditioning by the arrival of a new, sure and not contractive piece of information

(called the *product-based conditioning* rule) produces a normal revised possibility distribution. It can also be considered as the rule allowing to normalize Zadeh's revised possibility distribution by rescaling all possibility degrees upward to normality. Notice that this conditioning rule satisfies the property:

$$\pi_X(x) = \pi_{DE}(x/B).\Pi_Y(B)$$

which is similar to Bayesian conditioning.

These conditioning rules are illustrated in Fig. 3.15.

It is worthwhile to notice that in the case where the new *sure* piece of information B is such that $\Pi_X(B) = 1$, i.e., B is consistent with the possibilistic encoded knowledge π. In this case, the possibilistic revision (using different conditioning rules) is reduced to limiting the initial possibility distribution into the support of the information B and normalization has no reason to be conducted. This leads to consider:

$$\begin{aligned} \pi'(x) = min\,(\pi_X(x), \pi_Y(x)) \quad &= \quad \pi_X(x) \quad \text{if}\, x \in B \\ &= \quad 0 \qquad\quad \text{if}\, x \notin B \end{aligned}$$

Similarly, if $\mathcal{N}_X(B) = 1$ (i.e., $\Pi_X(B^C) = 0$), then no possibilistic revision is required and $\pi'(x) = \pi_X(x)$ for all $x \in \Omega$.

3.5.2 *Unreliable Crisp Event-Based Possibilistic Revision*

Let us consider a normal possibilistic variable (X, π_X) defined on the set of alternatives Ω and assume that a new arriving information is expressed as an unreliable crisp subset $(B \subseteq \Omega, \alpha)$ associated with an unreliability factor $\alpha \in [0, 1]$. The question arising is how should we revise the initial possibility distribution π_X consequently to this new, an unreliable subset $(B \subseteq \Omega, \alpha)$?

Dubois et al. [10] have suggested two approaches to consider the unreliability factor α. The first approach consists on considering α as a constraint that the resulting revised possibility distribution π' should satisfy. The second approach consists on considering α as the degree of certainty that B is the actual input information supplied by the external source of information.

3.5.2.1 Unreliability as a Constraint

This approach, allowing to deal with unreliability, consists on expressing the unreliability factor as a constraint given by $\mathcal{N}'(B) = \alpha$ on the targeted revised possibility distribution $\pi'(.) = \pi_X(./(B, \alpha))$. In other words, the resulting revised possibility distribution π' should produce:

- $\Pi'(B) = 1$ and
- $\Pi'(B^C) = 1 - \alpha$.

Dubois et al. [10] have suggested the following possibilistic revision rule:

$$\pi'(x) = \pi_{D\&P}(x/(B, \alpha)) = max\left[\pi(x/B), (1 - \alpha) * \pi\left(x/B^C\right)\right]$$

where "$*$" refers to the minimum or the product operator according as $\pi(./B)$ is the ordinal (i.e., Dubois' conditioning rule) or the scaling Dempster's product-based conditioning rule. $\pi(./B)$ and $\pi(./B^C)$ refer to one of the previously used conditioning rules in the case of new *sure* piece of information.

As illustrated in Fig. 3.16 (using the minimum operator and Dubois' conditioning rule), this approach conducts two possibilistic revisions in parallel.

A first revision, i.e., $\pi(x/B)$, is conducted using the event B as if it was a new and reliable piece of information.

The second revision, i.e., $\pi(x/B^C)$, is conducted using the complementary event B^C followed by a denormalization step (i.e., $(1 - \alpha) * \pi(x/B^C)$) so as to satisfy the constraint: $\Pi'(B^C) = 1 - \alpha$ for $\pi_{D\&P}(x/(B, \alpha))$.

It is worthwhile to notice that:

- If $\alpha = 1$, it is straightforward to show that $\pi_{D\&P}(x/(B, \alpha = 1)) = \pi_{D\&P}(x/B)$, that is the revision by conditioning is obtained
- If $\alpha = 0$, then, the resulting revised possibility distribution is less specific than π_X, such that: $\mathcal{N}'(B) = \mathcal{N}'\left(B^C\right) = 0$

Finally, the corresponding revised possibility distribution is given by:

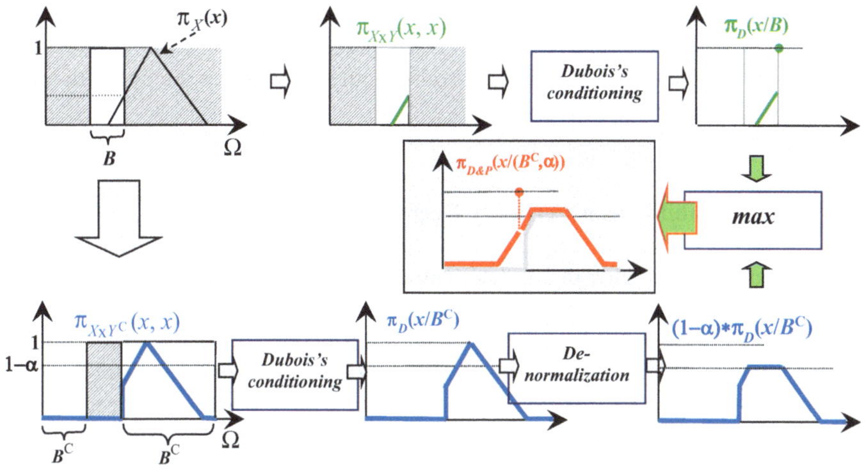

Fig. 3.16 Conditioning by the arrival of an unreliable new piece of information

$$\pi'(x) = \pi_{D\&P}(x/(B,\alpha)) = \begin{cases} \pi_X(x/B) & \text{if } x \in B \\ (1-\alpha) * \pi_X\left(x/B^C\right) & \text{if } x \in B^C \end{cases}$$

where $*$ is either the minimum or the product operator:

- The ***min-based conditioning rule*** ($*$ min operator):

$$\pi'(x) = \pi_{D\&P}(x/(B,\alpha)) = \begin{cases} 1 & \text{if } x \in B, \pi(x) = \Pi(B) \\ \pi(x) & \text{if } x \in B, \pi(x) < \Pi(B) \\ (1-\alpha) & \text{if } x \in B^C, \left(\pi(x) = \Pi\left(B^C\right) \text{ or } \pi(x) > 1-\alpha\right) \\ \pi(x) & \text{if } x \in B^C, \left(\pi(x) \neq \Pi\left(B^C\right) \text{ and } \pi(x) \leq 1-\alpha\right) \end{cases}$$

If $\mathcal{N}_X(B) \leq \alpha$, then all alternatives originally more plausible than $(1-\alpha)$ are forced to the conditional possibility degree level $(1-\alpha)$. This means that some strict ordering between alternatives from B^C may be lost.

If $\mathcal{N}_X(B) > \alpha$, then the best alternatives (i.e., for which $\pi_X(x) = \Pi'(B^C)$ or $\pi_X(x) > \alpha$) are raised to level $(1-\alpha)$.

- The ***product-based conditioning rule*** ($*$ Product operator):

$$\pi'(x) = \pi_{D\&P}(x/(B,\alpha)) = \begin{cases} \dfrac{\pi_X(x)}{\Pi_X(B)} & \text{if } x \in B \\ \dfrac{(1-\alpha)\,\pi_X(x)}{\Pi_X\left(B^C\right)} & \text{if } x \in B^C \end{cases}$$

If $\mathcal{N}_X(B) \leq \alpha$, then all plausibility levels are proportionally shifted down to the level $(1 - \alpha)$.

If $\mathcal{N}_X(B) > \alpha$, then the plausibility levels of other alternatives are proportionally shifted up (to level $1 - \alpha$).

3.5.2.2 Unreliability as a Certainty Degree

The second approach allowing to deal with a new but unreliable crisp information $(B \subseteq \Omega, \alpha)$ was proposed by *Dubois* et al. [10]. It consists on considering α as the degree of certainty that B is the actual input information supplied by the external source of information. In other words, $\mathcal{N}'(B) = \alpha$ is interpreted as an "extra piece of information" that may be useful, or not, to revise the current possibility distribution.

The unreliable information $(B \subseteq \Omega, \alpha)$ is, therefore, represented by the weighted nested pairs of subsets:

$$F = \{(B, 1), (\Omega, 1 - \alpha)\}$$

where the weights denote the degrees of possibility. The revised possibility distribution is, thus, defined as being the resulting disjunctive possibilistic fusion between both sets B and Ω:

$$\pi'(x) = \pi_{D\&P}(x/(B, \alpha)) = max\left[\pi(x/B), (1 - \alpha) * \pi_X(x)\right]$$

This means that $\pi(x/B)$ obtains the possibility 1 and π_X is kept with possibility $(1 - \alpha)$.

3.6 Conditioning and Possibilistic Medical Diagnosis

Medical diagnosis is the complex and challenging issue of determining a person's pathological statue and inferring diagnostic diseases from an available set of observations and findings. The physician (who has his/her own personal experience, knowledge, and mental endowment) is faced with a patient's signs, symptoms, medical history, laboratory findings, and physical examination and has to make a medical decision. The expected output of the medical diagnostic system is an "estimated degree" of presence of each potential disease. Such result may be expressed by a set of either "Yes/No" binary assertions, or, by a continuously graded ones (denoting disease "severity" from zero to, say, 1 or 100%). The major difficulty in medical diagnosis is due to the fact that:

- diagnosis is based on "indirect" evidences, the presence of symptoms, and the knowledge of medical reasoning mechanisms that relate presumed causes to observed effects;

- medical cause-effect relationships between diagnoses and their symptoms are hardly ever one-to-one (i.e., diagnoses share an overlapping range of symptoms);
- required observations can often not be made on continuous basis.

From an engineering point of view, the problem of medical diagnosis can be formalized as follows:

Let

1. D $=\{d_1, d_2, \ldots, d_M\}$ be an exhaustive set of M mutually exclusive possible diagnoses in the context of a certain medical problem (D is called the *set of diagnoses*).
2. S $=\{s_1, s_2, \ldots, s_N\}$ be an exhaustive set of N features, e.g., symptoms, clinical observations, measurements, findings, physiological parameters, test results, signal/image parameters or descriptors.... used for the description of an object (e.g., a patient, a suspicious region in a medical image...). S is called the *set of symptoms*.
3. Each symptom, s_n, $n = 1\ldots N$, is defined over a finite definition domain of L_n elements called feature's "n" modalities: $Dom_n = \{x_n(1), x_n(2)\ldots, x_n(L_n)\}$. The term symptom is used here in a generic sense.

An observed patient is, thus, described in the form of an N-dimensional observation vector

$$X = [f_1, f_2 \ldots, f_N] \in Dom_1 \text{x} \ldots Dom_N \text{x}.$$

Each symptom-modality pair: $(s_n, f_n \in Dom_n)$, i.e., each component of X, encodes the observed modality of the considered symptom and corresponds to a piece patient information, which we refer to as the symptom observed modality.

Moreover, in the context of a certain medical problem, several sources of knowledge may be used by the diagnostic process, Fig. 3.17:

- *Medical prior knowledge*: this knowledge source mainly concerns the available symptoms-diseases relationships.
- Patient-observed symptoms concerning the observations injected the physician.
- *Medical case base* formed by previous experience and encountered cases. This base is generally composed of a set of K diagnosed patients $\mathbb{P} = \{p_1, p_2 \ldots, p_K\}$, where each patient "$p$" is represented by its own symptoms observation vector X_p as well as the associated medical diagnose.

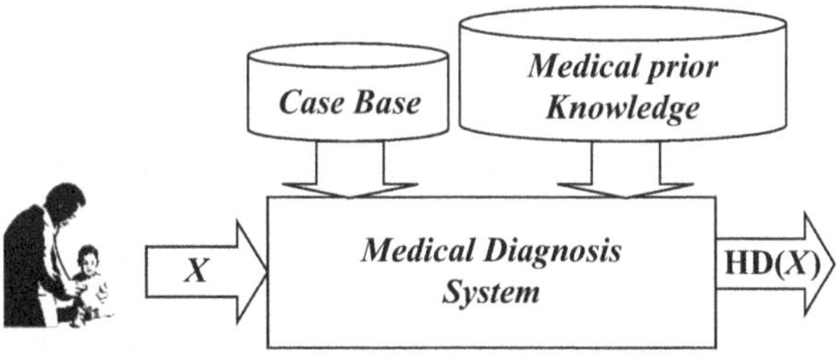

Fig. 3.17 Medical diagnosis system

The knowledge provided by these sources carries with it varying degrees of imperfections: uncertainty, ambiguity, errors, imprecision, incompleteness, missing data... For instance, the symptom s ≡ Patient's temperature may have an infinite set of modalities ([35 °C, 41 °C]), two modality values (No-Fever, Fever), etc. Faced with these imperfections concerning the observed symptoms and the relation of the symptoms to disease entities, it is nevertheless crucial for the physician to infer, from the symptoms displayed by the patient, a diagnosis label that will entail the appropriate therapeutic regimen.

The main target of medical diagnosis systems is to build a model which can accurately classify an undiagnosed patient into the diagnostic classes (i.e., diseases). A *crisp medical diagnosis* system is any mapping:

$$\begin{aligned} \mathrm{HD}: \quad & Dom_1 \mathrm{x} \ldots Dom_N \mathrm{x} \rightarrow \mathrm{D} \\ & X \rightarrow \mathrm{HD}(X) \in \mathrm{D} \end{aligned}$$

that is, for every "patient" X, the crisp medical diagnosis system specifies a single label which is interpreted as the associated diagnosis. *Possibilistic diagnosis* is characterized by the fact that the decision system relies on possibilistic information for solving the diagnostic problem. Possibilistic information can be encountered at different levels of design:

- *Possibilistic inputs*: instead of the original input values of observations and measurements, their possibility degrees, or linguistic variables represented as fuzzy constraints, can be the only available.
- *Possibilistic diagnoses*: instead of delivering a hard diagnosis, the system, for an observed object X, may produce a possibility degree to each possible diagnosis, i.e., producing a possibility distribution $\pi_X : \mathrm{D} \rightarrow [0, 1]$ (where $\pi_X(d_m)$ denotes the possibility degree that the observed object X is associated with the disease d_m, $m = 1, \ldots, M$).
- *Possibilistic reasoning*: the implementation of the reasoning mechanism is based on possibilistic reasoning concepts (e.g., inference, etc.).

References

1. L.A. Zadeh, Fuzzy sets as a basis for a theory of possibility. Fuzzy Sets Syst. **1**, 3–28 (1978)
2. E. Hisdal, Conditional possibilities independence and noninteraction. Fuzzy Sets Syst. **1**, 283–297 (1978)
3. D. Dubois, H. Prade, *Possibility Theory* (Plenum Press, New York and London, 2012)
4. A. Ramer, Conditional possibility measures. Cybern. Syst. **20**, 233–247 (1989)
5. D. Dubois, H. Prade, When upper probabilities are possibility measures. Fuzzy Sets Syst. **49**, 65–74 (1992)
6. A.P. Dempster, Upper and lower probabilities induced by a multivalued mapping, in *Classic Works of the Dempster-Shafer Theory of Belief Functions*, (Springer, Berlin, Heidelberg, 2008), pp. 57–72

7. H.T. Nguyen, On conditional possibility distributions. Fuzzy Sets Syst. **1**, 299–309 (1978)
8. B. Bouchon, Fuzzy inferences and conditional possibility distributions, in *Readings in Fuzzy Sets for Intelligent Systems*, (Elsevier, 1993), pp. 560–564
9. M.L.N. McAllister, Possibility theory: An approach to computerized processing of uncertainty (Didier Dubois and Henri Prade with the collaboration of Henri Farreny, Roger Martin-Clouaire, and Claudette Testemale; EF Handing, trans.). SIAM Rev. **34**, 147–148 (1992)
10. D. Dubois, H. Prade, A synthetic view of belief revision with uncertain inputs in the framework of possibility theory. Int. J. Approx. Reason. **17**, 295–324 (1997)

Chapter 4
Possibilistic Similarity Measures

4.1 Introduction

The concept of similarity plays an essential role in a wide range of application fields like pattern recognition, reasoning, data, and knowledge mining. Nevertheless, the formulation of a valid and general-purpose definition of the similarity concept cannot be easily and simply expressed by a formula and remains a challenging issue. N. Goodman [1] has even accused similarity, emphasizing its flexibility, its multi-aspect nature, and the absence of a general-purpose definition, of being a false friend, “often found where it doesn’t belong and professing powers it doesn’t possess.” An important concern, also expressed by Goodman, is that often similarity judgments are based on partial matching and a consideration of the whole structure of the compared objects is missing. Nevertheless, and despite these criticisms, the fundamental place that similarity holds in different theories of perception, knowledge representation, decision-making, and reasoning cannot be denied. Generally speaking, similarity allows assessing how two objects are alike, classifying patterns into different classes, inferring knowledge in order to “categorize” objects into a higher semantic level (classes, categories, etc.), helping a decision-maker to deal with a new encountered situation by comparing it with similar previously encountered ones, etc. From an engineering point of view, several similarity measures have been proposed in order to mathematically express and measure the similarity. All these measures are derived from a set of assumptions, are tied to particular applications, and are strongly related to different forms of knowledge representation and the available information. Before getting into specific details in measuring the similarity between possibility distributions, it is important to position things by responding to two fundamental questions:

- Measuring similarity between what and what?
- Measuring similarity for doing what?

B. Solaiman, É. Bossé, *Possibility Theory for the Design of Information Fusion Systems*, Information Fusion and Data Science,
https://doi.org/10.1007/978-3-030-32853-5_4

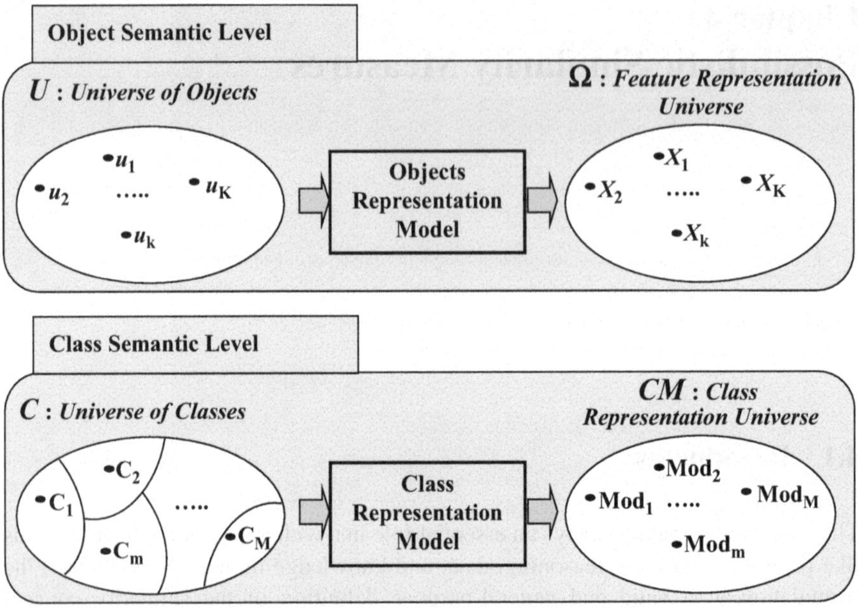

Fig. 4.1 The two semantic knowledge representation levels framework

To illustrate different application contexts of similarity measures, consider the very general two semantic knowledge representation levels framework, called here: the *Object Semantic Level* and the *Class Semantic Level* of a decision-making problem. This framework is illustrated in Fig. 4.1.

At the object semantic knowledge representation level, the *universe of objects*, $U = \{u_1, u_2, \ldots, u_K, ..\}$, is assumed to represent the set of all potential existing objects under consideration where the word *object* may be substituted (depending on the application context) by stimuli, pattern, instance, case, situation, event, etc. Using a predefined structure of an object knowledge representation, each object $u_K \in U$ is described (or encoded) as X_K, called the *object description*. X_K may be a simple vector of features (primitives, descriptors, formulas/rules in predicate logic, graph, etc.). Therefore, at this object semantic level, the universe of objects is "projected" into an observation universe $\Omega = \{X_1, X_2, .., X_K, ..\}$ called the *feature representation universe*. At the class semantic knowledge representation level, the universe of objects U is assumed to be "partitioned" into a set of disjoint and finite union of M subsets $C = \{C_1, C_2, \ldots, C_M\}$ called *Classes* (categories, labels, decisions, etc.). Each C_m is assumed to represent the class/category/label/decision shared by several objects from the universe U. Notice that this categorization constitutes the first intuitive level of similarity between objects since it consists in "grouping objects" assumed to "share a common property" or objects "considered as similar" with regard to the targeted decision. In a similar manner to the object semantic level, each class $C_m \in C$ is described (encoded or modeled) using a predefined class knowledge representation model Mod_m, called *class description model*. Mod_m

resumes the knowledge that we have (using the adopted representation model) of the class C_m. Representing each category by one or several "prototype" object(s) is the simplest class knowledge representation model (prototype, reference, or typical objects are considered as the most representative of their classes). Probabilistic, structural, knowledge-based, etc. models are also generally used. Therefore, at the class semantic representation level, the universe of classes is "projected" into the models universe $\mathrm{CM} = \{\mathrm{Mod}_1, \mathrm{Mod}_2, .., \mathrm{Mod}_M\}$ called the *class representation universe*.

To illustrate this object class framework in medical applications, for instance, we can consider the universe of objects as being the universe of all the patients where each patient is "described/observed/represented..." by a feature vector *X* containing some physical measures or observations; *X* may also be a medical image or even an electronic patient record (EPR). According to the targeted medical objective, the universe of patients is partitioned into subsets of patients' calluses/categories. It is worthwhile to notice, at this step, that this partitioning is not "unique" and that the object's description *X* is assumed to be consistent for the class description. Finally, classes modeling task should be conducted in a coherent manner with object's description allowing thus to obtain a medical diagnosis for an observed/examined patient. The interdependence between different universes, shown in this model, is behind the strong link for the joint development, in information systems, of the three following tasks: classes selection, objects description, and classes modeling. Moreover, this two-level semantic knowledge representation model allows to position three major tasks where the concept of similarity plays an essential role:

- Comparison of two objects in order to understand and to evaluate to which extent the two objects are close or similar to each other
- Comparison of a given object with a given class in order to evaluate the degree of similarity between the considered object and the given class
- Comparison of two classes in order to understand and to evaluate to which extent these classes overlap or are difficult to discriminate

Another interesting issue of the object—class semantic framework—is that it allows to distinguish between the two "similarity worlds": the human and the engineering worlds. This is a misleading point since the human concept of similarity is applied on the universes of objects and classes, whereas similarity measures that, we may define and evaluate, depend on the objects and classes adopted descriptions and models which are strongly related to the available means of measuring and computer-based knowledge representation. Therefore, they may not have, or cannot, capture the same discrimination capacities as we, humans, can conduct. This distinction between the human and the engineering world (Fig. 4.2) should be stressed in order to understand the performance and the limits of a given similarity measure (defined and applied in the engineering world) while compared to the human vision of the similarity concept.

In the "engineering world," where we as engineers are concerned, the most widely used mathematical definition of a similarity measure, *Sim*(., .), can be formulated as follows: let Ω be the objects feature universe. A similarity measure *Sim* is

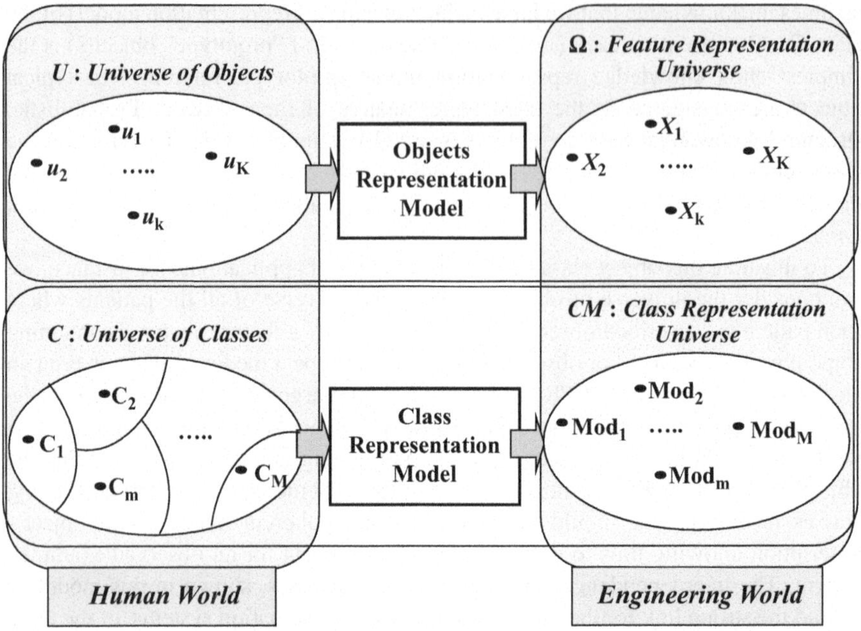

Fig. 4.2 The two similarity worlds

generally defined as a function $\Omega \times \Omega \rightarrow \mathbb{R}$ satisfying the following elementary mathematical properties:

- ***Positivity***: $Sim(X, Y) \geq 0$ for all $X, Y \in \Omega$
- ***Symmetry***: $Sim(X, Y) = Sim(Y, X)$ for all $X, Y \in \Omega$
- ***Maximality***: $Sim(X, X) \geq Sim(Y, X)$ for all $X, Y \in \Omega$

Moreover, *Sim* is called a ***normalized similarity measure*** if $Sim(X, Y) \in [0, 1]$ for all $X, Y \in \Omega$.

The similarity measure *Sim* is subjective to the taxonomic ad hoc representation of the object's features as well as the application target objective of using this similarity measure. The next section is dedicated to briefly resume the existing taxonomy of similarity measures.

4.2 Taxonomy of Similarity Measures

A careful analysis of different similarity measures, proposed in the literature, reveals that it is particularly impossible to single out a general and unique purpose similarity measure that works well for all applications. Several reasons stand behind this definition lack of a unique similarity measure. First, similarity measures are strongly related to the used objects' knowledge representation model (i.e., the way objects and categories from the universes U and C are represented in the feature space Ω and in the class representation universe CM) which also depends on the imperfection type affecting different features. Second, the similarity concept, by itself, is defined with respect to a predefined target objective. In other words, if the "application objective" changes, the "adopted" similarity measure becomes non-adapted. Finally, similarity measures are defined using some mathematical conditions where a huge amount of "mathematical similarity measures" may fulfill, but each measure has a different specificity.

It is worthwhile to notice that most of the existing measures target the first similarity measurement task (i.e., comparison between objects represented in the feature space Ω). Several reasons stand behind this situation. First, measuring the similarity between objects is certainly the most encountered situation in real applications. The second reason is that several class modeling approaches are based on representing each class by one or several representative objects (called reference objects, prototypes, centroids, etc.). Therefore, most class similarity measuring tasks are "somehow" reduced to measuring the similarity between objects.

The first representation model consists in considering objects as point X in a metric space. Different components of X are formed by relevant features extracted from different objects and on the basis of which we have to perform the required similarity measure. Similarity measuring to assess the agreement between two objects $u, v \in U$ is, thus, considered as a function of the distance between their feature vectors $X(u), Y(v) \in \Omega$. This metric representation model is considered through the first category of similarity measures called the ***metric-based similarity measures***.

An alternative way to compute the similarity between objects is considered when the object representation model consists in characterizing objects as subsets of binary features that an object possesses. This is, for instance, the case when each patient is characterized by the subset of symptoms out of the object representation space (considered in this case as the set of all observable symptoms). This category of measures, called the ***set-based similarity measures***, computes thus the similarity by taking into account both common and distinctive features of the objects being compared. These two categories of similarity measures are detailed in the following subsections.

4.2.1 Metric-Based Similarity Measures

4.2.1.1 Metric Distance Measures

The most obvious way of evaluating the similarity, or stimuli, between objects from the universe of objects U, is based on representing different objects as points in an N-dimensional metric feature representation universe Ω (Fig. 4.3).

The perceived similarity can be inversely associated with the distance leading to one-to-one correspondence between distance and similarity measures. In this way,

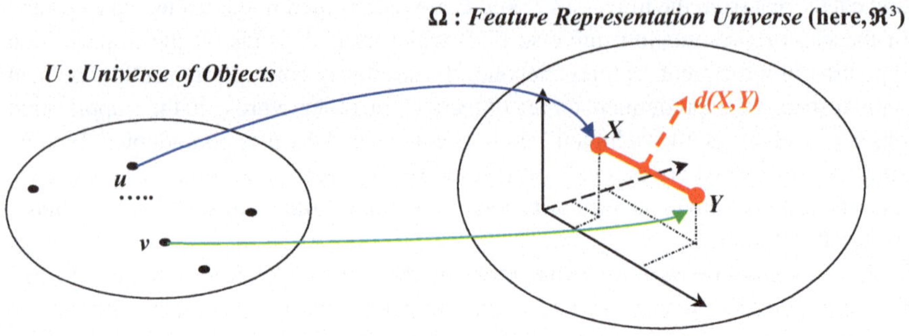

Fig. 4.3 Objects metric representation model

the similarity (or, more properly, dissimilarity) between two objects u, $v \in U$ is expressed as a distance between their corresponding feature vectors, or points, X, $Y \in \Omega$. The farther the points X, Y are from each other, the more dissimilar the objects are and vice versa.

Let $d(.,.)$ denote the distance function defined as follows:

$$\begin{array}{cccc} d: & \Omega \times \Omega & \rightarrow & \mathbb{R}^{+} \\ & (X, Y) & \rightarrow & d(X, Y) \end{array}$$

where $d(.,.)$ satisfies the following metric axioms:

- ***Constancy of self-similarity*** axiom: $d(X, X) = d(Y, Y)$, for all X and Y from Ω
- ***Minimality*** axiom: $d(X, X) \leq d(X, Y)$, for all X and Y from Ω
- ***Symmetry*** axiom: $d(X, Y) = d(Y, X)$, for all X and Y from Ω
- ***Triangle inequality*** axiom: $d(X, Y) \leq d(X, Z) + d(Z, Y)$, for all X, Y, and Z from Ω

In the technical literature, a huge amount of distance measures is available, each of which has its own advantages and its own lacks. In this section, the most widely used metric distance measures are resumed. Assume that the object u(resp. v) $\in U$ is represented as an N-valued feature vector $X = [x_1, x_2, .., x_N]$ (resp. $Y = [y_1, y_2, .., y_N]$), with x_n and y_n defined in the definition domain of the nth feature $\mathbb{D}_n$. A variety of distance functions are suggested in the literature. Here follows the most widely used Minkowski distance and Canberra distance.

Minkowski Distance

The most commonly employed and generic form of metric distance measures for multidimensional data ($\Omega = \mathfrak{R}^N$) is the r-metric Minkowski distance measure defined by:

$$d_{\mathrm{M},r}(X,Y) = \left[\sum_{n=1}^{N} |x_n - y_n|^r\right]^{1/r}, \quad r \geq 1$$

where "r" is called the *order* of Minkowski distance.

Three Minkowski distance functions corresponding to $r = 1, 2$, and ∞ are mostly used [10].

a_1. The *Hamming* distance ($r = 1$):

$$d_{\mathrm{H}}(X,Y) = \sum_{n=1}^{N} |x_n - y_n|$$

This distance measure is also called *Manhattan* or *City block* distance. A normalized Hamming distance is sometimes used:

$$d_{\mathrm{NH}}(X,Y) = \frac{1}{N} \sum_{n=1}^{N} |x_n - y_n|$$

a_2. The *Euclidean* distance ($r = 2$):

$$d_{\mathrm{E}}(X,Y) = \sqrt{\sum_{n=1}^{N} |x_n - y_n|^2}$$

A normalized Euclidean distance is sometimes used:

$$d_{\mathrm{NE}}(X,Y) = \sqrt{\frac{1}{N} \sum_{n=1}^{N} |x_n - y_n|^2}$$

a_3. The *Chebyshev* distance ($r = \infty$):

When $r \to \infty$, Minkowski distance is termed as *Chebyshev* distance or maximum distance. It represents the greatest distance between two vectors along any coordinate dimension. It is given as:

$$d_{\mathrm{Ch}}(X,Y) = max\, |x_n - y_n|, n = 1, 2, \ldots, N.$$

It is worthwhile to have a close look into the informational aspect of these distance measures. In fact, and as shown in Fig. 4.4a, objects from the universe

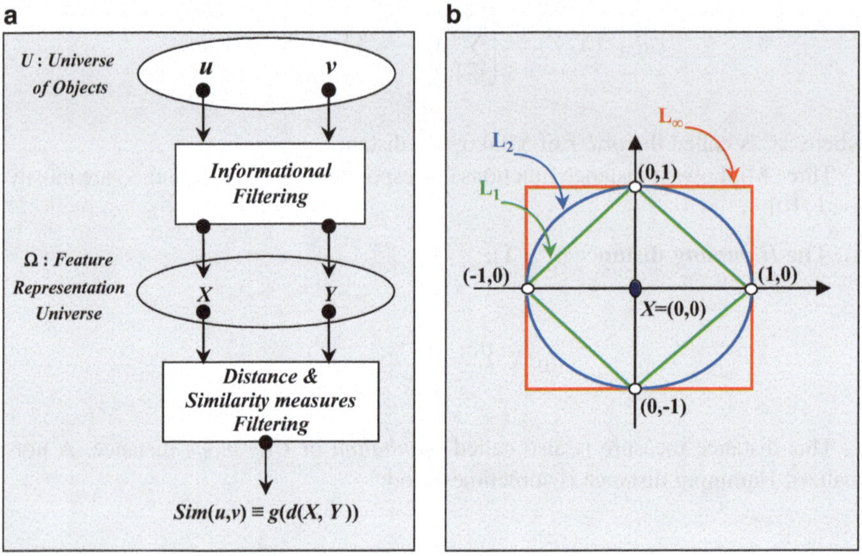

Fig. 4.4 (**a**) Information reduction aspect for similarity computation and (**b**) unit distance objects (from the origin) using Minkowski distances

U are rich in content and have a huge amount of informative facets; therefore, the object representation constitutes some "kind" of information filtering reducing the object informative facets into an N-valued feature vector in the universe $\Omega = \mathbb{R}^N$. Moreover, the similarity measure between two objects will produce a single measurement. As a consequent, given an object $u \in U$, several objects $v \in U$ will have the same similarity measure with u while being quite different. This is shown in Fig. 4.4b, where objects are represented in $\Omega = \mathbb{R}^2$ and where $X = (0, 0)$. Using Minkowski distance, objects for which the feature vector Y has a unit distance from $X = (0, 0)$ are positioned on the rectangle L_1 (Hamming distance, $r = 1$), on the rectangle L_∞ (Chebyshev distance, $r = \infty$) and on the unit circle L_2 (Euclidian distance, $r = 2$).

Canberra Distance

The Canberra distance [2] is similar to the Hamming (or Manhattan) distance. The distinction is that the absolute difference between feature values is divided by the sum of their absolute values *prior* to summing:

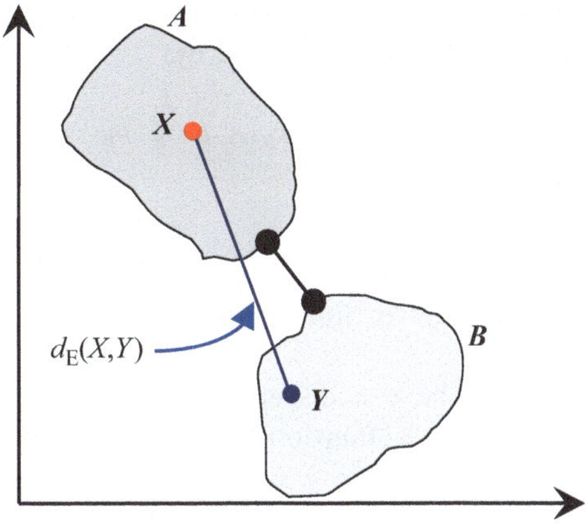

Fig. 4.5 Hausdorff distance

$$d_C(X,Y) = \sum_{n=1}^{N} \frac{|x_n - y_n|}{|x_n| + |y_n|}$$

The main advantage of using Canberra distance is that it leads to an upper distance bound which equals to N (i.e., number of used features): $0 \le d_C(X, Y) \le N$. Therefore, the use of the "normalized" version of Canberra distance:

$$d_{NC}(X,Y) = \frac{1}{N} \sum_{n=1}^{N} \frac{|x_n - y_n|}{|x_n| + |y_n|}$$

leads to a unit upper distance bound, i.e., $0 \le d_{NC}(X, Y) \le 1$.

Hausdorff Distance

The Hausdorff distance is a generalization of the distance between two points in a metric space to two compact nonempty subsets of the space. This is, for instance, the case of two observed regions in a satellite image or two observed organs on a medical image. If A and B are such compact nonempty subsets in $\mathbb{R}^2$, then the Hausdorff distance between A and B is defined by (Fig. 4.5):

$$d_{Hus}(A,B) = max\left\{ \underset{Y \in B}{max} \underset{X \in A}{inf} d_E(X,Y), \underset{X \in A}{max} \underset{Y \in B}{inf} d_E(X,Y) \right\}$$

It is worthwhile to notice that the universe of objects (on which Hausdorff distance is defined) is the universe of regions in an image. Nevertheless, the evaluation of Hausdorff distance between two such objects A and B is conducted using the Euclidian distances between different $\mathbb{R}^2$ (points) forming the two objects

A and B. In the case of real intervals, $A = [a_1, a_2]$ and $B = [b_1, b_2]$, the Hausdorff distance is simply expressed by:

$$d_{\text{Hus}}(A, B) = max\{|a_1 - b_1|, |a_2 - b_2|\}).$$

4.2.1.2 Metric Similarity Measures

Once we have a distance measure $d(.,.)$, a measurement of similarity can be expressed as a "transformation" of the distance measure:

$$Sim(u, v) = g[d(X, Y)]$$

where g is an order reversing one-to-one mapping (called distance-to-similarity transformation function):

$$\begin{array}{rcl} g : \text{Range}[distance] & \rightarrow & \text{Range}[Sim] \\ d(X, Y) & \rightarrow & Sim(u, v) \end{array}$$

such that $g[0] = 1$.

Here follows some popular transformation candidates for the distance-to-similarity transformation mapping g generally encountered in the associated literature:

1. **Richter's transformations** [3]:

$$Sim(u, v) \equiv Sim(X, Y) = \frac{d(X, Y)}{1 - d(X, Y)} \quad \text{for unbounded } d(X, Y);$$

$$Sim(u, v) \equiv Sim(X, Y) = 1 - \frac{d(X, Y)}{max_d} \quad \text{if } d(X, Y) \in [0, max_d]$$

2. **Santini's transformation** [4]:

$$Sim(u, v) \equiv Sim(X, Y) = 1 - d(X, Y) \quad \text{if } d(X, Y) \in [0, 1];$$

3. **Koczy's transformation** [5]:

$$Sim(u, v) \equiv Sim(X, Y) = \frac{1}{1 + d(X, Y)};$$

4. **Williams-Steele's transformation** [6]:

$$Sim(u, v) \equiv Sim(X, Y) = e^{-\alpha\, d(X,Y)} \quad (\alpha \text{ is the steepest measure}).$$

Assume that the object u(resp. v) $\in U$ is represented as an N-valued feature vector $X = [x_1, x_2, .., x_N]$ (resp. $Y = [y_1, y_2, .., y_N]$), with x_n and y_n defined in the definition domain $\mathbb{D}_n$. Then, using the abovementioned distance measures and distance-to-similarity transformation functions, the following measures can thus be defined in order to compute the similarity between objects represented in an N-valued feature vector representation universe:

- ***Manhattan*** (or *W-Similarity*) ***similarity measure*:**

This similarity measure is based on the use of the normalized Hamming distance and Santini's distance-to-similarity transformation [7]:

$$Sim_{\mathrm{M}}(u, v) \equiv Sim_{\mathrm{M}}(X, Y) = 1 - d_{\mathrm{NH}}(X, Y) = 1 - \frac{1}{N}\sum_{n=1}^{N} |x_n - y_n|$$

- ***Euclidean similarity measure***:

This similarity measure is based on the use of the normalized Euclidean distance and Santini's distance-to-similarity transformation:

$$Sim_{\mathrm{E}}(u, v) \equiv Sim_{\mathrm{E}}(X, Y) = 1 - d_{\mathrm{NE}}(X, Y) = 1 - \sqrt{\frac{1}{N}\sum_{n=1}^{N} (x_n - y_n)^2}$$

- ***Chebyshev*** (or *L-Similarity*) ***similarity measure***:

This similarity measure is based on the use of Chebyshev distance and Santini's distance-to-similarity transformation [8]:

$$Sim_{\mathrm{Ch}}(u, v) \equiv Sim_{\mathrm{Ch}}(X, Y) = 1 - d_{\mathrm{Ch}}(X, Y) = 1 - \max_{n=1...N} |x_n - y_n|$$

- ***Canberra similarity measure***:

This similarity measure is based on the use of Canberra distance and Santini's distance-to-similarity transformation [8]:

$$Sim_{\mathrm{S}}(u, v) \equiv Sim_{\mathrm{S}}(X, Y) = 1 - d_{\mathrm{C}}(X, Y) = 1 - \sum_{n=1}^{N} \frac{|x_n - y_n|}{|x_n| + |y_n|}$$

(to be taken 1 if $X = Y = 0$).

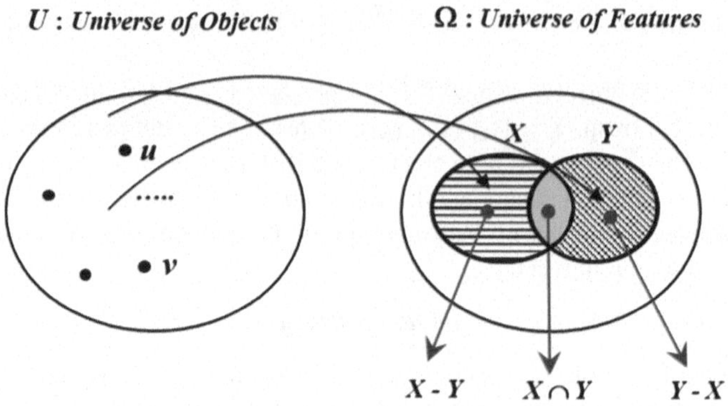

Fig. 4.6 Set-based object representation using a universe of binary features

As Sim_{M}, Sim_{E}, and Sim_{Ch} are based on metric distances which, by definition, verify the properties of positivity, symmetry, identity, and triangular inequality, hence, it is straightforward to verify that these similarity measures satisfy all the basic similarity measure properties (i.e., positivity, symmetry, and maximality).

4.2.2 Set-Based Similarity Measures

Let $\Omega = \{x_1, x_2, \ldots, x_N\}$ denote a finite set of *binary* features that an object $u \in U$ may possess. A feature may be a symptom (in a medical diagnosis situation), a characteristic feature (in a pattern recognition problem), etc. A nominal feature assuming "*K*" values can be transformed into "*K*" binary features where an object can possess one, and only one, of these "*K*" binary features. Using this set of binary features, an object can be represented as a "subset" of features characterizing the object as shown in Fig. 4.6.

X(resp. *Y*) is the subset of features characterizing the object *u*(resp. *v*).

Let *Sim*(.,.) be a similarity measure defined over the universe of objects as follows:

$$\begin{aligned} Sim: \; U \times U &\rightarrow \Re^+ \\ (u, v) &\rightarrow Sim(u, v) \end{aligned}$$

The assessment of similarity between two objects is thus considered as a feature matching process. It is formulated in terms of the set-theoretical notion of a matching function rather than in terms of the geometric concept of distance. That is, the similarity between two objects $u, v \in U$ increases with addition of common features and/or deletion of distinctive features. Let *X* (resp. *Y*) denote the subset of features characterizing the object *u* (resp. *v*). We denote:

- $X - Y$ the subset of features that belong to the object u but not to the object v
- $Y - X$ the subset of features that belong to the object v but not to the object u
- $X \cap Y$ the subset of features that are common to both objects u and v
- $X \cup Y$ the subset of features that are common to at least one of the two objects u and v

To compute the similarity between two objects $u, v \in U$, the focus would be on the intuition that the degree of similarity should take into account both the "amount" of overlap between the feature subsets X and Y (i.e., $X \cap Y$) and the "amount" of symmetric difference (i.e., $X - Y$ and $Y - X$).

The starting, and intuitive, point for the construction of set-based similarity measures defined on subsets from Ω is the measure of containment, also called *subsethood*, and the measure of *equality* between two subsets in terms of their union and intersection.

Let X and Y be two subsets from Ω, and then basic containment relation is defined as a *binary* relation:

$$(X \subseteq Y) = \begin{cases} 1 & \text{if } x \in X \Rightarrow x \in Y \ \ \forall x \in \Omega \\ 0 & \text{in other case} \end{cases}$$

As it is well known in algebraic set theory that $X \subseteq Y$ if and only if $X \cap Y = X$, therefore a reasonable *soft* measure of the degree to which X is a subset of Y (i.e., *subsethood degree*) can be expressed using the following ratio:

$$\text{Degree}(X \subseteq Y) = \frac{|X \cap Y|}{|X|} = \frac{\text{Number of features in the set } X \cap Y}{\text{Number of features in the set } X}$$

where |.| denotes the cardinality of subsets. Figure 4.7 illustrates this definition of the subsethood degree.

Similarly, the *degree of equality* between X and Y is defined as a *binary* relation:

$$(X = Y) = \begin{cases} 1 & (X \cap Y) = (X \cup Y) \\ 0 & \text{in other case} \end{cases}$$

Therefore, to check whether X is equal to Y, it seems natural to consider the ratio:

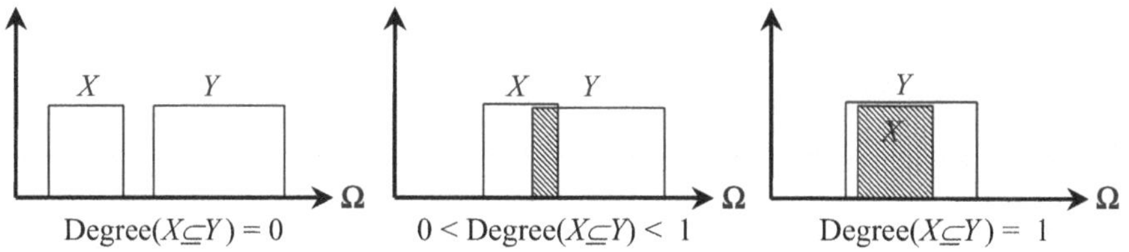

Fig. 4.7 Subsethood degree illustrated when $\Omega = \mathbb{R}$

$$\text{Degree}(X = Y) = \frac{|X \cap Y|}{|X \cup Y|} = \frac{\text{Number of fetures in the set } X \cap Y}{\text{Number of fetures in the set } X \cup Y}$$

as a reasonable degree to which X equals to Y.

Considering X (resp. Y) as being the subset of features characterizing the object u (resp. v), then degree ($X = Y$) can be considered as a potential similarity measure between the two objects u and v. Notice that this ratio produces a normalized similarity measure, i.e., ranging in the unit interval [0, 1], with $Sim(u, v) = 0$ when the feature vectors have no common elements and $Sim(u, v) = 1$ when the two feature subsets X and Y are equal.

In 1977, Tversky [9] has proposed an interesting set-theoretic similarity measuring approach. Assuming X (resp. Y, P, and Q) as referring to the feature subset representing the object u (resp. v, p, and q), Tversky suggested the two following assumptions to be verified by a similarity measure:

1. ***Matching* assumption**

The similarity measure is expressed as a function $\mathcal{F}$ of three arguments: $X \cap Y$, $X - Y$, and $Y - X$:

$$Sim(u, v) = \mathcal{F}(X \cap Y, X - Y \text{ and } Y - X)$$

2. ***Monotonicity* assumption**: If

$$X \cap P \subseteq X \cap Y$$
$$X - Y \subseteq X \cap P$$
$$Y - X \subseteq P \cap X$$

then $Sim(u, v) > Sim(u, p)$.

A function $\mathcal{F}$ that satisfies the matching and the monotonicity assumptions is called a *matching function*. The pairs of objects (u, v) and (p, q) are said to agree on one (two or three) component whenever one (two or three) of the following properties hold:

$$X \cap Y \cong P \cap Q$$
$$X - Y \cong P - Q$$
$$Y - X \cong Q - P$$

Based on this definition, Tverski postulates a third assumption:

3. ***Independence* assumption**

If the pairs of objects (u, v) and (p, q) as well as the pairs of objects $(u^{'}, v^{'})$ and $(p^{'}, q^{'})$ agree on the same two components, while the pairs of objects (u, v) and $(u^{'}, v^{'})$ as well as (p, q) and $(p^{'}, q^{'})$ agree on the remaining third component, then:

$$Sim(u, v) \geq Sim\left(u^{'}, v^{'}\right) \Leftrightarrow Sim(p, q) \geq Sim\left(p^{'}, q^{'}\right)$$

In other words, this assumption simply states that if (u, v) are "closer" than (p, q), then $(u^{'}, v^{'})$ are "closer" than $(p^{'}, q^{'})$.

Tversky [9] suggested two families of matching functions satisfying the three abovementioned assumptions:

- ***The contrast model:*** The matching function is defined as *a linear combination* of a "certain set measure" $f(.)$ of their common, i.e., $X \cap Y$, and distinctive features, i.e., $X - Y$ and $Y - X$:

$$\begin{aligned} Sim_{\text{Tversky}}(u, v) &= \mathrm{F}(X \cap Y, X - Y \text{ and } Y - X) \\ &= \theta f(X \cap Y) - \alpha f(X - Y) - \beta f(Y - X) \end{aligned}$$

According to this model, common features increase similarity, whereas distinctive features that are unique to the two objects decrease similarity. Notice that if $\theta = 0$ and $\alpha = \beta = 1$, then

$$Sim_{\text{Tversky}}(u, v) = f(X - Y) + f(Y - X)$$

which is the dissimilarity between the two feature subsets X and Y.

- ***The ratio model:*** The matching function is defined as *ratio*:

$$\begin{aligned} Sim_{\text{Tversky}}(u, v) &= \mathrm{F}(X \cap Y, X - Y \text{ and } Y - X) \\ &= \frac{f(X \cap Y)}{f(X \cap Y) + \alpha f(X - Y) + \beta f(Y - X)} \end{aligned}$$

where $\theta, \alpha, \beta \geq 0$, and $f(.)$ is "usually" the cardinality $Card(.)$, or $|.|$, of the considered subset, but may be any function that satisfies additivity (i.e., $f(X \cup Y) = f(X) + f(Y)$, whenever X and Y are disjoint).

Notice that the similarity values produced by Tversky's contrast model are not bounded between 0 and 1 making, thus, the result interpretation as a difficult task. Moreover, the more unique features an object possesses, the lower the similarity measure will be obtained.

These difficulties are overcome using Tversky's ratio model which is defined as a normalized ratio ranging in the unit interval [0,1] (i.e., $0 \leq Sim_{\text{Tversky}}(u, v) \leq 1$). Also, the ratio model is irrespective of the size of the features being compared. Therefore, the Tversky's ratio model seems more appropriate to be used in practice.

It is worthwhile to notice that Tversky's ratio model does not define a single similarity measure but rather a family of measures characterized by different values of the parameters α and β and by the function $f(.)$. For instance, if $f(.)$ is the cardinality function and $\alpha = \beta = 1$, Tversky's similarity measure turns to the subset equality degree:

$$Sim_{\text{Tversky}}(u, v)\big|_{\alpha,\beta=1} = \frac{f(X \cap Y)}{f(X \cup Y)} = \frac{|X \cap Y|}{|X \cup Y|} = \text{Degree}(X = Y)$$

(this similarity measure is called *Jaccard Index*). Also, when $\alpha = 1$ and $\beta = 0$ (resp. $\alpha = 0$ and $\beta = 1$), Taversky's similarity measure is reduced to the subsethood measure (called *inclusion index*):

$$Sim_{\text{Tversky}}(u, v)\big|_{\alpha=1,\beta=0} = \frac{f(X \cap Y)}{f(X)}$$

$$Sim_{\text{Tversky}}(u, v)\big|_{\alpha=1,\beta=1} = \frac{f(X \cap Y)}{f(Y)}$$

In other words, α (resp. β) represents a "weighting" factor of distinctive features possessed by the object u (resp. v). Therefore, in order to ensure the symmetric property of a similarity measure (i.e., $Sim(u, v) = Sim(v, u)$), both factors α and β should be considered as equal. The most used Tversky's ratio similarity measures are resumed in Table 4.1.

4.3 Fuzzy Sets Theory and Similarity Measures

In real world, objects similarity is considered as a naturally given property, whereas designing similarity-based systems in the "engineering world" starts with the underlying purposes and, then, features selection, and the relations establishing similarities are chosen corresponding to the relationships of actual interest. Before getting into the existing approaches and methods developed in order to measure the similarity

Table 4.1 Tversky's ratio similarity measures for different weight values of distinctive features

Tversky's ratio similarity measure	$\alpha = \beta$	Measure
Sneath & Sokal Measure	2	$\frac{f(X \cap Y)}{f(X \cap Y)+2[f(X)+f(Y)]}$
Jaccard Index	1	$\frac{f(X \cap Y)}{f(X \cup Y)}$
Jaccard Measure (Eisler measure)	1/2	$\frac{2\,f(X \cap Y)}{f(X)+f(Y)}$
Sorennsen Measure	1/4	$\frac{4\,f(X \cap Y)}{2\,f(X \cap Y)+f(X)+f(Y)}$
Anderberg Measure	1/8	$\frac{8\,f(X \cap Y)}{6\,f(X \cap Y)+f(X)+f(Y)}$

between fuzzy sets, it is important to stress the fact that fuzzy sets theory and similarity measures are closely related concepts.

Let us consider the pattern recognition framework, where objects (or patterns) from the universe of objects, U, are "observed" through sensors so that each object $u \in U$ is represented by a feature vector:

$$X = [x_1, x_2, \ldots x_N] \in \Omega = \Omega_1 \times \Omega_2 \times \ldots . \times \Omega_N$$

where Ω denotes the universe of features and Ω_n denotes the nth feature's, x_n, universe. Based on the use of the feature vector $X(u)$, the objective of the pattern recognition system is to assign a label from the class universe $C = \{C_1, C_2, \ldots, C_M\}$ to a newly observed object $u \in U$. This label assignment is based on the use of different forms of knowledge and models $\{\text{Mod}_1, \text{Mod}_2, \ldots, \text{Mod}_M\}$ used to represent different classes.
Here follow different situations where similarity measures and fuzzy sets theory are closely related:

- ***A similarity measure leading to a fuzzy set***

When a normalized similarity measure is considered (i.e., when the obtained similarity values are bounded in the interval [0,1]), in order to compute the similarity degree between an object and the universe of categories, then the similarity measure defines, in a natural way, a fuzzy set over the universe of ***classes*** (Fig. 4.8). Using this "basic vision" of a similarity measure, the use of fuzzy sets fusion operators can be applied in order *to merge similarity measures issued from different universes of features* as, for instance, in multisensory data fusion applications.

- ***Classes ambiguous knowledge representation model***

Class's knowledge is often expressed, by human experts, in terms of linguistic variables defined over different features. As shown in Fig. 4.9, the relationship between each class C_m and each *binary* feature x_n is expressed by the expert as a membership degree $a_{n,m} \in [0, 1]$ representing how much the feature x_n is involved (adequate, affected, related, etc.) by the class C_m (we may think about the relationship between diseases and observed symptoms, for instance). Each class C_m is, thus,

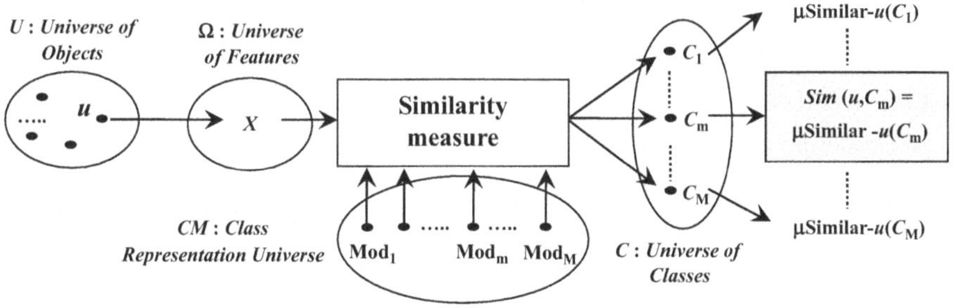

Fig. 4.8 Similarity measure defining a fuzzy set over the universe of classes

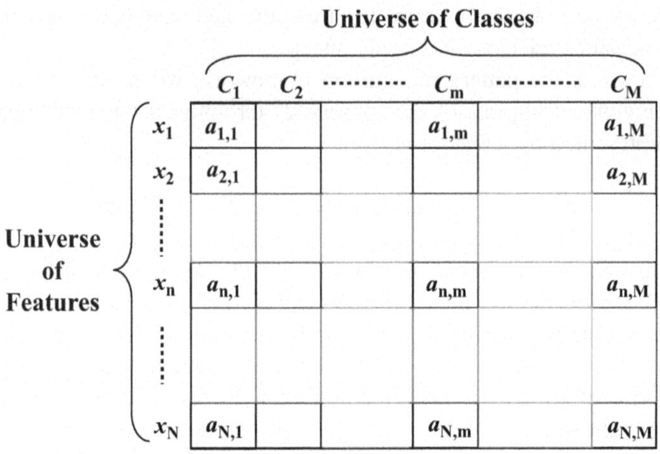

Fig. 4.9 Ambiguous knowledge modeling different classes

represented as a fuzzy set A_{C_m} over the set of the binary features: $\mu_{A_{C_m}}(x_n) = a_{n,m} \in [0, 1], n = 1, \ldots, N$.

Notice that this is the same situation when the expert category C_m knowledge is expressed through a linguistic parameter over a single feature x_n (for the category, the observed feature is around, or close to, a given value x). Considering this situation, a natural question raises:

To what extent two categories C_{m1} and C_{m2} are similar?

The answer to this question is obtained by measuring the similarity between the two fuzzy sets A_{Cm1} and A_{Cm2}.

- ***Feature vague knowledge representation model***

A third practical situation is encountered when the knowledge about features is expressed by human experts (while observing an object) in terms of linguistic variables defined over different features.

As shown in Fig. 4.10, the relationship between each feature x_n and each objet u_k is expressed by the expert as a membership degree $b_{k,\,n} \in [0, 1]$ representing how much the feature x_n is involved (adequate, affected, related, etc.) by the object u_k (we may think about the degree of "activation" between a given symptom, x_n, and an observed patient u_k).

Similarly to the previous situation, each object u_k is, thus, represented as a fuzzy set B_{u_k} over the universe of features: $\mu_{B_{u_k}}(x_n) = b_{k,n}, n = 1, \ldots, N$.

Considering this situation, a natural question raises:

To what extent two objects (patients, cases, patterns, etc.) u_{k1} and u_{k2} are similar?

The answer to this question is obtained by measuring the similarity between the two fuzzy sets $B_{u_{k1}}$ and $B_{u_{k2}}$. As an example of this situation, let us consider the following example in order to characterize hotels in a given town (i.e., U is the set of

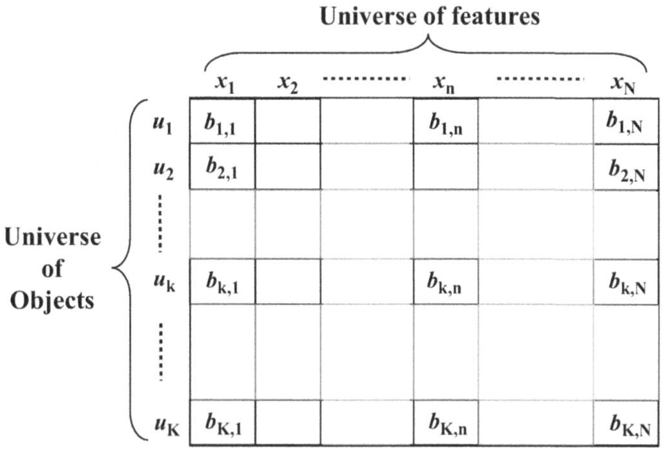

Fig. 4.10 Feature vague knowledge

all hotels in the town). The first feature, x_1, is considered to be *close to downtown*; feature, x_2, is considered as *comfortable rooms*; and feature, x_3, is considered as *not expensive*. The universe of features is, thus, defined by:

$$\Omega = \{x_1 \equiv \text{"close to downtown"}, x_2 \equiv \text{"comfortable rooms"}, x_3 \equiv \text{"not expensive"}\}.$$

Each hotel $u \in U$ is then represented as a fuzzy set A_u defined over $\Omega : X = [\mu_{A_u}(x_1)\mu_{A_u}(x_2)\mu_{A_u}(x_3)]$, where $\mu_{A_u}(x_n) \in [0, 1]$ denotes the membership of the feature x_n, $n = 1, 2, 3$, to the fuzzy set representing the hotel $u \in U$.

Given the universe of features $\Omega = \{x_1, x_2, x_3\}$, an intuitive and immediate question, related to the need of computing the similarity between fuzzy sets, can be formulated as to what extent two hotels are similar?

As shown in these real cases, we are frequently faced with the situation of evaluating the similarity between two fuzzy sets A, B defined on the same universe Ω (μ_A and μ_B denote their respective membership functions).

To do so, let Ω denote a universe of features, $\mathscr{F}(\Omega)$ the set of all fuzzy sets defined on the universe Ω.

A real function *Sim*: $\mathscr{F}(\Omega) \times \mathscr{F}(\Omega) \rightarrow [0, 1]$ is called a *similarity measure between fuzzy sets*, if it satisfies the following properties:

1. $Sim(A, B) = Sim(B, A)$ for all A, $B \in \mathscr{F}(\Omega)$;
2. $Sim(A, A) = 1$ for all $A \in \mathscr{F}(\Omega)$;
3. ***Inclusion Transitive property***:

$$\text{If } A \subseteq B \subseteq C: \quad Sim\,(A, B) \geq Sim\,(A, C) \text{ and}$$
$$Sim\,(B, C) \geq Sim\,(A, C) \text{ for all } A, B, C \in \mathscr{F}(\Omega).$$

Fig. 4.11 Fuzzy sets as points in the metric space $[0, 1]^2$, $\Omega = \{x_1, x_2\}$

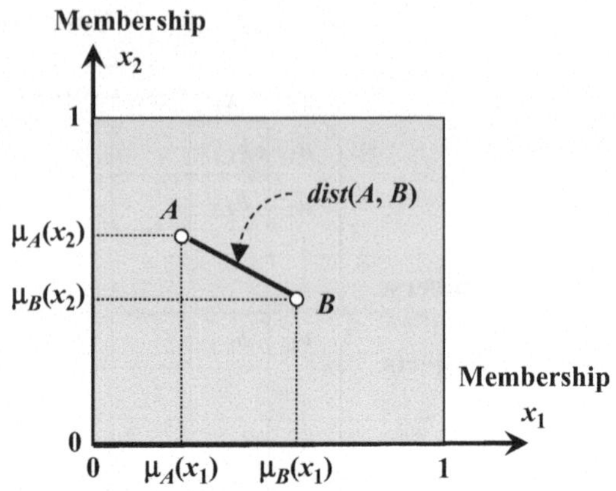

A huge amount of similarity measures between fuzzy sets, and satisfying the previous properties, has been proposed in the literature. These measures can be divided into three major groups:

- Measures obtained through metric considerations (called *metric-based similarity measures*)
- Measures obtained through subsethood considerations (called *set-based similarity measures*)
- Measures based on approximate reasoning logical implications consideration (called *implication-based similarity measures*)

4.3.1 Metric-Based Similarity Measures of Fuzzy Sets

In order to define a similarity measure between two fuzzy sets defined over a finite universe of features $\Omega = \{x_1, x_2, \ldots., x_N\}$, the metric-based approach consists in considering each fuzzy set $A \in \mathcal{F}(\Omega)$ as a point from the metric space $[0, 1]^N$, called *membership metric space* (see Fig. 4.11, in the two dimensional case, i.e., when the universe of features contains only two features). This means that the object representation model is "simply" an N-dimensional vector $X = [\mu_A(x_1), \mu_A(x_2)\ldots\mu_A(x_N)] \in [0, 1]^N$ where the nth component, $n = 1, 2, \ldots, N$, represents the membership of the nth feature to the fuzzy set A.

Using this metric vision, a similarity measure $Sim(A, B)$ between two fuzzy sets A, $B \in \mathcal{F}(\Omega)$ is then obtained as a "transformation" $g(.)$ of a distance measure $dist(A, B)$ between the two fuzzy sets A, B.

Recall that a bounded distance $dist(.,.)$, or a bounded metric, defined on $\mathcal{F}(\Omega)$ is a function:

$$\begin{array}{rcl} dist: \mathscr{F}(\Omega) \times \mathscr{F}(\Omega) & \to & [0,1] \\ (A,B) & \to & dist(A,B) \end{array}$$

satisfying the following properties:

- *Non-negativity:* $dist(A,B) \geq 0$;
- *Identity of indiscernibles:* $dist(A,B) = 0$ if and only if $A = B$;
- *Symmetry:* $dist(A,B) = dist(B,A) \quad \forall\, A, B \in \mathscr{F}(\Omega)$;
- *Triangular inequality:* $dist(A,C) \leq dist(A,B) + dist(B,C) \quad \forall\, A, B, C \in \mathscr{F}(\Omega)$.

($\mathscr{F}(\Omega)$, *dist*) is called a ***metric space***.

Relaxing some properties gives rise to various notions of generalized metric spaces:

- ($\mathscr{F}(\Omega)$, *dist*) is called an ***ultrametric*** space if it satisfies the following stronger version of the triangular inequality: $dist(A,C) \leq \max\,(dist(A,B), dist(B,C))$.
- ($\mathscr{F}(\Omega)$, *dist*) is called a ***pseudometric*** space if the second property (i.e., identity of indiscernables) is "softened" such that $dist(A,B) = 0$ for some distanced fuzzy sets A and B.
- ($\mathscr{F}(\Omega)$, d) is called a ***quasimetric*** space if the third property (i.e., symmetry) is dropped: $dist(A,B) \neq dist(B,A)$.

A variety of distance functions and distance-to-similarity transformation functions lead to a huge amount of possibilistic similarity measures.

Concerning the distance functions, the use of the r-metric Minkowski distance measure defined by:

$$d_{\mathrm{M},r}(A,B) = \left[\sum_{n=1}^{N} |\mu_A(x_n) - \mu_B(x_n)|^r\right]^{1/r}, r \geq 1$$

leads to:

- *The Hamming distance* ($r = 1$) between two fuzzy sets $A, B \in \mathscr{F}(\Omega)$, [10]:

$$d_{\mathrm{H}}(A,B) = \sum_{n=1}^{N} |\mu_A(x_n) - \mu_B(x_n)|$$

(the normalized Hamming distance is given by: $d_{\mathrm{NH}}(A,B) = \frac{1}{N}\sum_{n=1}^{N} |\mu_A(x_n) - \mu_B(x_n)|$);

- The *Euclidean distance* ($r = 2$) between two fuzzy sets $A, B \in \mathscr{F}(\Omega)$, [10]:

$$d_{\mathrm{E}}(A,B) = \sqrt{\sum_{n=1}^{N} |\mu_A(x_n) - \mu_B(x_n)|^2}$$

(the normalized Euclidean distance is given by: $d_{\mathrm{NE}}(A,B) = \sqrt{\frac{1}{N}\sum_{n=1}^{N} |\mu_A(x_n) - \mu_B(x_n)|^2}$);

- The *Chebyshev distance* ($r = \infty$) between two fuzzy sets A, $B \in \mathscr{F}(\Omega)$, [10]:

$$d_{\mathrm{Ch}}(A,B) = \max_{n=1,\ldots,N} |\mu_A(x_n) - \mu_B(x_n)|$$

Also, the use of Canberra distance leads to the following expression of Canberra distance between two fuzzy sets A, $B \in \mathscr{F}(\Omega)$, [2]:

$$d_{\mathrm{C}}(A,B) = \sum_{n=1}^{N} \frac{|\mu_A(x_n) - \mu_B(x_n)|}{\mu_A(x_n) + \mu_B(x_n)}$$

recall that $0 \leq d_{\mathrm{C}}(A,B) \leq N$.

Using this metric vision, a similarity measure $Sim(A,B)$ between two fuzzy sets A, $B \in \mathscr{F}(\Omega)$ is finally obtained as a "transformation" $g(.)$ of a distance measure $dist(A,B)$ between the two fuzzy sets A, B:

$$Sim(A,B) = g(dist(A,B))$$

Table 4.2 resumes some proposed similarity measures using Santini's distance-to-similarity transformation (i.e., $Sim(A,B) = 1 - dist(A,B)$):

As $Sim_{\mathrm{H}}(A,B)$, $Sim_{\mathrm{NE}}(A,B)$, and $Sim_{\mathrm{Ch}}(A,B)$ are based on metric distances which, by definition, satisfies the properties of non-negativity, symmetry, identity, and triangular inequality, hence, it is straightforward to check that these similarity measures satisfy all the basic similarity measure properties.

Richter's and Koczy's distance-to-similarity transformations can also be used to obtain other similarity measures between fuzzy sets.

4.3.2 Set-Based Similarity Measures of Fuzzy Sets

As previously mentioned, the concept of similarity can be interpreted in different ways depending on the context and the nature of the available information in order to compute the similarity between objects. An important family of similarity measures

Table 4.2 Santini's based similarity measure using different distance functions

Distance function	Santini's based similarity measure
Normalized Hamming distance	*Manhattan* (or ***W-Similarity***) *similarity measure* [7] $Sim_{\mathrm{H}}(A,B) = 1 - \frac{1}{N}\sum_{n=1}^{N} \lvert\mu_A(x_n) - \mu_B(x_n)\rvert$
Normalized Euclidean distance	*Euclidean similarity measure* $Sim_{\mathrm{NE}}(A,B) = 1 - \sqrt{\frac{1}{N}\sum_{n=1}^{N} \lvert\mu_A(x_n) - \mu_B(x_n)\rvert^2}$
Chebyshev distance	*Chebyshev* (or ***L-Similarity***) *similarity measure* [8] $Sim_{\mathrm{Ch}}(A,B) = 1 - \max_{n=1,\ldots,N} \lvert\mu_A(x_n) - \mu_B(x_n)\rvert$
Canberra distance	*Canberra* (or ***S-Similarity***) *similarity measure* [8] $Sim_{\mathrm{C}}(A,B) = 1 - \sum_{n=1}^{N} \frac{\lvert\mu_A(x_n) - \mu_B(x_n)\rvert}{\mu_A(x_n) + \mu_B(x_n)}$ (to be taken 1 if $A = B = \Phi$)

between two fuzzy sets is based on defining the similarity as the degree to which these fuzzy sets are equal. The focus would be on the intuition that the degree of similarity should take into account both the amount of overlap between the sets and the amount of symmetric difference.

The starting point for the construction of set-based similarity measures defined on $\mathcal{F}(\Omega)$ is the measure of containment and the measure of equality between two membership functions. In fact, if A and B are two fuzzy sets from Ω where $\mu_A(x)$ and $\mu_B(x)$ denote their respective membership functions, then the fuzzy sets containment relation is defined as a *binary relation*:

$$(A \subseteq B) = \begin{cases} 1 & \text{if } \mu_A(x) \leq \mu_B(x) \;\; \forall x \in \Omega \\ 0 & \text{in other case} \end{cases}$$

Similarly, the equality between the two fuzzy sets A and B is defined, as a binary relation, in the two equivalent ways:

$$(A = B) = \begin{cases} 1 & (A \subseteq B) \text{ and } (B \subseteq A) \\ 0 & \text{in other case} \end{cases}$$
$$= \begin{cases} 1 & \text{if } \mu_A(x) = \mu_B(x) \;\; \forall x \in \Omega \\ 0 & \text{in other case} \end{cases}$$

or

$$(A = B) = \begin{cases} 1 & (A \cap B) = (A \cup B) \\ 0 & \text{in other case} \end{cases}$$
$$= \begin{cases} 1 & \text{if } \min(\mu_A(x), \mu_B(x)) = \max(\mu_A(x), \mu_B(x)) \ \forall x \in \Omega \\ 0 & \text{in other case} \end{cases}$$

In order to relax the containment crispness issue (i.e., the binary aspect) and allow having "soft" degrees of containment and of equality, several alternative methods are proposed. In the same way that fuzzy sets allow for gradual transition between full membership and total nonmembership, a similarity measure is appealed to capture a gradual transition between equality and non-equality.

The following approach expressing the containment, also called *subsethood*, and the equality in terms of union and intersection is probably the most frequently used. In fact, it is well known in algebraic set theory that $A \subseteq B$ if and only if $A \cap B = A$. The degree to which A is a subset of B (i.e., subsethood degree) proposed in the case of crisp sets:

$$\text{Degree}(A \subseteq B) = \frac{|A \cap B|}{|A|} = \frac{\text{Number of elements in the set } A \cap B}{\text{Number of elements in the set } A}$$

can be extended to the case of two fuzzy sets A and B defined on the finite set of alternatives $\Omega = \{x_1, x_2 \ldots, x_N\}$, as the ratio [11]:

$$\text{Degree}(A \subseteq B) = \frac{|A \cap B|}{|A|} = \frac{\sum_{n=1}^{N} \min(\mu_A(x_n), \mu_B(x_n))}{\sum_{n=1}^{N} \mu_A(x_n)}$$

where $\mu_A(x)$ and $\mu_B(x)$ denote the membership functions of the two fuzzy sets A and B. In the case where $A = \Phi$ (i.e., the empty set), we consider Degree($A \subseteq B$) = 1.

Degree($A \subseteq B$) measures how well A and B satisfy the inequality $\mu_A \leq \mu_B$, relative to the size of A:

- If $\mu_A(x) \leq \mu_B(x)$ for all $x \in \Omega$, then Degree($A \subseteq B$) fully fulfills Zadeh's fuzzy sets containment definition (i.e., Degree($A \subseteq B$) = 1).
- If $A \cap B = \Phi$ then Degree($A \subseteq B$) = 0 (i.e., when the supports of the two fuzzy sets A and B do not intersect, then their similarity is null).
- If $A \cap B \neq \Phi$ then 0 <Degree($A \subseteq B$) ≤ 1 (i.e., partial support intersection leads to partial similarity between the two fuzzy sets A and B).

The set-based similarity measures defined on $\mathscr{F}(\Omega)$ are mainly based on the use of equality measures between fuzzy sets. Here follow the main approaches proposed in the literature:

A. ***Mutual Subsethood Similarity Measure***

Using the following definition of the binary relation of crisp set equality:

$$(A = B) = \begin{cases} 1 & (A \cap B) = (A \cup B) \\ 0 & \text{in other case} \end{cases}$$

several authors [10–13], etc. proposed to consider the Jaccard Index as a potential equality measure:

$$\text{Degree}(A = B) = \frac{|A \cap B|}{|A \cup B|}$$

where |.| denotes the cardinality function. The smaller the ratio, the more there are elements from $A \cup B$ which are not part of the intersection $A \cap B$.

The generalization of this similarity measure into fuzzy sets is straightforward:

$$Sim_{\text{M}}(A,B) = \text{Degree}(A = B) = \frac{|A \cap B|}{|A \cup B|} = \frac{\sum_{n=1}^{N} min\left(\mu_A(x_n), \mu_B(x_n)\right)}{\sum_{n=1}^{N} max\left(\mu_A(x_n), \mu_B(x_n)\right)}$$

Notice that $Sim_{\text{M}}(A,B)$ is symmetric (i.e., $Sim_{\text{M}}(A,B) = Sim_{\text{M}}(B,A)$) and produces unit self-similarity (i.e., $Sim_{\text{M}}(A,A) = 1$).

A "normalized" version of this fuzzy set similarity measure is also proposed [14]:

$$Sim_{\text{NM}}(A,B) = \frac{1}{N}\frac{|A \cap B|}{|A \cup B|} = \frac{1}{N}\frac{\sum_{n=1}^{N} min\left(\mu_A(x_n), \mu_B(x_n)\right)}{\sum_{n=1}^{N} max\left(\mu_A(x_n), \mu_B(x_n)\right)}$$

This normalized version represents the "average" of the similarity degree of A and B on all elements of Ω.

B. ***Tversky's Subsethood Similarity Measure***

As previously mentioned, Tversky suggested the so-called contrast model, where similarity is computed between two crisp sets. This model describes a family of equality degree functions (considered as potential similarity measures [9]) in a ratio model:

$$Sim_{\text{Tversky} \mid \alpha_1,\alpha_2}(A,B) = \text{Degree}(A = B) = \frac{f(A \cap B)}{f(A \cap B) + \alpha_1 f(A - \text{B}) + \alpha_2 f(B - A)}$$

for some α_1, $\alpha_2 \geq 0$, and the function $f(.)$ is any function satisfying the additivity property:

Table 4.3 Tversky's ratio similarity measures for different weight values of distinctive features

Tversky's ratio similarity measure	$\alpha = \beta$	Measure
Sneath & Sokal fuzzy Measure	2	$\frac{\sum_{n=1}^{N} min(\mu_A(x_n), \mu_B(x_n))}{\sum_{n=1}^{N} min(\mu_A(x_n), \mu_B(x_n)) + 2\left[\sum_{n=1}^{N} \mu_A(x_n) + \sum_{n=1}^{N} \mu_B(x_n)\right]}$
Jaccard Index	1	$\frac{\sum_{n=1}^{N} min(\mu_A(x_n), \mu_B(x_n))}{\sum_{n=1}^{N} max(\mu_A(x_n), \mu_B(x_n))}$
Jaccard Measure (Eisler measure)	1/2	$\frac{2\sum_{n=1}^{N} min(\mu_A(x_n), \mu_B(x_n))}{\sum_{n=1}^{N} \mu_A(x_n) + \sum_{n=1}^{N} \mu_B(x_n)}$
Sorennsen Measure	1/4	$\frac{4\sum_{n=1}^{N} min(\mu_A(x_n), \mu_B(x_n))}{2\left[\sum_{n=1}^{N} min(\mu_A(x_n), \mu_B(x_n))\right] + \sum_{n=1}^{N} \mu_A(x_n) + \sum_{n=1}^{N} \mu_B(x_n)}$
Anderberg Measure	1/8	$\frac{8\sum_{n=1}^{N} min(\mu_A(x_n), \mu_B(x_n))}{6\left[\sum_{n=1}^{N} min(\mu_A(x_n), \mu_B(x_n))\right] + \sum_{n=1}^{N} \mu_A(x_n) + \sum_{n=1}^{N} \mu_B(x_n)}$

$$f(\mathrm{A} \cup \mathrm{B}) = f(A) + f(B)$$

The mostly used function is *Card*(.).

In order to ensure the symmetric property of a similarity measure (i.e., $Sim(A, B) = Sim(B, A)$), both factors α and β should be considered as equal. The most used Tversky's ratio similarity measures are resumed in Table 4.3 (with $f(A) = Card(A)$).

Notice that for $\alpha = \beta = 1$, then Tversky's ratio similarity measure is reduced to the mutual subsethood similarity measure.

C. ***Degree of Sameness***

In algebraic set theory, several equivalent approaches are used in order to define crisp set equality. One of these approaches, called *mutual containment*, is resumed by the following binary relation defining crisp set equality:

$$(A = B) = \begin{cases} 1 & (A \subseteq B) \text{ and } (B \subseteq A) \\ 0 & \text{in other case} \end{cases}$$

Using this approach, two crisp, or fuzzy, sets are considered as similar if:

$$A \text{ and } B \text{ are similar} \Leftrightarrow (A \subseteq B) \text{ and } (B \subseteq A)$$

This leads to the definition of a fuzzy sets similarity measure, called *Degree of Sameness*, assumed to measure the degree to which A and B are the same [15]:

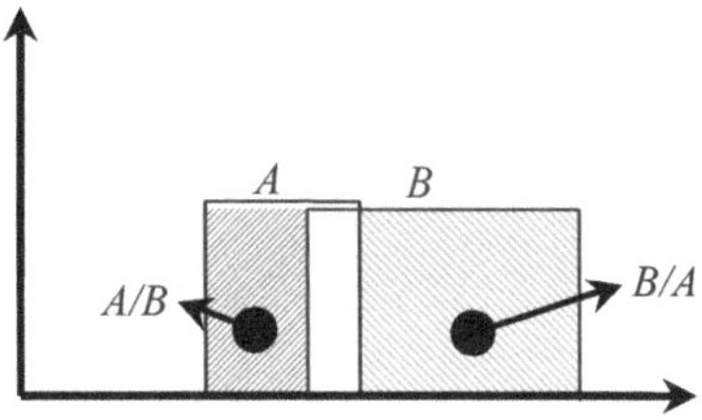

Fig. 4.12 Restle's set equality approach

$$
\begin{aligned}
Sim_{\mathrm{BK}}(A,B) &= \mathrm{Degree}(A = B)\\
&= min\,(A \subseteq B, B \subseteq A)\\
&= min\,[\mathrm{Degree}(A \subseteq B), \mathrm{Degree}(B \subseteq A)]
\end{aligned}
$$

where $A \subseteq B$ (resp. $B \subseteq A$) indicates the degree to which A (resp. B) is included in B (resp. A). This leads to the following expression of the degree of sameness measure of similarity:

$$
\begin{aligned}
Sim_{\mathrm{BK}}(A,B) &= min\left(\frac{\sum_{n=1}^{N} min\,(\mu_A(x_n), \mu_B(x_n))}{\sum_{n=1}^{N} \mu_A(x_n)}, \frac{\sum_{n=1}^{N} min\,(\mu_A(x_n), \mu_B(x_n))}{\sum_{n=1}^{N} \mu_B(x_n)}\right)\\
&= \left[\sum_{n=1}^{N} min\,(\mu_A(x_n), \mu_B(x_n))\right] \cdot \left[max\left(\sum_{n=1}^{N} \mu_A(x_n), \sum_{n=1}^{N} \mu_B(x_n)\right)\right]
\end{aligned}
$$

D. ***Restle's Similarity Measure***

A third approach allowing to define crisp set equality is proposed by Restle [10]. In fact, from a set-theoretic point of view, two crisp sets A and B are considered as equal, if the two difference subsets A/B and B/A are "empty" (Fig. 4.12).

Using the fuzzy sets definitions, $A/B = \{x \in A \,\mathrm{and}\, x \notin B\}$ and $B/A = \{x \in B \,\mathrm{and}\, x \notin A\}$, two membership functions, called *difference membership functions*, can be defined as follows:

$$
\begin{aligned}
\mu_{A/B}(x) &= min\,[\mu_A(x), 1 - \mu_B(x)]\\
\mu_{B/A}(x) &= min\,[\mu_B(x), 1 - \mu_A(x)]
\end{aligned}
$$

where $\mu_{A/B}(.)$ (resp. $\mu_{B/A}(.)$) denotes the membership function of alternatives that approximately belong to A (resp. B) and not to B (resp. A). From this point of view, Restle's similarity measure between the two fuzzy sets A and B is defined as:

$$
\begin{aligned}
Sim_{\mathrm{R}}(A,B) &= \text{Degree}(A = B) \\
&= \text{Degree}[((A/B) \cup (B/A)) = \Phi] \\
&= 1 - \mu((A/B) \cup (B/A)) \\
&= 1 - \sum_{n=1}^{N} max\left(\mu_{A/B}(x_n), \mu_{B/A}(x_n)\right) \\
&= 1 - \sum_{n=1}^{N} max\{ min(\mu_A(x_n), 1 - \mu_B(x_n)), min(\mu_B(x_n), 1 - \mu_A(x_n))\}
\end{aligned}
$$

E. ***Degree of Nearness***

Let us consider two fuzzy sets A and B defined on the finite set of alternatives $\Omega = \{x_1, x_2 \ldots x_N\}$. Given an alternative $x \in \Omega$, if $max(\mu_A(x), \mu_B(x)) \neq 0$, then the ratio

$$0 \leq \frac{max(\mu_A(x), \mu_B(x)) \cdot min(\mu_A(x), \mu_B(x))}{[max(\mu_A(x), \mu_B(x))]^2} \leq 1$$

constitutes a potential similarity measure between the two fuzzy degrees $\mu_A(x)$ and $\mu_B(x)$. Using this basic idea, X.J.Tong [16] proposed to consider the following similarity measure between the two fuzzy sets A and B, called the *degree of nearness*:

$$Sim_{\mathrm{DN}}(A,B) = \frac{\sum_{n=1}^{N} [max(\mu_A(x_n), \mu_B(x_n)) \cdot min(\mu_A(x_n), \mu_B(x_n))]}{\sum_{n=1}^{N} [max(\mu_A(x_n), \mu_B(x_n))]^2}$$

It can be easily proved that $Sim_{\mathrm{DN}}(A,B)$ is a similarity measure.
Also, it is straightforward to show that:

- $Sim_{\mathrm{DN}}(A,B) = Sim_{\mathrm{DN}}(B,A)$ for all A, $B \in \mathcal{F}(\Omega)$;
- $Sim_{\mathrm{DN}}(A,A) = 1$ for all $A \in \mathcal{F}(\Omega)$;
- If A is a crisp subset of Ω, then $Sim_{\mathrm{DN}}(A,A^{c}) = 0$;
- If $A \subseteq B \subseteq C$, then $A \cup B = \mathrm{B}$, $A \cap B = \mathrm{A}$, $A \cup C = \mathrm{C}$, $A \cap C = \mathrm{A}$, hence:
  - $Sim_{\mathrm{DN}}(A,B) \geq Sim_{\mathrm{DN}}(A,C)$
  - $Sim_{\mathrm{DN}}(B,C) \geq Sim_{\mathrm{DN}}(A,C)$ for all A, B, $C \in \mathcal{F}(\Omega)$.

The degree of nearness similarity measure shows, also, two interesting properties:

- $Sim_{\mathrm{DN}}(A,B) = Sim_{\mathrm{DN}}(A \cup B, A \cap B)$;
- $Sim_{\mathrm{DN}}(A,B) \neq Sim_{\mathrm{DN}}(A^{c}, B^{c})$.

4.3.3 Implication-Based Similarity Measures of Fuzzy Sets

Implication-based similarity measures of fuzzy sets are constructed following the basic idea that the more two membership functions imply each other, the more they are similar. This approach is general in the sense that by using different implication operators, one gets the distance-based as well as set-based similarity measures previously developed. The basic idea behind the implication-based similarity measures is to use the implication operation in order to define fuzzy sets containment and the logical equivalence operation for the definition of fuzzy set equality. Therefore, the two fundamental concepts used in this approach are the implication operation and the equality indices.

1. ***Implication Operator***

The implication operator ($\Rightarrow$) is defined as an application:

$$\begin{aligned} Impl: [0,1]\times[0,1] &\rightarrow [0,1] \\ (a,b) &\rightarrow Impl(a,b) = (a \Rightarrow b) \end{aligned}$$

such that:

1. $Impl(a,b)$ is a decreasing function of a, (i.e., if $a_1 \le a_2$ then $Impl(a_1,b) \ge Impl(a_2,b)$);
2. $Impl(a,b)$ is an increasing function of b, (i.e., if $b_1 \le b_2$ then $Impl(a,b_1) \le Impl(a,b_2)$);
3. $Impl(0,1) = Impl(0,0) = Impl(1,1) = 1$ and $Impl(1,0) = 0$.

$Impl(a,b) = (a \Rightarrow b)$, read: a entails b, means that it can never happen that a is true and b is not true. The full interpretation of the material implication $a \Rightarrow b$ is that the degree of truth of $a \Rightarrow b$ quantifies to what extend b is at least as true as a.

There are basically two major families of implication operators: S-Implication and R-Implication operators.

S - Implication Operators

This family of implication operators, called *Strong*, results from using the classical definition of the material implication:

$$\begin{aligned} Impl(a,b) = (a \Rightarrow b) &\leftrightarrow [(\text{Not } a).\text{OR}.b] \\ &\leftrightarrow C(a^C, b) \end{aligned}$$

where C is a triangular conorm (allowing to compute the logical operator OR) and a^C denotes the negation (or the complement, of a).

The two widely used strong implication operators are:

Kleen-Dienes implication operator: $Impl_{KD}(a,b) = max\,(1 - a, b)$

$$\left(a^{\mathrm{C}} = 1 - a \text{ and } C(\alpha,\beta) = max\,(\alpha,\beta)\right)$$

and

Reichenbach implication operator: $Impl_{\mathrm{R}}(a, b) = 1 - a + a.\, b$

$$\left(a^{\mathrm{C}} = 1 - a \text{ and } C(\alpha,\beta) = \alpha + \beta - \alpha\beta\right)$$

R - Implication Operators

This family of implication operators, called *residuated*, results from the deduction process where $a \Rightarrow b$ means that b is at least as true as a. Two basic R-implication operators are:

- ***Gödel implication operator*:** $Impl_{\mathrm{G}}(a, b) = 1$ if $a \leq b$ and b otherwise.
- ***Goguen implication operator***: $Impl_{\mathrm{Gg}}(a,b) = \begin{cases} 1 & \text{if } \; a = 0 \\ min\left(\frac{b}{a}, 1\right) & \text{otherwise} \end{cases}$

Notice that Lukasiewicz implication operator $Impl_{\mathrm{L}}(a, b) = min\,(1 - a + b, 1)$ is both strong and residuated.

2. ***Equality Indices***

Equality indices [17] are viewed as a generic implication-based definitions of a similarity relation between two truth values $a, b \in [0, 1]$.

Two equality indices can be defined:

- *Truth values co-implication equality index*:

$$[a \equiv b]_1 = T(a \Rightarrow b, b \Rightarrow a);$$

- *Truth values equivalence equality index*:

$$[a \equiv b]_2 = \frac{1}{2}\left\{T(a \Rightarrow b, b \Rightarrow a), T\left(a^{\mathrm{C}} \Rightarrow b^{\mathrm{C}}, b^{\mathrm{C}} \Rightarrow a^{\mathrm{C}}\right)\right\};$$

where T denotes a triangular norm (t-norm), $a^{\mathrm{C}} = 1 - a$ and $b^{\mathrm{C}} = 1 - b$.

$[a \equiv b]_1$ measures through the t-norm T the "strength of co-implication" between the two truth values a, b. This strength is considered as a degree of "equality" between a and b. Notice that $[a \equiv b]_1$ verifies the properties of a similarity measure [18]. On the other hand, $[a \equiv b]_2$ measures the mean value of $[a \equiv b]_1$ and $[a^{\mathrm{C}} \equiv b^{\mathrm{C}}]_1$.

When the considered t-norm (T) is the *min* conjunctive operation, here follows some equality indices ($[a \equiv b]_1$) depending on the used implication operator:

- ***Lukasiewicz implication operator***: $[a \equiv b]_1 = 1 - |a - b|$;
- ***Kleen-Dienes implication operator***: $[a \equiv b]_1 = min\{max(1 - a, b), max(1 - b, a)\}$;
- ***Gödel implication operator***: $[a \equiv b]_1 = 1$ if $a = b$ and $min(a, b)$ otherwise.

Example If we consider the Lukasiewicz implication operator $Impl_L(a, b) = min(1 - a + b, 1)$ and the min operator t-norm, then $[a \equiv b]_2 = 1 - |a - b|$.

This result is obtained as follows:

$$\begin{aligned}
[a \equiv b]_2 &= \frac{1}{2}\{T(a \Rightarrow b, b \Rightarrow a), T(a^C \Rightarrow b^C, b^C \Rightarrow a^C)\} \\
&= \frac{1}{2}\{min(a \Rightarrow b, b \Rightarrow a), min(a^C \Rightarrow b^C, b^C \Rightarrow a^C)\} \\
&= \frac{1}{2}\{min(min(1 - a + b, 1)), min(1 - b + a, 1)), \\
&\quad min(min(1 - [1 - a][1 - b], 1), min(1 - [1 - b] + [1 - a], 1))\} \\
&= \frac{1}{2}\{min(min(1 - a + b, 1)), min(1 - b + a, 1)), min(min(1 - b + a, 1), min(1 - a + b, 1))\} \\
&= min(min(1 - a + b, 1), min(1 - b + a, 1)) \\
&= 1 - |a - b|
\end{aligned}$$

since if $a < b$, then:

$$min(1 - a + b, 1) = 1 \text{ and } min(1 - b + a, 1) = 1 - b + a \leq 1, \text{thus} : [a \equiv b]_2 = 1 - b + a;$$

and if $b < a$, then:

$$min(1 - a + b, 1) = 1 - a + b \text{ and } min(1 - b + a, 1) = 1 \leq 1, \text{thus} : [a \equiv b]_1 = 1 - b + a;$$

Therefore, both results arc resumed as: $[a \equiv b]_2 = 1 - |a - b|$.

Theorem $[a \equiv b]_1 = [a \equiv b]_2$ whenever an S-implication operator is substituted for implication.

Proof Let us consider an S-implication operator $a \Rightarrow b = C(a^C, b)$, and let T denote the dual t-norm of C: $T(a, b) = 1 - C(a^C, b^C)$.

$\Rightarrow$ is an S-Implication operator means that $a \Rightarrow b = C(a^C, b)$, therefore,

$$\begin{aligned}
[a \equiv b]_{2\text{-S-Impl}} &= \frac{1}{2}\{T(C(a^C, b), C(b^C, a)), T(C(a, b^C), C(b, a^C))\} \\
&= T(a \Rightarrow b, b \Rightarrow a) = [a \equiv b]_1
\end{aligned}$$

since $C(b, a^C) = C(a^C, b)$ and $C(a, b^C) = C(b^C, a)$, commutative character of triangular conforms.

It is important to notice that the agreement between truth values co-implication and equality indices does not always hold when R-implication operators are used. The application of these concepts into the determination of implication-based similarity measures between fuzzy sets is conducted at two levels:

- The alternative similarity level
- The global similarity level

Let us consider two fuzzy sets A and B defined over the finite set of alternatives $\Omega = \{x_1, x_2 \ldots x_N\}$.

- ***Alternative similarity level***

Given an alternative $x_n \in \Omega$, this level aims at defining the similarity between the two membership values $\mu_A(x_n)$ and $\mu_B(x_n)$. If we consider $a = \mu_A(x_n)$ and $b = \mu_B(x_n)$, then equality indices $[a \equiv b]_1$ and $[a \equiv b]_2$ can be applied at this level leading to:

$$\begin{aligned} \textit{Similarit-Degree}_1(\mu_A(x_n), \mu_B(x_n)) &= [\mu_A(x_n) \equiv \mu_B(x_n)]_1 \\ &= T(\mu_A(x_n) \Rightarrow \mu_B(x_n), \mu_B(x_n) \Rightarrow \mu_A(x_n)) \end{aligned}$$

and

$$\begin{aligned} &\textit{Similarit-Degree}_2(\mu_A(x_n), \mu_B(x_n)) = [\mu_A(x_n) \equiv \mu_B(x_n)]_2 \\ &= \frac{1}{2}\left\{T(a \Rightarrow b, b \Rightarrow a), T\left(a^{\mathrm{C}} \Rightarrow b^{\mathrm{C}}, b^{\mathrm{C}} \Rightarrow a^{\mathrm{C}}\right)\right\} \\ &= \frac{1}{2}\left\{T(\mu_A(x_n) \Rightarrow b, b \Rightarrow \mu_A(x_n)), T\left([\mu_A(x_n)]^{\mathrm{C}} \Rightarrow [\mu_B(x_n)]^{\mathrm{C}}, [\mu_B(x_n)]^{\mathrm{C}} \Rightarrow [\mu_A(x_n)]^{\mathrm{C}}\right)\right\} \end{aligned}$$

Recall that this level depends on the adopted definition of the t-norm T and the implication operator $\Rightarrow$.

Table 4.4 shows some examples using several implication operators when the considered t-norm (T) is the *min* conjunctive operation.

- ***Global similarity level***

Given the similarity between the two membership degrees $\mu_A(x_n)$ and $\mu_B(x_n)$, for all $x_n \in \Omega$, this level produces a "global" similarity degree between both fuzzy sets A and B at a global level. In other words, this level is regarded as a synthetic function of μ_A and μ_B around the whole universe of alternatives Ω:

$$Sim(A, B) = h(\textit{Similarity-Degree}(\mu_A(x_n), \mu_B(x_n), x_n \in \Omega)$$

where $h(.)$, called the truth combining function, is a scalar truth evaluator "resuming/synthesizing" the similarity degree of all the alternatives $x_n \in \Omega$. Notice that the use of truth combining functions as *Inf* or *Sup* allows the values of $\mu_A(x_n)$ and $\mu_B(x_n)$ at one single point to represent the global similarity between the two fuzzy sets A and

Table 4.4 Truth values co-implication equality at the alternative similarity level ($[\mu_A(x_n) \equiv \mu_B(x_n)]_1$)

	***t*-norm (*T*): The *min* conjunctive operation**
Lukasiewicz implication operator	$1 - \mid \mu_A(x_n) - \mu_B(x_n) \mid$
Kleen-Dienes implication operator	$min\{ max(1 - \mu_A(x_n), \mu_B(x_n)), max(1 - \mu_B(x_n), \mu_A(x_n))\}$
Gödel implication operator	1 If $\mu_A(x_n) = \mu_B(x_n)$ $min(\mu_A(x_n), \mu_B(x_n))$ Otherwise
Reichenbach implication operator	$1 - \mu_A(x_n) + \mu_A(x_n)\mu_B(x_n)$ If $\mu_A(x_n) > \mu_B(x_n)$ $1 - \mu_B(x_n) + \mu_A(x_n)\mu_B(x_n)$ If $\mu_A(x_n) < \mu_B(x_n)$
Mamdani implication operator	$min(\mu_A(x_n), \mu_B(x_n))$

B. On the other hand, other truth combining functions (as the relative cardinality) produce a similarity measure taking into account all alternatives from the universe Ω.

Here follows some encountered implication-based similarity measures:

- ***Beg's Similarity Measure*** [19]

This implication-based similarity measure is based on the use of *Similarity-Degree*$_1$ (in order to measure the similarity between the two fuzzy degrees $\mu_A(x_n)$ and $\mu_B(x_n)$) as well as the truth combining function $h(.) \equiv Inf_x \in \Omega$:

$$Sim_{\text{Beg}}(A, B) = Inf_{x\in\Omega}\, [\mu_A(x) \equiv \mu_B(x)]_1$$
$$Sim_{\text{Beg}}(A, B) = Inf_{x\in\Omega}\, T(\mu_A(x) \Rightarrow \mu_B(x), \mu_B(x) \Rightarrow \mu_A(x))$$

Using the *Inf* function is considered as the t-norm, and using Kleen-Dienes implication operator, Beg's implication-based similarity measure is given by:

$$\begin{aligned} Sim_{\text{Beg}}(A, B) &= Inf_{x\in\Omega}\; min\,(\, max\,(1 - \mu_A(x), \mu_B(x)),\; max\,(1 - (\mu_B(x), \mu_A(x))) \\ &= Inf_{x\in\Omega}\; min\,(1 - \; min\,(\mu_A(x), 1 - \mu_B(x)), 1 - \; min\,(\mu_B(x), 1 - \mu_A(x)))) \end{aligned}$$

Finally, we get:

$$Sim_{\text{Beg}}(A, B) = Inf_{x\in\Omega}\, [1 - \; max\,(\, min\,(\mu_A(x), 1 - \mu_B(x)),\; min\,(\mu_B(x), 1 - \mu_A(x))))]$$

- ***Manhattan Similarity Measure*** [7]

An example of the generic aspect of the implication-based similarity measures is given when relative cardinality function is used at the global similarity level, the Lukasiewicz implication operator and the *min* operator *t*-norm, then $[a \equiv b]_1 = [a \equiv b]_2 = 1 - \mid a - b \mid$. In this case, the resulting implication-based similarity measure is the Manhattan (or W-Similarity) measure already obtained using the metric distance similarity measure:

$$Sim_{\mathrm{M}}(A,B) = 1 - \frac{1}{N}\sum_{n=1}^{N} |\mu_A(x_n) - \mu_B(x_n)|$$

- ***Degree of Matching*** (or *Consistency Index* [20])

This implication-based similarity measure is based on the use of *Similarity-Degree*$_1$, Mamdani implication operator with the *min* operator *t*-norm as well as the truth combining function $h(.) \equiv \max_{x \in \Omega}$. The resulting similarity measure, also called *Consistency index*, describes the highest degree to which both fuzzy sets overlap:

$$Sim_{\mathrm{DM}}(A,B) = max_{n=1,2\ldots,N}(\, min\,(\mu_A(x_n), \mu_B(x_n)))$$

4.4 Possibility Distributions Similarity Measures

Assessing agreement between possibility distributions is a very important issue in the application of possibility theory. For example, in possibilistic knowledge-based, expert, or database systems, the degree of compatibility of an input possibilistic information with a stored possibilistic information in the system is often used to indicate the degree to which the stored information (rule, knowledge, or simply data) is applicable to the context of the input information. The mathematical approaches for assessing the agreement between two possibility distributions are based on the concept of similarity.

Measuring the similarity between two possibilistic information elements from different information sources is an important issue. This problem is encountered in most real-world problems and has attracted a lot of attention for a long time. Most encountered studies consider possibility distributions "as if they were" membership functions of fuzzy sets, and thus, a possibilistic measure of similarity is "reduced" to the estimation of the similarity degree between fuzzy sets.

This "transposition" of similarity measures between fuzzy sets into similarity measures between possibility distributions uses the fact that membership functions and possibility distributions are mathematically defined in a similar manner, i.e., both produce measures in the unit interval over the universe of discourse. Nevertheless, this "vision" neglects the fundamental difference in terms of the semantic meaning between the two concepts of membership functions and possibility distributions.

To clarify this point, let us consider the example of three experts estimating the distance between two objects. In this example (see Fig. 4.13), each expert gives an

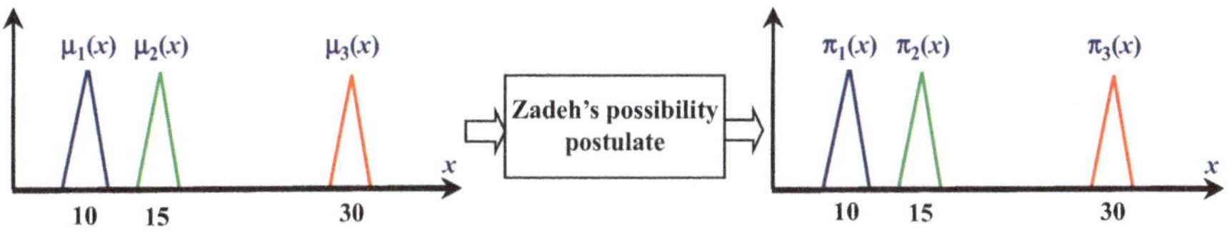

Fig. 4.13 Similarity measure between fuzzy sets versus possibility distributions

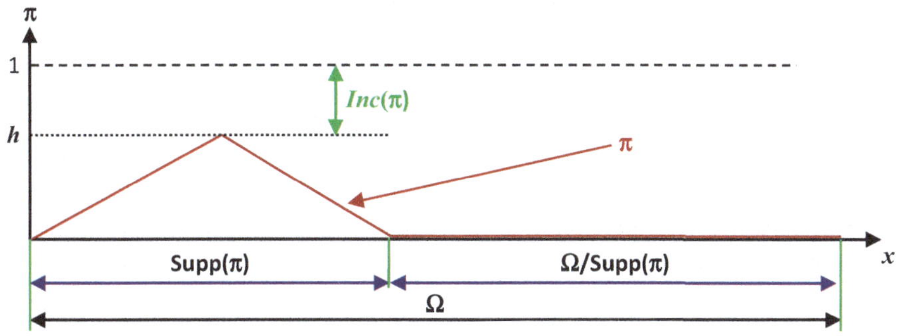

Fig. 4.14 Informative knowledge elements encapsulated in a possibility distribution

estimation of the target distance such that Expert1, the distance is *close to* 10 meters; Expert2, the distance is *close to* 15 meters; and Expert3, the distance is *close to* 30 meters. The "ambiguous" knowledge expressed by the three experts is represented using simple triangular membership functions to the three corresponding fuzzy sets representing the linguistic variables: *Close to*, 10, 15, and 30 meters (defined over the universe of distances). Using this fuzzy sets representation, all similarity measures between fuzzy sets will produce a zero similarity value between any two considered fuzzy sets. Knowing that the experts are expressing their own ambiguous pieces of knowledge about a unique (but unknown) distance value (i.e., we are facing three possibilistic information), then the projection of the membership functions into possibility distributions (using Zadeh's possibility postulate) is natural, leading thus to three possibility distributions, one per expert.

The transposition of similarity measures on fuzzy sets into possibility distributions will lead to a zero similarity between any couple of experts. Taking into account the semantic meaning of possibility distributions (trying to "discover" the unknown distance value) will intuitively show that Expert1 and Expert2 are somehow similar to each other much more that any of them when compared to the third expert. Therefore, the semantic meaning difference between fuzzy sets and possibility distributions is not completely "captured" through the direct transposition of fuzzy sets similarity measures into possibilistic similarity measures.

This section intends to present and to discuss the major issue of similarity measures between possibility distributions. It is worthwhile to recall, at this level, the main informative knowledge elements encapsulated in a possibility distribution.

As it is shown in Fig. 4.14, a possibility distribution vehicles three main informative knowledge elements:

- ***The domain of possible***, Supp(π): The alternatives universe Ω is partitioned into two domains, the domain of possible containing alternatives having a possibility degree different from zero (i.e., $\text{Supp}(\pi) = \{x \in \Omega \text{ such that } \pi(x) \neq 0\}$) and the domain of impossible $\Omega/\text{Supp}(\pi) = \{x \in \Omega \text{ such that } \pi(x) = 0\}$.
- ***Possibilistic ordering***: The second informative source of knowledge is related to relative possibility degrees attributed to different alternatives. In fact, consider two alternatives x_1 and $x_2 \in \Omega$ for which the possibilistic source of information has attributed the two possibility degrees $\pi(x_1)$ and $\pi(x_2)$. Besides the importance of the possibility degrees attributed to both alternatives, the fact that $\pi(x_1) > \pi(x_2)$, for instance, encapsulates an important informative aspect for which the possibilistic source of knowledge "considers" the occurrence of the alternative x_1 is more credible than the alternative x_2.
- ***Inconsistency***: The third informative source of knowledge encapsulated in a possibility distribution is its degree of inconsistency defined as $Inc(\pi) = 1 - h(\pi)$ ($Inc(\pi) \in [0, 1]$), where $h(\pi)$ indicates the height of the possibility distribution (i.e., the highest possibility degree). In fact, this element reflects the degree to which the possibility distribution can confirm, or not, if at least one of the alternatives is totally possible to occur. Unfortunately, this important informative knowledge source element simply "disappears" when a possibility distribution is normalized (forcing, thus, to have at least one alternative to be totally possible).

In the following subsections, the general conditions allowing to define a possibilistic similarity measure will be detailed. The three major categories of possibilistic similarity measures will then be detailed. The first one is the metric-based possibilistic similarity measures category; the second is the set-based possibilistic similarity measures category. Finally, the third category, called informational-based possibilistic similarity measures category, groups some similarity measures that do not necessarily transpose similarity measures between fuzzy sets into possibility distributions but try to exploit some of the informative facets of possibility distributions, integrating somehow the semantic meaning difference between membership functions and possibility distributions.

4.4.1 Definition, Possibilistic Similarity Measures

As previously mentioned, a similarity measure is a function that assigns a nonnegative real number to each pair of patterns, objects, images, functions, or possibility distributions, defining a notion of resemblance. Let $\Omega = \{x_1, x_2 \ldots x_N\}$ denote a finite set of alternatives and $\pi(\Omega)$ denote the set of all possibility distributions defined on Ω (we will use the notation $\pi * (\Omega)$ to indicate the set of all normalized possibility distributions $\pi * (\Omega) \subset \pi(\Omega)$).

A *possibilistic similarity measure*, denoted by *Sim*, is defined as a function:

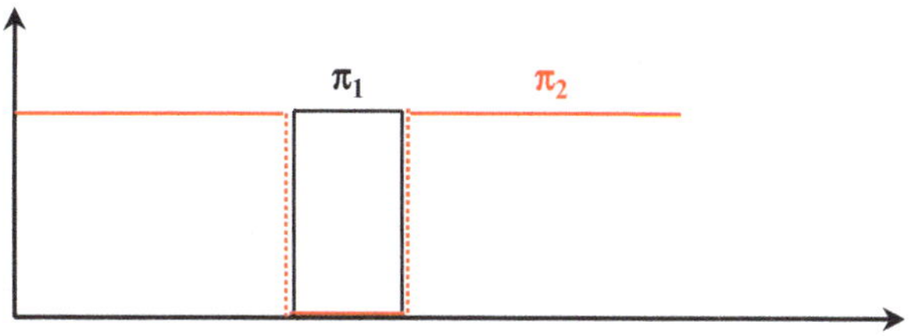

Fig. 4.15 Strong degeneracy property

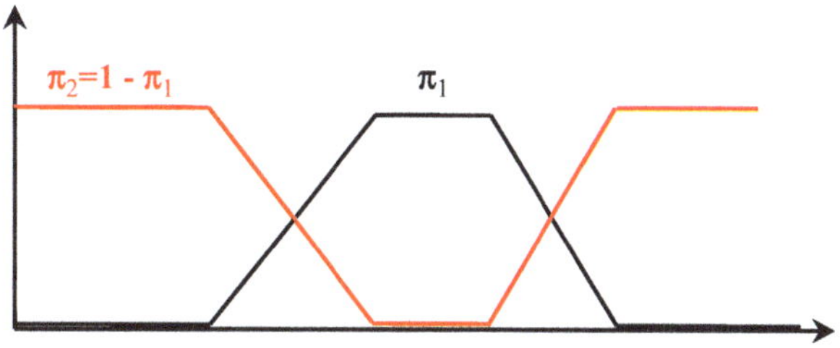

Fig. 4.16 Soft version of the strong degeneracy property

$$\begin{array}{rccc} Sim: & \pi(\Omega) \times \pi(\Omega) & \rightarrow & [0,1] \\ & (\pi_1, \pi_2) & \rightarrow & Sim(\pi_1, \pi_2) \end{array}$$

and satisfying some, or all, of the following properties of a similarity measure [21, 22], and [23]:

P1. *Nonnegative* and *Upper Bound Property*

$$0 \leq Sim(\pi_1, \pi_2) \leq 1 \forall \pi_1 \text{ and } \pi_2 \in \boldsymbol{\pi}(\Omega)$$

Maximum similarity is assumed to be obtained for two equal possibility distributions (i.e., $Sim(\pi, \pi) = 1$, $\forall \pi \in \pi(\Omega)$). In other words, identity implies full similarity. The reverse is not guaranteed, i.e., two different possibility distributions may have the full unit similarity without, necessarily, being identical.

P2.a *Strong Non-degeneracy Property*

Non-degeneracy fixes the conditions under which the similarity measure between two possibility distributions is null (i.e., the lower bound of the similarity value). This bound is assumed reached for two maximally contradictory possibility distributions. The strong non-degeneracy property stipulates that the null similarity is only obtained for binary-valued complementary possibility distributions; see Fig. 4.15:

$Sim(\pi_1, \pi_2) = 0$ if and only if:

- $\pi_1(x)$ and $\pi_2(x) \in \{0, 1\} \; \forall \; x \in \Omega$
- $\pi_1(x) = 1 - \pi_2(x) \; \forall \; x \in \Omega$

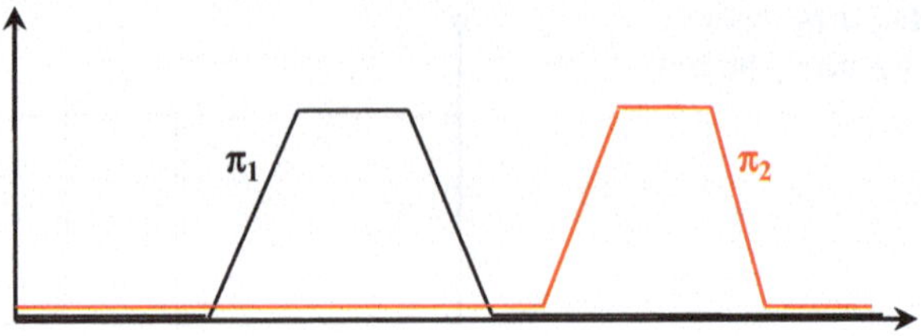

Fig. 4.17 Weak degeneracy property

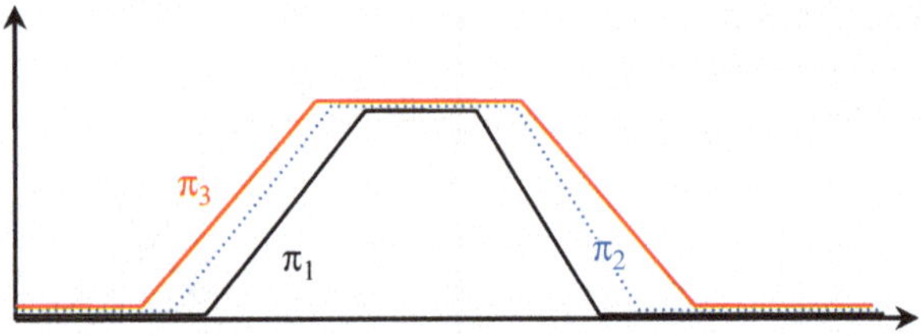

Fig. 4.18 Large inclusion property

This maximal contradiction concept is sometimes "softened" (Fig. 4.16) by removing the "binary" restriction of the considered possibility distributions (i.e., $\pi_1(x)$ and $\pi_2(x) \in [0, 1] \quad \forall\, x \in \Omega$).

P2.b ***Weak Non-degeneracy Property***

The weak non-degeneracy property does not impose binary-valued and complementary possibility distributions to attain the null similarity measure. This property stipulates non-degeneracy when the supports of the two possibility distributions have no intersection (i.e., $\mathrm{Supp}(\pi_1) \cap \mathrm{Supp}(\pi_2) = \Phi$) (Fig. 4.17):

$Sim(\pi_1, \pi_2) = 0$ if and only if $min(\pi_1(x), \pi_2(x)) = 0 \; \forall\, x \in \Omega$.

Notice that similarity measures satisfying any of the two non-degeneracy conditions will not capture the whole semantic significance of possibility distributions (discussed in the example developed in Fig. 4.13).

P3. ***Symmetric Property***:

$Sim(\pi_1, \pi_2) = Sim(\pi_2, \pi_1), \; \forall\, \pi_1, \pi_2 \in \boldsymbol{\pi}(\Omega)$

This intuitive property states that the degree of similarity of π_1 to π_2 is the same as the degree of similarity of π_2 to π_1.

P4.a ***Inclusion Transitive Property***

$\forall \pi_1, \pi_2, \pi_3 \in \pi(\Omega)$, where π_1 is more specific than π_1 (i.e., $\pi_1(x) \leq \pi_2(x), \forall x \in \Omega$) which is in turn more specific than π_3 (i.e., $\pi_1 \leq \pi_2 \leq \pi_3$), then:

$$Sim(\pi_1, \pi_3) \leq \; min\,[Sim(\pi_1, \pi_2), Sim(\pi_2, \pi_3)]$$

This property (also referred to as the *Large Inclusion* property) means that if π_2 "lies between" π_1 and π_3, then the degree of similarity π_1 to π_3 is at most equal to both the degrees of similarity of π_1 to π_2 and π_2 to π_3 (Fig. 4.18).

P4.b ***Strict Inclusion Property***

$\forall \pi_1, \pi_2, \pi_3 \in \pi(\Omega)$, where $\pi_1 \neq \pi_2 \neq \pi_3$. If $\pi_1 \leq \pi_2 \leq \pi_3$, then $Sim(\pi_1, \pi_3) < Sim(\pi_1, \pi_2)$.

It is worthwhile to notice that $\pi_1 \neq \pi_2$ (resp. $\pi_2 \neq \pi_3$) and $\pi_1 \leq \pi_2$ (resp.$\pi_2 \leq \pi_3$) implies the strict specificity (i.e., $\pi_1 < \pi_2$, and $\pi_2 < \pi_3$; see Fig. 4.18).

P5. ***Permutation Invariant Property***

Let π_1, π_2, π_3 and $\pi_4 \in \pi(\Omega)$ such that $Sim(\pi_1, \pi_2) > Sim(\pi_3, \pi_4)$. Suppose that for any x_n, $x_m \in \Omega$ and that $\forall j = 1...4$, we have $\pi'_j(x_n) = \pi_j(x_m)$, $\pi'_j(x_m) = \pi_j(x_n)$, and $\forall x_k \neq x_n, x_m\ \pi'_j(x_k) = \pi_j(x_k)$. Then:

$$Sim(\pi'_1, \pi'_2) > Sim(\pi'_3, \pi'_4)$$

4.4.2 Metric-Based Possibilistic Similarity Measures

The first category of possibilistic similarity measures is based on "projecting" the metric-based similarity measures of fuzzy sets into the possibilistic domain by representing possibility distributions as "points" in a metric space, and the similarity between these distributions is, then, regarded as a kind of an inverse of their distance in this metric space. Therefore, similarity measures, from this category, are conducted in two steps. First, the distance between possibility distributions is obtained using a distance measure; second, the degree of similarity is evaluated by converting the distance measure into a similarity measure using one of the existing distance-to-similarity transformations.

Let us consider two possibility distributions π_1 and $\pi_2 \in \pi(\Omega)$ defined on the finite set of alternatives $\Omega = \{x_1, x_2 ... x_N\}$. Table 4.5 resumes some proposed metric-based possibilistic similarity measures using Santini's distance-to-similarity transformation (i.e., $Sim(\pi_1, \pi_2) = 1 - dist(\pi_1, \pi_2)$).

Notice that among these metric-based measures, only the Chebyshev possibilistic similarity measure produces a similarity value related to only one alternative: the alternative showing the highest possibility difference. As $Sim_{\mathrm{NH}}(\pi_1, \pi_2)$, $Sim_{\mathrm{NE}}(\pi_1, \pi_2)$, and $Sim_{\mathrm{Ch}}(\pi_1, \pi_2)$ are based on metric distances which, by definition, satisfy the properties of non-negativity, symmetry, identity, and triangular inequality, hence, it is straightforward to check that these similarity measures satisfy all the basic similarity measure properties.

Table 4.5 Metric-based possibilistic similarity measures (using Santini's distance-to-similarity transformation)

Distance function	Metric-based possibilistic similarity measures
Normalized Hamming distance	*Manhattan* (or ***W-Similarity***) *similarity measure* [21] $Sim_{NH}(\pi_1,\pi_2) = 1 - \frac{1}{N}\sum_{n=1}^{N}\lvert\pi_1(x_n) - \pi_2(x_n)\rvert$
Normalized Euclidean distance	*Euclidean similarity measure* $Sim_{NE}(\pi_1,\pi_2) = 1 - \sqrt{\frac{1}{N}\sum_{n=1}^{N}\lvert\pi_1(x_n) - \pi_2(x_n)\rvert^2}$
Chebyshev distance	*Chebyshev* (or ***L-Similarity***) *similarity measure* [8] $Sim_{Ch}(\pi_1,\pi_2) = 1 - \max_{n=1,\ldots,N}\lvert\pi_1(x_n) - \pi_2(x_n)\rvert$
Canberra distance	*Canberra* (or ***S-Similarity***) *similarity measure* [8] $Sim_C(\pi_1,\pi_2) = 1 - \sum_{n=1}^{N}\frac{\lvert\pi_1(x_n)-\pi_2(x_n)\rvert}{\pi_1(x_n)+\pi_2(x_n)}$

4.4.3 Set-Based Possibilistic Similarity Measures

As previously stated, set-based similarity measures between fuzzy sets are the generalization of the classical set-theoretic similarity measures heavily relying on the definitions of cardinality and subsethood. This generalization is semantically consistent since membership functions define subsets with fuzzy boarders for which the use of an operator, like the α-cut operator, will "reduce" a fuzzy set into a classical crisp set for which different alternatives belong, or not, to this crisp set. Consequently, the generalization of the classical set-theoretic similarity measures towards fuzzy sets sounds natural and can be easily justified and interpreted. The important similarity, in terms of mathematical functions, between the two concepts of a membership function and a possibility distribution and the "extensive use" of Zadeh's possibility postulate (allowing to "numerically transform" a membership function into a possibility distribution) have encouraged researchers to extend the application of set-based similarity measures between fuzzy sets into the possibilistic context by a simple "duplication" of the existing measures while substituting membership functions by possibility distributions. Let us consider two possibility distributions π_1 and $\pi_2 \in \pi(\Omega)$ defined on the finite set of alternatives $\Omega = \{x_1, x_2 \ldots x_N\}$. Table 4.6 resumes major set-based possibilistic similarity measures proposed and used in the literature.

At this level, it is important to recall the interpretation and the semantic difference between the two concepts of membership functions and possibility distributions. In fact, let us consider a membership function μ_A:

$$\begin{aligned} \mu_A : \quad \Omega &\rightarrow [0,1] \\ x &\rightarrow \mu_A(x) \end{aligned}$$

Table 4.6 Major set-based possibilistic similarity measures

Possibilistic mutual subsethood $Sim_{\mathrm{M}}(\pi_1, \pi_2)$	$\frac{\lvert\pi_1 \cap \pi_2\rvert}{\lvert\pi_1 \cup \pi_2\rvert} = \frac{\sum_{n=1}^{N} min\,(\pi_1(x_n), \pi_2(x_n))}{\sum_{n=1}^{N} max\,(\pi_1(x_n), \pi_2(x_n))}$
Normalized possibilistic mutual subsethood $Sim_{\mathrm{NM}}(\pi_1, \pi_2)$	$\frac{1}{N} \frac{\sum_{n=1}^{N} min\,(\pi_1(x_n), \pi_2(x_n))}{\sum_{n=1}^{N} max\,(\pi_1(x_n), \pi_2(x_n))}$
Jaccard possibilistic index $Sim_{\mathrm{J}}(\pi_1, \pi_2)$	$\frac{\sum_{n=1}^{N} min\,(\pi_1(x_n), \pi_2(x_n))}{\sum_{n=1}^{N} max\,(\pi_1(x_n), \pi_2(x_n))}$
Eisler's possibilistic measure $Sim_{\mathrm{Ei}}(\pi_1, \pi_2)$	$\frac{2\sum_{n=1}^{N} min\,(\pi_1(x_n), \pi_2(x_n))}{\sum_{n=1}^{N} \pi_1(x_n) + \sum_{n=1}^{N} \pi_2(x_n)}$
Possibilistic degree of sameness $Sim_{\mathrm{BK}}(\pi_1, \pi_2)$	$min\left(\frac{\sum_{n=1}^{N} min\,(\pi_1(x_n), \pi_2(x_n))}{\sum_{n=1}^{N} \pi_1(x_n)}, \frac{\sum_{n=1}^{N} min\,(\pi_1(x_n), \pi_2(x_n))}{\sum_{n=1}^{N} \pi_2(x_n)}\right) = \left[\sum_{n=1}^{N} min\,(\pi_1(x_n), \pi_2(x_n))\right] \cdot \left[max\left(\sum_{n=1}^{N} \pi_1(x_n), \sum_{n=1}^{N} \pi_2(x_n)\right)\right]$
Possibilistic Restle's similarity $Sim_{\mathrm{R}}(\pi_1, \pi_2)$	$1 - \sum_{n=1}^{N} max\,(min\,(\pi_1(x_n), 1 - \pi_2(x_n)), min\,(\pi_2(x_n), 1 - \pi_1(x_n)))$
Possibilistic degree of nearness $Sim_{\mathrm{DN}}(\pi_1, \pi_2)$	$\frac{\sum_{n=1}^{N} [max\,(\pi_1(x_n), \pi_2(x_n)).\, min\,(\pi_1(x_n), \pi_2(x_n))]}{\sum_{n=1}^{N} [max\,(\pi_1(x_n), \pi_2(x_n))]^2}$

μ_A encapsulates the ambiguous knowledge defining a fuzzy set A, where all alternatives $x \in \Omega$ "physically" belong to the fuzzy set A with total, partial, or null membership degrees. Therefore, if two fuzzy sets A and B are defined on the universe Ω (with μ_A and μ_B as their respective membership functions), then computing the similarity degree between these two fuzzy sets, i.e., $Sim(A, B)$, in terms of subsethood and set equality is fully meaningful. On the other hand, a possibility distribution π defined on the universe of alternatives Ω:

$$\begin{aligned} \pi:\ \Omega &\rightarrow [0, 1] \\ x &\rightarrow \pi(x) \end{aligned}$$

encapsulates an epistemic knowledge at the level of each alternative $x \in \Omega$ in terms of its possibility degree of being the unique occurring alternative (i.e., represents an epistemic knowledge in the framework of uncertainty). This means that one and only one alternative from the universe Ω occurs and our state of knowledge concerning the unique occurring alternative is expressed through possibility degrees associated to all alternatives from Ω. Therefore, the extension use of a set-based similarity measure between fuzzy sets into a similarity measure between possibility distributions means that the possibility distributions are considered as if they "represent" physical fuzzy sets defined on the universe Ω. In other words, computing the similarity degree $Sim(\pi_1, \pi_2)$ between two possibility distributions π_1 and π_2 as being numerically equal to $Sim(A, B)$ where $\mu_A(x) = \pi_1(x)$ and $\mu_B(x) = \pi_2(x)$ means that the possibility distributions are used as inputs at an inverse transformation of Zadeh's possibility postulate in order to obtain fuzzy sets defined on Ω. This means that the two equivalent fuzzy sets A and B have no physical existence and their membership functions only reflect the possibility degrees of occurrence considered as virtually representing fuzzy sets. Hence, the focus of set-based possibilistic similarity measures is to be positioned on the intuition that the degree of similarity should take into account both the amount of overlap and the amount of symmetric difference between the "virtual" sets of possibility degrees in the domain of possible.

4.4.4 Informational-Based Possibilistic Similarity Measures

In the previous sections, possibilistic similarity measures based on "transforming" the existing metric-based and set-based similarity measures in order to be applied for measuring the similarity between possibility distributions were detailed. In this section, some similarity measures developed in the framework of possibility theory and taking into consideration some possibilistic informational facets are detailed.

4.4.4.1 Possibilistic Degree of Matching

The first and simplest, informational-based possibilistic similarity measure, called the *possibilistic degree of matching* [20], or the *consistency index*, considers the highest degree to which two possibility distributions $\pi_1$and π_2 (defined on the universe Ω) overlap: $h(\pi_1 \cap \pi_2) = 1 - Inc(\pi_1 \cap \pi_2)$, as the similarity measure between any two possibility distributions (Fig. 4.19).

$$Sim_{\mathrm{DM}}(\pi_1, \pi_2) = max_{n=1,2\ldots,N}(min(\pi_1(x_n), \pi_2(x_n)))$$

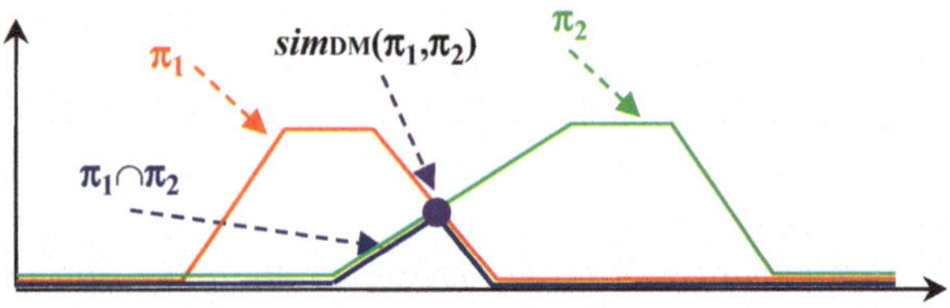

Fig. 4.19 Possibilistic degree of matching

It is worthwhile to notice that besides the fact that the possibilistic degree of matching captures an important informative facet of the joint possibility distribution $\pi_1 \cap \pi_2$, it produces a similarity degree related to only one alternative having the highest possibility degree in the overlap domain of possibility of both distributions $\pi_1$and π_2. Moreover, it is straightforward to prove that this possibilistic similarity measure satisfies the following properties:

- *Nonnegative* and *upper bound property*: $0 \leq \quad Sim_{\mathrm{DM}}(\pi_1, \pi_2) \leq 1$, $\forall \pi_1$ and $\pi_2 \in \pi * (\Omega)$;
- *Weak non-degeneracy property:* $Sim_{\mathrm{DM}}(\pi_1, \pi_2) = 0$ if and only if *min* $(\pi_1(x), \pi_2(x)) = 0 \; \forall \; x \in \Omega$;
- *Symmetric property:*
$Sim_{\mathrm{DM}}(\pi_1, \pi_2) = Sim_{\mathrm{DM}}(\pi_2, \pi_1) \leq 1$, $\forall \; \pi_1$ and $\pi_2 \in \pi * (\Omega)$;
- *Inclusion transitive property:* $\forall \pi_1$, π_2, $\pi_3 \in \pi * (\Omega)$, where $\pi_1 \leq \pi_2 \leq \pi_3$, then:

$$Sim_{\mathrm{DM}}(\pi_1, \pi_3) \leq \; min \left[Sim_{\mathrm{DM}}(\pi_1, \pi_2), Sim_{\mathrm{DM}}(\pi_2, \pi_3)\right];$$

- *Strict inclusion property:* $\forall \pi_1$, π_2, $\pi_3 \in \pi * (\Omega)$, where $\pi_1 \leq \pi_2 \leq \pi_3$, and $\pi_1 \neq \pi_2 \neq \pi_3$ then:

$$Sim_{\mathrm{DM}}(\pi_1, \pi_3) < Sim_{\mathrm{DM}}(\pi_1, \pi_2).$$

4.4.4.2 Information Closeness Index

The information closeness index is a possibilistic measure of similarity based on the use of the *U-uncertainty* measure proposed by Higashi and Klir [24, 25]. In this section, the definition of the *U-uncertainty* measure is recalled, and the use of this measure in order to obtain another measure called the *gain of information* is detailed. Finally, the information closeness distance measure and the information closeness index are defined as a possibilistic similarity measure.

Let $\Omega = \{x_1, x_2, x_3 \ldots x_N\}$ denote a finite set of alternatives, and π: $\Omega \rightarrow [0, 1]$ an ordered possibility distribution over Ω (i.e., different values $\pi(x_n)$ are arranged in a decreasing order such that $1 = \pi(x_1) \geq \pi(x_2) \geq \pi(x_3) \geq \ldots.. \geq \pi(x_N)$).

The U-uncertainty is a non-specificity measure expressed by the formula:

$$U(\pi) = \sum_{n=1}^{N} \left(\pi(x_n) - \pi(x_{n+1})\right) \log_2(n) + \left[1 - \pi(x_1)\right] \log_2(N)$$

where $\pi(x_{N+1}) = 0$ by convention. This U-uncertainty measures the "amount" of imprecision, uncertainty, or non-specificity related to the piece of ambiguous

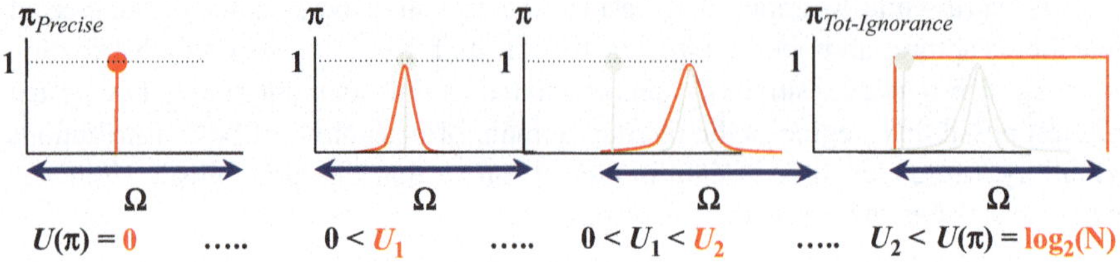

Fig. 4.20 The U-uncertainty measure of non-specificity

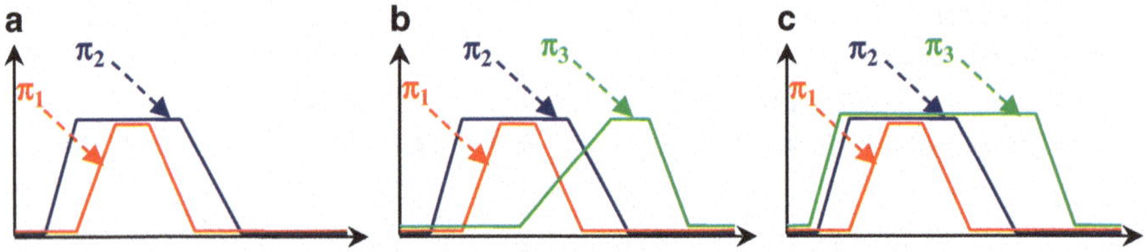

Fig. 4.21 The gain of information concept

information expressed by the possibility distribution π. In fact, as illustrated in Fig. 4.20, the range of $U(\pi)$ is $[0, \log_2(N)]$:

$$0 \leq U(\pi) \leq \log_2(N)$$

As can be noticed, the minimal U-uncertainty value (i.e., $U(\pi) = 0$) is obtained in the case of *Precise Knowledge* (i.e., when there is only one alternative $x_n \in \Omega$ for which $\pi_{\text{Precise}}(x_n) = 1$, and $\pi_{\text{Precise}}(x_m) = 0$ for all $m \neq n$); and the maximal U-uncertainty value (i.e., $U(\pi) = \log_2(N)$) is obtained in the case of *Total Ignorance* (i.e., $\pi_{\text{Ignorance}}(x_n) = 1$, for all n). Thus, $U(\pi)$ measures the effect of a passage from $\pi_{\text{Ignorance}}$ to the actual more specific possibility distribution π. In some applications, a normalized version of this measure can be used:

$$\widehat{U}(\pi) = U(\pi)/\log_2(N)$$

to obtain the property $0 \leq \widehat{U}(\pi) \leq 1$.

If two possibility distributions π_1 and π_2 are considered where π_1 is more specific than π_2: $\pi_1 \leq \pi_2$ (i.e., $\pi_1(x_n) \leq \pi_2(x_n)$, $n = 1 \ldots N$), then it can be easily shown that $U(\pi_1) \leq U(\pi_2)$. The quantity:

$$g(\pi_1, \pi_2) = U(\pi_2) - U(\pi_1)$$

is called the *gain of information* when π_2 is replaced by π_1.

Here follows some interesting properties of the gain of information function:

(a) If two possibility distributions π_1 and π_2 are considered where π_1 is more specific than π_2: $\pi_1 \leq \pi_2$ (Fig. 4.21a), then: $g(\pi_1, \pi_2) \geq 0$. The equality holds (i.e., $g(\pi_1, \pi_2) = 0$) if and only if $\pi_1 = \pi_2$.
(b) If two possibility distributions π_1 and π_2 are considered where π_1 is more specific than π_2: $\pi_1 \leq \pi_2$, then for any possibility distribution π_3 defined on the same universe (Fig. 4.21b), we have:

$$g(\pi_1, \pi_1 \cup \pi_3) \geq g(\pi_2, \pi_2 \cup \pi_3)$$

This means that the gain of information when $\pi_1 \cup \pi_3$ is replaced by π_1 is greater than the gain of information when $\pi_2 \cup \pi_3$ is replaced by π_2.

(c) If three possibility distributions π_1, π_2, and π_3 are considered where $\pi_1 \leq \pi_2 \leq \pi_3$ (Fig. 4.21c) then:

- $g(\pi_1, \pi_2) \leq g(\pi_1, \pi_3)$. The equality holds if and only if $\pi_2 = \pi_3$;
- $g(\pi_1, \pi_3) = g(\pi_1, \pi_2) + g(\pi_3, \pi_3)$. This means that the gain of information is additive when more specific possibility distributions are considered.

(d) Let π_1, π_2, and π_3 be three possibility distributions defined on the same universe Ω, then:

$$g(\pi_1, \pi_1 \cup \pi_2) + g(\pi_2, \pi_1 \cup \pi_2) \leq g(\pi_1 \cap \pi_2, \pi_1) + g(\pi_1 \cap \pi_2, \pi_2)$$

The information closeness, denoted by G, was proposed by Higashi and Klir [25] as a proximity measure (i.e., behaving as a distance measure) between two possibility distributions π_1 and π_2 defined on the same universe Ω. It is defined as follows:

$$G(\pi_1, \pi_2) = 2.U(\pi_1 \cup \pi_2) - U(\pi_1) - U(\pi_2)$$

where $U(\pi)$ is the U-uncertainty measure and $\cup$ is taken as the maximum operator.

Notice that $G(\pi_1, \pi_2)$ can be written as:

$$\begin{aligned} G(\pi_1, \pi_2) &= [U(\pi_1 \cup \pi_2) - U(\pi_1)] + [U(\pi_1 \cup \pi_2) - U(\pi_2)] \\ &= g(\pi_1, \pi_1 \cup \pi_2) + g(\pi_2, \pi_1 \cup \pi_2) \end{aligned}$$

In other terms, $G(\pi_1, \pi_2)$ measures the cumulative gain of information when $\pi_1 \cup \pi_2$ is replaced by π_1 and when $\pi_1 \cup \pi_2$ is replaced by π_2.

Here follows some important characteristics of this proximity measure:

- $G(\pi_1, \pi_2) \leq 2.\log_2(N) - \log_2(N-1)$;
- If $\pi_1 \leq \pi_2$ (i.e., π_1 is more specific than π_2), then $G(\pi_1, \pi_2) = U(\pi_2) - U(\pi_1)$ measures the gain of information if π_2 is replaced by π_1;
- *Non-degeneracy property:* $G(\pi_1, \pi_2) = 0$ if and only if $\pi_1 = \pi_2$;

- If $\pi_1 = \pi_2$, then $\pi_1 \cup \pi_2 = \pi_2 \cup \pi_1 = \pi_1 = \pi_2$. Therefore, $G(\pi_1, \pi_2) = 2.\ U(\pi_1 \cup \pi_2) - U(\pi_1) - U(\pi_2) = 0$;
- If $G(\pi_1, \pi_2) = 0$, then as $G(\pi_1, \pi_2) = g(\pi_1, \pi_1 \cup \pi_2) + g(\pi_2, \pi_1 \cup \pi_2)$ and $g(\pi_1, \pi_1 \cup \pi_2) \geq 0$ and $g(\pi_2, \pi_1 \cup \pi_2) \geq 0$; this implies that $g(\pi_1, \pi_1 \cup \pi_2) = 0$ and $g(\pi_2, \pi_1 \cup \pi_2) = 0$. Consequently, $\pi_1 = \pi_1 \cup \pi_2$ and $\pi_2 = \pi_1 \cup \pi_2$, this leads to $\pi_1 = \pi_2$;
- *Symmetry property:* $G(\pi_1, \pi_2) = G(\pi_2, \pi_1)$ (trivial);
- *Triangle property:* Let π_1,π_2, and π_3 be three possibility distributions defined on the same universe Ω, then $G(\pi_1, \pi_3) \leq G(\pi_1, \pi_2) + G(\pi_2, \pi_3)$ where the equality holds if and only if $\pi_2 \leq \pi_1 \cup \pi_3$;
- If $\pi_1 \cup \pi_2 = \Omega$, then $G(\pi_1, \pi_2) = 2\log_2(N) - U(\pi_1) - U(\pi_2) = 2.\ \log_2(N) - \log_2(N-1)$

(this can be easily shown using the definition of $U(\pi_1) + U(\pi_2)$ and the fact that $\pi_2 = 1 - \pi_1$). In other words, the maximal value of $G(\pi_1, \pi_2)$ is reached for two maximally contradictory possibility distributions.

From the previous properties, it is straightforward to show that the information closeness is a metric distance in $\pi(\Omega)$, the set of all possibility distributions defined on the universe Ω.

Example Consider the following possibility distributions: π_1, π_2, π_3, and π_4 defined on $\Omega = \{x_1, x_2, x_3, x_4\}$ by:

$$\pi_1 = [\pi_1(x_1), \pi_1(x_2), \pi_1(x_3), \pi_1(x_4)] = [1, 0.5, 0.3, 0.7]$$

$$\pi_2 = [\pi_2(x_1), \pi_2(x_2), \pi_2(x_3), \pi_2(x_4)] = [1, 0, 0, 0]$$

$$\pi_3 = [\pi_3(x_1), \pi_3(x_2), \pi_3(x_3), \pi_3(x_4)] = [0.9, 1, 0.3, 0.7]$$

$$\pi_4 = [\pi_4(x_1), \pi_4(x_2), \pi_4(x_3), \pi_4(x_4)] = [0, 1, 0.3, 0.7]$$

We have $G(\pi_1, \pi_2) = 1.12$, $G(\pi_1, \pi_3) = 0.52$, $G(\pi_1, \pi_4) = 1.08$. Thus, according to G, π_3 is the closest to π_1 and π_4 is closer to π_1 than π_2.

Finally, the use of Richter's transformations [3] allows transforming the information closeness distance measure into a possibilistic similarity measure called information closeness index:

$$Sim_{\mathrm{ICI}}(\pi_1, \pi_2) = 1 - \frac{G(\pi_1, \pi_2)}{max_G}$$

where $G(\pi_1, \pi_2) \in [0, max_G]$, for all π_1, $\pi_2 \in \pi(\Omega)$, and $max_G = [2.\ \log_2(N) - \log_2(N-1)]$ is the maximal information closeness distance.

It is worthwhile to notice that $Sim_{\mathrm{ICI}}(\pi_1, \pi_2) \in [0, 1]$ for all π_1, $\pi_2 \in \pi(\Omega)$.

4.4.4.3 Information Affinity Index

Let $\Omega = \{x_1, x_2..., x_N\}$ denote a finite universe; π_1 and π_2 are two normal possibility distributions defined on Ω (i.e., $\pi_1, \pi_2 \in \pi * (\Omega)$). Jenhani et al. [23] have proposed a possibilistic similarity measure, called *information affinity index* and denoted *InfoAff* (π_1,π_2). It takes into account the Manhattan distance along with the inconsistency measure as follows:

$$InfoAff(\pi_1, \pi_2) = 1 - \frac{d(\pi_1, \pi_2) + Inc(\pi_1 \cap \pi_2)}{2}$$

where $d(\pi_1, \pi_2) = \frac{1}{N} \sum_{n=1}^{N} |\pi_1(x_n) - \pi_2(x_n)|$ represents the Manhattan distance between the two possibility distributions π_1 and π_2 and

$$Inc(\pi_1 \cap \pi_2) = 1 - \max_{x_n \in \Omega} \left(\min \left[\pi_1(x_n), \pi_2(x_n) \right] \right)$$

is the consistency degree of the conjunctive possibility distribution $\pi_1 \cap \pi_2$. Notice that $d(\pi_1, \pi_2)$ computes the "possibilistic" Manhattan distance (i.e., between possibility degrees) and $Inc(\pi_1 \cap \pi_2)$ measures the degree of "conflict" between the two possibility distributions.

Two possibility distributions, π_1 and π_2, are said to have a *strong affinity* if *InfoAff* $(\pi_1, \pi_2) = 1$. This case is obtained when $d(\pi_1, \pi_2) = 0$ meaning that $\pi_1 = \pi_2$ (as we are dealing with normal possibility distributions, this will lead to have *Inc* $(\pi_1 \cap \pi_1) = 0$).

Similarly, two distributions, π_1 and π_2, are said to have a *weak affinity* if *InfoAff* $(\pi_1, \pi_2) = 0$.

Information affinity index satisfies the following properties:

- *InfoAff*(.,.) is a ***nonnegative, upper,*** and ***lower bounded*** similarity measure.
  This can be proved by noticing that $0 \leq d(\pi_1, \pi_2) \leq 1$ as well as $0 \leq Inc(\pi_1 \cap \pi_2) \leq 1$ (leading to $0 \leq InfoAff(\pi_1, \pi_2) \leq 1$).
- *InfoAff*(.,.) is a ***symmetric*** similarity measure: $InfoAff(\pi_1, \pi_2) = InfoAff(\pi_2, \pi_1)$.
- *InfoAff*(.,.) is a ***non-degenerate*** similarity measure ($InfoAff(\pi_1, \pi_2) = 0$ if and only if $\pi_2 = \pi_1$).
- *InfoAff*(.,.) is an ***inclusion transitive*** similarity measure.

If π_1 is more specific than π_2 (i.e., $\pi_1 \leq \pi_2$) which is in turn more specific than π_3 (i.e., $\pi_2 \leq \pi_3$), then π_1, π_2, and π_3 are fully consistent with each other (i.e., $\exists x \in \Omega$ for which $\pi_1(x) = \pi_2(x) = \pi_3(x) = 1$). This leads to the fact that $Inc(\pi_1 \cap \pi_2) = Inc(\pi_1 \cap \pi_3) = Inc(\pi_2 \cap \pi_3) = 1$. Moreover, it is straightforward to see that $d(\pi_1, \pi_2) \leq d(\pi_1, \pi_3)$.

Therefore,

$$1 - \frac{d(\pi_1, \pi_2) + 1}{2} \leq 1 - \frac{d(\pi_1, \pi_3) + 1}{2}$$

and thus *InfoAff*$(\pi_1, \pi_2) \geq$ *InfoAff*(π_1, π_3).

4.4.4.4 Possibilistic Similarity Index

Let $\Omega = \{x_1, x_2 \ldots x_N\}$ denote a finite universe; π_1 and π_2 are two possibility distributions defined on Ω. The *similarity index* of the pair (π_1, π_2) [26], denoted by $SI(\pi_1, \pi_2)$, is defined by:

$$SI(\pi_1, \pi_1) = min\{\alpha(\pi_1, \pi_2), \alpha(1 - \pi_1, 1 - \pi_2)\}$$

where

$$\alpha(\pi_1, \pi_2) = \frac{\sum_{n=1}^{N} \pi_1(x_n) \cdot \pi_2(x_n)}{\sum_{n=1}^{N} \{max(\pi_1(x_n)\pi_2(x_n))\}^2}$$

In case where $\sum_{n=1}^{N} \{max(\pi_1(x_n), \pi_2(x_n))\}^2 = 0$, we set $\alpha(\pi_1, \pi_2) = 1 = SI(\pi_1, \pi_2)$.

Notice that $\alpha(\pi_1, \pi_2)$ (similar to Jaccard Index for which the conjunctive operator is the product operation and the disjunctive operator is the maximum operation) measure the similarity between the two possibility distributions π_1, $\pi_2 \in \pi(\Omega)$. Therefore, the similarity index is based on evaluating both similarity degrees $\alpha(\pi_1, \pi_2)$ between π_1 and π_2 as well as $\alpha(1 - \pi_1, 1 - \pi_2)$ between their contradictory possibility distributions $1 - \pi_1$ and $1 - \pi_2$ and choosing the smallest value as being the adopted possibilistic similarity degree between $\pi_1$and π_2.

Example Consider the following possibility distributions π_1 and π_2 defined on $\Omega = \{x_1, x_2, x_3, x_4\}$ by:

$$\pi_1 = [\pi_1(x_1), \pi_1(x_2), \pi_1(x_3), \pi_1(x_4)] = [0.3, 0.5, 0.7, 1]$$
$$\pi_2 = [\pi_2(x_1), \pi_2(x_2), \pi_2(x_3), \pi_2(x_4)] = [0.1, 0.3, 0.1, 1]$$

Then:

$$\sum_{n=1}^{N} \pi_1(x_n) \cdot \pi_2(x_n) = 0.3 \times 0.1 + 0.5 \times 0.3 + 0.7 \times 0.1 + 1 \times 1$$

$$= 0.03 + 0.15 + 0.07 + 1 = 1.25$$

$$\sum_{n=1}^{N} \{max(\pi_1(x_n), \pi_2(x_n))\}^2 = (0.3)^2 + (0.5)^2 + (0.7)^2 + (1)^2$$

$$= 0.09 + 0.25 + 0.49 + 1 = 1.83$$

$$\sum_{n=1}^{N} [1 - \pi_1(x_n)] \cdot [1 - \pi_2(x_n)] = 0.7 \times 0.9 + 0.5 \times 0.7 + 0.3 \times 0.9 + 0 \times 0$$

$$= 0.63 + 0.35 + 0.27 = 1.25$$

$$\sum_{n=1}^{N} \{max([1 - \pi_1(x_n)], [1 - \pi_2(x_n)])\}^2 = (0.9)^2 + (0.7)^2 + (0.9)^2$$

$$= 0.81 + 0.49 + 0.81 = 2.11$$

Thus, $\alpha(\pi_1, \pi_2) = 1.25/1.83 = 0.68$, $\alpha(1 - \pi_1, 1 - \pi_2) = 1.25/2.11 = 0.59$, and finally:

$$SI(\pi_1, \pi_2) = min\{\alpha(\pi_1, \pi_2), \alpha(1 - \pi_1, 1 - \pi_2)\} = min\{0.68, 0.59\} = 0.59.$$

The *similarity index* $SI(\pi_1, \pi_2)$ satisfies the following properties:

(a) $SI(\pi_1, \pi_2)$ is a ***nonnegative*** measure, ***upper bonded, and non-degenerate*** ($0 \leq SI(\pi_1, \pi_2) \leq 1$) with $SI(\pi, \pi) = 1$, for all possibility distributions π defined on Ω;

Proof If $\sum_{n-1}^{N} \{max(\pi_1(x_n), \pi_2(x_n))\}^2 = 0$, then $\forall x_n \in \Omega$, $\pi_1(x_n) = \pi_2(x_n) = 0$ and, therefore, $\alpha(\pi_1, \pi_2) = 1$. Again, $1 - \pi_1(x_n) = 1 - \pi_2(x_n) = 1$, so that $\alpha(1 - \pi_1, 1 - \pi_2) = 1$. Hence, $SI(\pi_1, \pi_2) = 1$. Otherwise, $\exists x \in \Omega$ such that $max(\pi_1(x), \pi_2(x)) > 0$. In this case, $\alpha(\pi_1, \pi_2) \geq 0$. Similarly, $\alpha(1 - \pi_1, 1 - \pi_2) \geq 0$; therefore, $SI(\pi_1, \pi_2) \geq 0$. Again, $\forall x_n \in \Omega : \pi_1(x_n) \leq max(\pi_1(x), \pi_2(x))$ and $\pi_2(x_n) \leq max(\pi_1(x), \pi_2(x))$ will imply at once that $\sum_{n=1}^{N} \pi_1(x_n) \times \pi_2(x_n) \leq \sum_{n=1}^{N} \{max(\pi_1(x_n), \pi_2(x_n))\}^2$, i.e., $\alpha(\pi_1, \pi_2) \leq 1$.

Similarly, $\alpha(1 - \pi_1, 1 - \pi_2) \leq 1$ and hence $SI(\pi_1, \pi_2) \leq 1$. Thus, $0 \leq SI(\pi_1, \pi_2) \leq 1$.

(b) $SI(\pi_1, \pi_2)$ is a ***symmetric*** measure: (i.e., $SI(\pi_1, \pi_2) = SI(\pi_2, \pi_1)$).
(c) For all π_1 and π_2, $SI(\pi_1, \pi_2) = SI(1 - \pi_1, 1 - \pi_2)$.
(d) Two possibility distributions π_1 and π_2 are equal if and only if $SI(\pi_1, \pi_2) = 1$.

Proof If $\pi_1 = \pi_2$ then from the definition, it follows that $SI(\pi_1, \pi_2) = 1$. Let us first suppose that $SI(\pi_1, \pi_2) = 1$, i.e., $\min\{\alpha(\pi_1, \pi_2), \alpha(1 - \pi_1, 1 - \pi_2)\} = 1$. Now, since $\alpha(\pi_1, \pi_2) \leq 1$ and $\alpha(1 - \pi_1, 1 - \pi_2) \leq 1$, it follows that

$\alpha(\pi_1, \pi_2) = \alpha(1 - \pi_1, 1 - \pi_2) = 1$. Therefore:

$$\sum_{n=1}^{N} \pi_1(x_n) \times \pi_2(x_n) = \sum_{n=1}^{N} \{max(\pi_1(x_n), \pi_2(x_n))\}^2$$

Again, we have:

$\pi_1(x_n) \leq max(\pi_1(x_n), \pi_2(x_n))$ and $\pi_2(x_n) \leq max(\pi_1(x_n), \pi_2(x_n))$ leading to:

$$\pi_1(x_n) \cdot \pi_2(x_n) \leq [max(\pi_1(x_n), \pi_2(x_n))]^2 \forall x_n \in \Omega.$$

Given that $\sum_{n=1}^{N} \pi_1(x_n).\pi_2(x_n) = \sum_{n=1}^{N} \{max(\pi_1(x_n), \pi_2(x_n))\}^2$, we must have:

$\pi_1(x_n).\ \pi_2(x_n) = [max(\pi_1(x_n), \pi_2(x_n))]^2\ \forall\ x_n \in \Omega$ (otherwise, $\alpha(\pi_1, \pi_2)$ would become less than unity, which is untrue). Now, for a given $x_n \in \Omega$ if we assume $\pi_1(x_n) \geq \pi_2(x_n)$, then to obtain

$$\pi_1(x_n).\pi_2(x_n) = [max(\pi_1(x_n), \pi_2(x_n))]^2$$

either $\pi_1(x_n) = \pi_2(x_n) \neq 0$, or $\pi_1(x_n) = \pi_2(x_n) = 0$. The same reasoning can be conducted if $\pi_1(x_n) \leq \pi_2(x_n)$. So finally $\pi_1 = \pi_2$.

(e) If $SI(\pi_1, \pi_2) = 0$, then either:

- $\pi_1 \cap \pi_2 = 0$, i.e., $min(\pi_1(x_n), \pi_2(x_n)) = 0\ \forall x_n \in \Omega$;
- $\pi_1 \cup \pi_2 = \Omega$, i.e., $max(\pi_1(x_n), \pi_2(x_n)) = 1\ \forall x_n \in \Omega$;
- $\pi_2 = 1 - \pi_1$, i.e., π_2 is the complementary possibility distribution of π_1.

Proof $SI(\pi_1, \pi_2) = 0 = min\ \{\alpha(\pi_1, \pi_2), \alpha(1 - \pi_1, 1 - \pi_2)\}$.

- If $\alpha(\pi_1, \pi_2) = 0$, then $\sum_{n=1}^{N} \pi_1(x_n).\pi_2(x_n) = 0$. Since $\pi_1(x_n).\ \pi_2(x_n) \geq 0$ for all $x_n \in \Omega$, therefore, $min(\pi_1(x_n), \pi_2(x_n)) = 0$ meaning that $\pi_1 \cap \pi_2 = 0$.
- If $\alpha(1 - \pi_1, 1 - \pi_2) = 0$, then $\sum_{n=1}^{N} (1 - \pi_1(x_n)).(1 - \pi_2(x_n)) = 0$, therefore,

$$min(1 - \pi_1(x_n), 1 - \pi_2(x_n)) = 0$$

meaning that $\pi_1 \cup \pi_2 = \Omega$.

- If $\alpha(\pi_1, \pi_2) = 0$ and $\alpha(1 - \pi_1, 1 - \pi_2) = 0$, then both equalities:

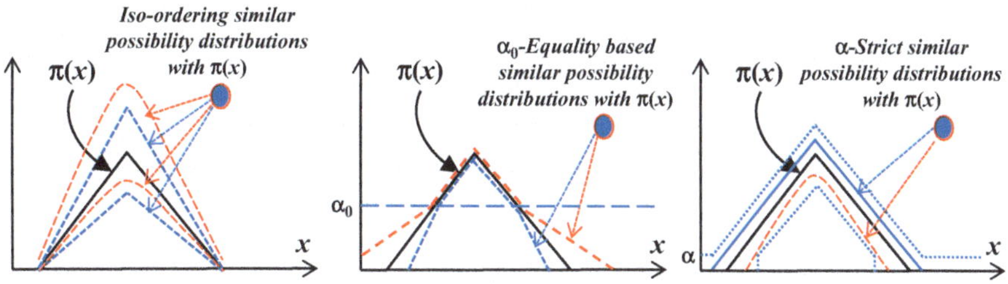

Fig. 4.22 Examples of Huete binary possibility

$$\pi_1(x_n).\pi_2(x_n) = 0 \text{ and } (1 - \pi_1(x_n)).(1 - \pi_2(x_n)) = 0$$

are verified for all $x_n \in \Omega$. Therefore, if $\pi_1(x_n) = 0(\text{resp}.\pi_2(x_n) = 0)$, then $1 - \pi_2(x_n) = 0$ (resp. $1 - \pi_1(x_n) = 0$). This implies that $\pi_2 = 1 - \pi_1$.

4.4.4.5 Huete Similarity Measures

An important "family" of similarity measures between possibility distributions has been proposed by J. F. Huete [27]. These measures produce a binary value whether two possibility distributions are similar or not. Let $\Omega = \{x_1, x_2 .. .x_N\}$ denote a finite reference set; π_1 and π_2 are two possibility distributions defined on Ω. Three binary similarity measures are then proposed by J. F. Huete [27] (Fig. 4.22):

- ***Iso-ordering similarity measure*:** π_1 and π_2 are considered as iso-ordering similar if and only if both distributions exhibit the same "ordering" of possibility degrees for all singletons, i.e.,

$$\text{Sim}_{\text{Iso-ordering}}(\pi_1, \pi_2) = 1 \text{ if and only if :}$$
$$\forall x, x^{'} \in \Omega, \left[\pi_1(x) < \pi_1\left(x^{'}\right) \Leftrightarrow \pi_2(x) < \pi_2\left(x^{'}\right)\right]$$

- ***α_0-Equality-based similarity measure*:** π_1 and π_2 are considered as α_0-Equality-based similar if $\pi_1(x) = \pi_2(x)$ for all $x \in \Omega$ and all values of $\pi_1(x)$, $\pi_2(x)$ that are greater than a fixed possibility threshold value α_0: $\text{Sim}_{\alpha 0\text{ - }Equality\ -\ based}(\pi_1, \pi_2) = 1$ if and only if α - Cut$(\pi_1) = \alpha$ - Cut(π_2), $\forall\ \alpha \geq \alpha_0$, where the α-Cut is the crisp set $\{x/\pi_1(x) \geq \alpha\}$.
- ***α-Strict similitude measure*:** π_1 and π_2 are considered as α-Strict-similitude if the values for each $x \in \Omega$ for $\pi_1(x)$ and $\pi_2(x)$ differ in less than a given predefined possibility threshold value α:

$$Sim_{\alpha\text{-Strict similitude}}\ (\pi_1, \pi_2) = 1 \text{ if and only if : } \forall x \in \Omega, |\pi_1(x) - \pi_2(x)| \leq \alpha$$

These possibilistic similarity measures are illustrated in the Fig. 4.22:

Notice that the three possibilistic similarity measures proposed by Huete produce binary values instead of graded similarity degrees (i.e., Sim $(\pi_1, \pi_2) \in \{0, 1\}$); thus, they are extremely fragile. In fact, if only one singleton, which may be marginal, does not obey the similarity measure definition, then the considered possibility distributions will be considered as being not similar.

In order to remediate to the binary aspect of these measures, R. Sangüesa et al. [28] have extended Huete's measures as follows:

Given a possibilistic difference threshold value $\alpha \in [0, 1]$, two possibility distributions π_1 and π_2, defined over the same domain Ω, will be considered similar, if for most domain's singletons, then the possibility values $\pi_1(x)$ and $\pi_2(x)$ differ by less than α. Sangüesa's degree of similarity is defined as follows:

$$Sim_{\text{Sangüesa}}(\pi_1, \pi_2, \alpha) = 1 - \frac{\sum\limits_{\substack{n = 1,..N \\ |\pi_{1(x_n)} - \pi_2(x_n)| \geq \alpha}} |\pi_1(x_n) - \pi_2(x_n)|}{\sum\limits_{n=1,..N} |\pi_1(x_n) - \pi_2(x_n)|}$$

Notice that $Sim_{\text{Sangüesa}}(\pi_1, \pi_2, \alpha) = 1$ corresponds to total similarity using Huete's α-Strict-similitude measure. In this case π_1 and π_2 are said to be identical at α level.

References

1. N. Goodman, Seven strictures on similarity (1972)
2. E. Diday, Recent progress in distance and similarity measures in pattern recognition, Proc. Second IJCPR, vol 534, (1974)
3. M.M. Richter, Classification and learning of similarity measures, in *Information and Classification*, (Springer-Verlag Berlin Heidelberg, 1993), pp. 323–334
4. S. Santini, R. Jain, Similarity is a geometer. Multimed. Tools Appl. **5**, 277–306 (1997)
5. L.T. Kóczy, D. Tikk, Fuzzy rendszerek, *TypoTEX, Budapest* (2000)
6. J. Williams, N. Steele, Difference, distance and similarity as a basis for fuzzy decision support based on prototypical decision classes. Fuzzy Sets Syst. **131**, 35–46 (2002)
7. S.-M. Chen, M.-S. Yeh, P.-Y. Hsiao, A comparison of similarity measures of fuzzy values. Fuzzy Sets Syst. **72**, 79–89 (1995)
8. C.P. Pappis, N.I. Karacapilidis, A comparative assessment of measures of similarity of fuzzy values. Fuzzy Sets Syst. **56**, 171–174 (1993)
9. A. Tversky, Features of similarity. Psychol. Rev. **84**, 327 (1977)
10. R. Zwick, E. Carlstein, D.V. Budescu, Measures of similarity among fuzzy concepts: A comparative analysis. Int. J. Approx. Reason. **1**, 221–242 (1987)
11. B. Kosko, Fuzziness vs. probability. Int. J. Gen. Syst. **17**, 211–240 (1990)
12. R.A.M. Gregson, *Psychometrics of Similarity* (Academic Press, 1975)
13. D.J. Dubois, *Fuzzy Sets and Systems: Theory and Applications*, vol 144 (Academic Press, 1980)
14. W. Jia, W. Zhenyuan, Using neural networks to determine Sugeno measures by statistics. Neural Netw. **10**, 183–195 (1997)
15. W. Bandler, L. Kohout, Fuzzy power sets and fuzzy implication operators. Fuzzy Sets Syst. **4**, 13–30 (1980)

16. X.-J. Tong, S.-M. Zhang, L. Zhou, Q.-M. Huang, Similarity and nearness of fuzzy sets, in *2005 International Conference on Machine Learning and Cybernetics*, (2005), pp. 2668–2670
17. K. Hirota, W. Pedrycz, Handling fuzziness and randomness in process of matching fuzzy data, in *Proceedings of the Third IFSA Congress*, (1989), pp. 97–100
18. A. Cain, G. Harrison, An analysis of the taxonomist's judgment of affinity, in *Proceedings of the Zoological Society of London*, (1958), pp. 85–98
19. I. Beg, S. Ashraf, Similarity measures for fuzzy sets. Appl. Comput. Math **8**, 192–202 (2009)
20. H. Lee-Kwang, Y.-S. Song, K.-M. Lee, Similarity measure between fuzzy sets and between elements. Fuzzy Sets Syst. **62**, 291–293 (1994)
21. L. Xuecheng, Entropy, distance measure and similarity measure of fuzzy sets and their relations. Fuzzy Sets Syst. **52**, 305–318 (1992)
22. H. Bustince, Indicator of inclusion grade for interval-valued fuzzy sets. Application to approximate reasoning based on interval-valued fuzzy sets. Int. J. Approx. Reason. **23**, 137–209 (2000)
23. I. Jenhani, N.B. Amor, Z. Elouedi, S. Benferhat, K. Mellouli, Information affinity: A new similarity measure for possibilistic uncertain information, in *European Conference on Symbolic and Quantitative Approaches to Reasoning and Uncertainty*, (2007), pp. 840–852
24. M. Higashi, G.J. Klir, Measures of uncertainty and information based on possibility distributions. Int. J. Gen. Syst. **9**, 43–58 (1982)
25. M. Higashi, G.J. Klir, On the notion of distance representing information closeness: Possibility and probability distributions. Int. J. Gen. Syst. **9**, 103–115 (1983)
26. S. Raha, K.S. Ray, Reasoning with vague truth. Fuzzy Sets Syst. **105**, 385–399 (1999)
27. J. Huete, Aprendizaje de redes de creencia mediante la detecci on de independencias: modelos no probabilisticos, Ph. D. Thesis, Universidad de Granada, 1995
28. R. Sangüesa, J. Cabós, U. Cortes, Possibilistic conditional independence: A similarity-based measure and its application to causal network learning. Int. J. Approx. Reason. **18**, 145–167 (1998)

Chapter 5
The Interrelated Uncertainty Modeling Theories

5.1 Introduction

Uncertainty is one of the greatest challenges for engineering design. It is at the heart of decision-making processes in most real-world application areas. Two major and distinct reasons are behind the fact of encountering uncertainty. The first reason concerns the natural random property of "variability" in real-life observed phenomena (e.g., a measured value, the age of a person, the label of observed objects, etc.). The second reason is due to incomplete information and the lack of knowledge to achieve the decision-making process.

Let Ω denote a finite set of *mutually exclusive* alternatives. A useful interpretation of uncertainty consists of considering different alternatives, decisions, assumptions, or hypotheses, from Ω as propositions where precisely one and only one of which is true. The unique occurring alternative, called the *true alternative*, is unknown. This, in fact, resumes the uncertainty type of information imperfection where we face the problem of "discovering" or identifying the identity of the true occurring alternative. A real-world decision situation involving uncertainty can be formalized in different ways. Each formalization constitutes a mathematical model representation with its own semantics.

The very first mathematical framework for representing and dealing with uncertainty is based on the classical set theory. In fact, an ill-known quantity is represented as a subset of mutually exclusive alternatives, one of which is the true one. In this framework, incomplete and uncertain information is expressed in the form of an interval, or a subset, of alternatives assigned to the unknown quantity. This modeling process is referred to as *imprecision modeling*. For instance, in a physical measurements, the real observed value x_0 is frequently given as $x_0 \pm \Delta x$ (i.e., the precise and unknown true measurement x_0 is assumed to take one and only one value in the interval $[x_0 - \Delta x, x_0 + \Delta x]$, but we do not know which one it is). This imprecision modeling is not limited to numerical imprecisions. For instance, in semantic

B. Solaiman, É. Bossé, *Possibility Theory for the Design of Information Fusion Systems*, Information Fusion and Data Science,
https://doi.org/10.1007/978-3-030-32853-5_5

imprecision the unknown label of an observed object may be expressed as belonging to the subset of labels $\{C_2, C_4, C_5\}$, but we cannot tell precisely which one it is. Notice that imprecision modeling mainly considers the incomplete information source of uncertainty (i.e., the true alternative identity suffers from the lack of knowledge, and it is not necessarily random).

Nevertheless, since the eighteenth century and up to the middle of the twentieth century, most theoretical advances dealing with uncertainty modeling and uncertainty processing were devoted to the theory of probabilities. The basic probabilistic modeling approach consists, first, on normalizing the "total certainty" of alternatives' occurrence to the unit value and, then, attributing to each probable alternative a fraction of belief/certainty, called its probability, or chance, of occurrence. In fact, let $\Omega = \{x_1, \ldots, x_N\}$ denote a finite set of *mutually exclusive* alternatives, where the unique occurring alternative, i.e., the true alternative, is unknown. Therefore, the probabilistic modeling assigns to each alternative x_n, $n = 1, 2, \ldots, N$, a partial belief $P\{x_n\} \in [0, 1]$, such that:

$$\sum_{n=1}^{N} P\{x_n\} = 1$$

Partial beliefs $P\{x_n\}$, $n = 1, 2, \ldots, N$ are called *probability density values*. This point-function vision for the definition of probability values (i.e., probability density function) was then "enlarged," or enriched, by considering the definition of the probability measure, Pr, which is a set function defined on the power set 2^Ω (i.e., the set of all subsets of Ω) through the following axioms:

- $\Pr(\Phi) = 0$, $\Pr(\Omega) = 1$;
- $\Pr(A \cup B) = \Pr(A) + \Pr(B)$, for all $A, B \in 2^\Omega$ such that $A \cap B = \Phi$.

where Φ stands for the empty set.

In the early 1980s, two more mathematical formalisms appeared: evidence theory (also called belief functions theory, BFT, or *Dempster-Shafer theory*, DST) and possibility theory. DST is, somehow, a generalization of probability theory where the normalized total certainty unit value if fractioned and distributed through a *b*asic *p*robability *a*ssignment, *b.p.a*, function (also called a *mass function*) $m(.)$ over elements from the power set 2^Ω instead of the set of alternatives Ω. The basic axioms defining the *b.p.a* function are as follows:

- $m(\Phi) = 0$;
- $\sum_{A \subseteq \Omega} m(A) = 1$.

Moreover, the *b.p.a* function allows determining two dual nonadditive measures, *belief* and *plausibility measures*, which constitute a bracketing of the probability measure of different subsets of Ω. It is worthwhile to notice that DST has brought a real flexibility to probability theory by allowing to represent, in a logical framework, both imprecision and uncertainty. In fact, if we only know that the true alternative belongs to the subset $\{x_1, x_2\}$ without any other knowledge allowing to distinguish these two alternatives, then the probabilistic modeling cannot capture this kind of imprecise knowledge. In fact, $P\{x_1\} = P\{x_2\} = 1/2$ means that both alternatives have the same probability of occurrence and that the available knowledge does not allow distinguishing between these two alternatives. On the contrary, evidential modeling can express this type of knowledge simply by the following *b.p.a* function: $m(\{x_1, x_2\}) = 1$, $m(\{x_1\}) = 0$ and $m(\{x_2\}) = 0$.

Finally, possibility theory captures another facet of uncertainty where the available knowledge is epistemic. In this case, the total certainty is normalized to the unit at the level of each individual alternative x_n, $n = 1, 2, \ldots, N$, and not at the level of the whole set Ω. The possibilistic modeling assigns to each alternative x_n, $n = 1, 2, \ldots, N$, a possibility degree $\pi(x_n) \in [0, 1]$. Two confidence dual set functions, namely, *possibility* and *necessity measures*, are deduced from the basic possibilistic modeling.

As it can be witnessed, probability, evidence, and possibility theories share the same idea of normalizing the total certainty, or belief, and their fundamental difference should be positioned on the way this total certainty is distributed. In probability theory, this total certainty is fractioned between different alternatives, whereas in evidence theory this total certainty is fractioned between all subsets of alternatives. Finally, in possibility theory, the total certainty is not fractioned but is positioned at the level of each alternative. This difference between these three mathematical frameworks (in terms of total uncertainty distribution) vanishes when we have to deal with the ultimate target of decision-making. In fact, probability, evidence, and possibility theories have to transform the basic certainty fractioning elements (i.e., probability density functions, *b.p.a* functions, and possibility distributions) into decisional set functions, called measures: probability, belief, plausibility, possibility, and necessity measures. These measures are assumed to satisfy some specific properties and bring a meaningful knowledge in order to characterize the uncertainty of each subset of alternatives, and this, for the ultimate decision process.

Different representation formalisms of uncertainty utilized by probability, evidence, and possibility theories are best compared and contrasted when considered as special cases of the more general fuzzy measure theory[1] (to be referred as nonadditive or monotonic measures according to Wang and Klir [1]). In fact, the five set-function measures (i.e., probability, belief, plausibility, possibility, and necessity measures) are all shown to be nonadditive subset measures.

[1]From the preface of the (2009) Wang and Klir's book entitled *Generalized Measure Theory*, here is why we will be using *monotone measures* rather than *fuzzy measures* in the rest of this chapter. "...However, it should be emphasized that we made a deliberate decision to abandon the central term of our previous book (fuzzy measure theory), the term 'fuzzy measure.' We judge this term to be highly misleading. Indeed, the so-called fuzzy measures do not involve any fuzziness. They are just special set functions that are defined on specified classes of classical sets, not on classes of fuzzy sets. Since the primary characteristic of such functions is monotonicity, we deemed it reasonable to call these set functions monotone measures rather than fuzzy measures."

In the next section, a brief review of the fundamental definitions and concepts used in monotone measures theory is presented. A brief recall of probability, evidence, and possibility theories is then given. Finally, major bridges allowing to transform one uncertainty model into other models are detailed.

5.2 The Monotone Measures Theory

The concept of measures is one of the most important concepts in information processing, decision-making, and data mining. Generally speaking, an uncertainty measure consists of assigning to each event (i.e., a subset of the universe of discourse, set of alternatives Ω) a number in the unit interval and evaluating the likelihood (confidence, trust...) that the unknown true alternative (for which we are uncertain about its identity) belongs to the considered event. In this section, a *trust measure*, called a *capacity measure* by G. Choquet [2] or a *nonadditive measure* by Sugeno [3], is presented.

Let Ω denote a finite universe of discourse composed of exhaustive and mutually exclusive alternatives, and let $\mathcal{P}(\Omega)$ denote the power set of Ω (i.e., the set of all the subsets $A \subseteq \Omega$).

Definition, *σ-field*, $\mathcal{B}$ A *σ-field* (also called *σ-algebra*) $\mathcal{B}$ is a subset of the power set $\mathcal{P}(\Omega)$ such that:

1. $\Phi \in \mathcal{B}$ and $\Omega \in \mathcal{B}$;
2. If $A \in \mathcal{B}$, then $A^C \in \mathcal{B}$ (where A^C denotes the complement set of A);
3. If $A_n \in \mathcal{B}$, $n = 1, 2, \ldots$ (any sequence of sets in $\mathcal{B}$) then $\bigcup_{n=1}^{\infty} A_n \in \mathcal{B}$.

The couple $(\Omega, \mathcal{B})$ is called a *measurable space*.

5.2.1 Sugeno Monotone Measure Definition

Consider an exhaustive and finite set of mutually exclusive alternatives Ω, a σ-field $\mathcal{B}$ from the power set $\mathcal{P}(\Omega) = 2^{\Omega}$ and let $g : \mathcal{B} \rightarrow [0, 1]$ denote a function defined on the σ-field $\mathcal{B}$ and assuming values from the unit interval. In an uncertainty context framework on Ω, composed of exhaustive and mutually exclusive alternatives (i.e., Ω contains all possible alternatives that may occur and that one and only one of these alternatives occurs at a time, i.e., the true alternative), the function g is called a *Sugeno monotone measure* [4] (simply, monotone measure), if it satisfies the following axioms:

1. *Boundary conditions*: $g(\Phi) = 0$ and $g(\Omega) = 1$ (i.e., normality);
2. *Monotony*: $A, B \in \mathcal{B}$, if $A \subseteq B$, then $g(A) \leq g(B)$.

The issue is that we don't know this unique occurring alternative. The Sugeno monotone measure of a given event $A \subseteq \Omega, g(A)$, represents the amount of "belief," "*trust*," "certainty," or "*confidence*" that we have in the occurrence of A (i.e., that the true alternative belongs to A).

In fact,

- $g(\Phi) = 0$ means that no trust is placed on the empty set (i.e., the true alternative definitely does not belong to the empty set).
- $g(\Omega) = 1$ means that our global trust is placed on the universe of discourse Ω (i.e., the true alternative definitely does belong to the universe of discourse). For this, Ω is frequently called the *certain set*.
- If $A \subseteq B$ (i.e., the occurrence of A implies the occurrence of B), then it is natural that our confidence in the occurring of B is greater than our confidence in the occurring of A: $g(A) \leq g(B)$. In other words, the belief that the true alternative belongs to a given set must be at least as great as the belief that the true alternative belongs to any subset of the given set.

One of the reasons behind calling $g(.)$ as a *Sugeno monotone measure* is that it attributes values in the unit interval [0, 1] to each subset of Ω (in a similar way to fuzzy membership functions but with the difference that $g(.)$ "describes" subsets and not the elementary alternatives). Sugeno monotone measures should have been called: *trust or confidence measure functions*.

Two important properties of Sugeno monotone measures can be witnessed:

- Since $A \subseteq A \cup B$, $B \subseteq A \cup B$ and using the monotonicity axiom, we have:

$$g(A \cup B) \geq max\ [g(A), g(B)]$$

This means that the trust attributed to the occurrence of $A \cup B$ is greater than the maximum trust attributed to A and B.

- Since $A \cap B \subseteq A$, $A \cap B \subseteq B$ and using the monotonicity axiom, we have:

$$g(A \cap B) \leq min\ [g(A), g(B)]$$

This means that the trust attributed to the occurrence of $A \cap B$ is less than the minimum trust attributed to A and B.

Moreover, given a monotone measure g defined on Ω, a *dual monotone measure* g^{d} is defined as follows:

$$\begin{aligned} g^{\mathrm{d}}: \quad & \mathcal{B} \rightarrow [0, 1] \\ & A \rightarrow g^{\mathrm{d}}(A) = 1 - g(A^{\mathrm{C}}) \end{aligned}$$

where $\mathcal{B}$ is a σ-field from the power set $\mathcal{P}(\Omega)$ and A^{C} is the complement subset of A.

5.2.2 Distinguished Classes of Monotone Measures

In the framework of monotone measures theory, some measures may have particularly interesting additional properties. In fact, a monotone measure g defined on Ω is called:

- *Sub-additive*, whenever $\forall A, B \in \mathcal{P}(\Omega) : g(A \cup B) \leq g(A) + g(B)$.
- *Superadditive*, whenever $\forall A, B \in \mathcal{P}(\Omega) : g(A \cup B) \geq g(A) + g(B)$.
- *Possibility*: Recall that all monotone measures verify the property,

$$\forall A, B \in \mathcal{P}(\Omega) : g(A \cup B) \geq \ max\,[g(A), g(B)]$$

  The most "pessimistic" monotone measure, i.e., verifying:

$$g(A \cup B) = \ max\,[g(A), g(B)] \quad \forall A, B \in \mathcal{P}(\Omega)$$

  is called a *possibility measure*.

- *Necessity*, whenever $\forall A, B \in \mathcal{P}(\Omega) : g(A \cap B) = \ min\,[g(A), g(B)]$. Recalling that all monotone measures verify the property, $\forall A,\ B \in \mathcal{P}(\Omega) : g(A \cap B) \leq min\,[g(A), g(B)]$, means that a necessity measure is the most optimistic trust measure.

Moreover, here follows two distinguished classes of monotone measures:

A. ***Belief Measures***

A monotone measure which satisfies the additional following superadditivity axiom:

$$g(A_1 \cup A_2 \cup \ldots \cup A_N) \geq \sum\nolimits_n g(A_n) - \sum\nolimits_{i<k} g(A_i \cap A_k) + \ldots$$
$$+ (-1)^{n+1} g(A_1 \cap A_2 \cap \ldots \cap A_N)$$

for all families of subsets A_n of Ω is called a *belief measure* [5]. In the case of two events A and B from Ω, this inequality becomes:

$$g(A \cup B) \geq g(A) + g(B) - g(A \cap B)$$

From this definition, it is straightforward to show that:

- $g(A) + g(A^C) \leq 1$.
- Each belief measure is superadditive (i.e., $g(A \cup B) \geq g(A) + g(B)$, since $g(A \cap B) \geq 0$).
- If $A \cap B = \Phi$, then $g(A \cup B) \geq g(A) + g(B)$ means that the degree of belief associated with the union of two events is not smaller than the sum of the degrees of belief pertaining to the individual events.

B. ***Plausibility Measures***

A monotone measure which satisfies the additional following sub-additivity axiom:

$$g(A_1 \cap A_2 \cap \ldots \cap A_N) \leq \sum_n g(A_n) - \sum_{i<k} g(A_i \cup A_k) + \ldots$$
$$+ (-1)^{n+1} g(A_1 \cup A_2 \cup \ldots \cup A_N)$$

is called a *plausibility measure* [5]. In the case of two events A and B from Ω, this inequality becomes:

$$g(A \cap B) \leq g(A) + g(B) - g(A \cup B)$$

Moreover, it is easy to show that for a plausibility measure $g : g(A) + g(A^C) \geq 1$.

Given a belief measure, g, then the set function:

$$g^d : \quad \mathscr{B} \rightarrow [0, 1]$$
$$A \rightarrow 1 - g(A^C)$$

is a plausibility measure. The couple $(g, g^d = 1 - g(.^C))$ is called the *dual belief/plausibility measures*. In the particular case of dual belief/plausibility measures, we can easily prove that $g(A) \leq g^d(A)$ for all subsets $A \subseteq \Omega$ (since, $g^d(A) + g^d(A^C) \geq 1$, then $g^d(A) + 1 - g(A) \geq 1$ leading to $g(A) \leq g^d(A)$).

5.3 Uncertainty Theories in the Framework of Monotone Measures Theory

Let $\Omega = \{x_1, \ldots, x_N\}$ denote a finite set of *mutually exclusive* alternatives, where the unique occurring alternative, i.e., the true alternative, is unknown. This resumes the uncertainty type of information imperfection where we face the problem of "discovering" or identifying the identity of the true occurring alternative. In this section, the three theories dealing with uncertainty (i.e., DST, probability theory, and possibility theory) will be synthetized as particular cases of the monotone measures theory.

Dempster-Shafer Theory (DST)

Independently from monotone measures theory, Dempster-Shafer theory, DST, is considered as one of the most important uncertainty modeling theories. The DST is based on a seminal work of Dempster [6] and Shafer [7] that can be considered as a "generalization" of probability theory where probability densities are assigned to subsets and not only to the mutually exclusive alternatives.

Let Ω denote a finite universe of discourse (defined as a set of exhaustive and mutually exclusive alternatives) and assume that we are facing the uncertainty form of information imperfection (i.e., only one alternative from Ω occurs and that we do not know the identity of this true alternative). Uncertainty representation model used by DST is based on two fundamental assumptions:

(i) The total "certainty" is normalized to one, and,
(ii) Depending on our state of knowledge, the total unit certainty is "distributed" on the power set elements $\mathcal{P}(\Omega) = 2^{\Omega}$. This certainty distribution is, thus, defined through a *basic probability assignment* (*b.p.a*), or *mass function* m: $\mathcal{P}(\Omega) \to [0, 1]$, such that $m(\Phi) = 0$ and,

$$\sum_{A \subseteq \Omega} m(A) = 1$$

For each event (subset) $A \subseteq \Omega$, $m(A)$ expresses the proportion of all available relevant evidence that supports the claim that the true alternative belongs to the subset A but to no particular subset of A (i.e., the value $m(A)$ represents the support for the elements of A that cannot be apportioned to proper subsets of A).

Moreover, the support assigned to the entire set of alternatives Ω, i.e., $m(\Omega)$, represents the uncommitted belief (or ignorance). Every subset $A \subseteq \Omega$ for which $m(A) \neq 0$ is called a *focal element*. It is worthwhile to notice that it is not required that $m(\Omega) = 1$.

Given the *b.p.a* function, $m(.)$, two dual set functions *Bel* and *Pl* are then uniquely determined for all $A \in \mathcal{P}(\Omega)$ as follows:

- The *belief function* $Bel(A)$: $\mathcal{P}(\Omega) \to [0, 1]$ is defined as $Bel(A) = \sum_{B \subseteq A} m(B)$ where $Bel(A)$ defines our belief in the occurrence of the event $A \in \mathcal{P}(\Omega)$ as the sum of all the masse assignments that support its constituents. In other words, $Bel(A)$ is assumed to represent the evidence that the true alternative belongs to A and to subsets of A;
- The *dual plausibility function* defined as the summation of the masse assignments of the subsets B that intersect with the set of interest A, i.e., $B \cap A \neq \Phi$: $Pl(A) = \sum_{B \cap A \neq \Phi} m(B)$. $Pl(A)$ computes the masse assignments that are not supporting the complement of A (i.e., A^C). In other words, $Pl(A)$ represents the evidence that the true alternative belongs to A, to subsets of A as well as to sets that overlap with A.

These two set functions *Bel* and *Pl* give, respectively, the lower and upper trust bounds in the occurrence of each event, i.e., $Bel(A) \leq Pl(A)$ for all events $A \in \mathcal{P}(\Omega)$. They are mutually dual in the sense that one of them can be uniquely transformed by the other using the equation: $Pl(A) = 1 - Bel(A^C)$. It is worthwhile to notice that $Bel(A) + Bel(A^C) \leq 1$ (this inequality is called the *fundamental property of belief measures*). The value $(1 - Bel(A) + Bel(A^C))$ is, somehow, a measure of uncertainty, Fig. 5.1, concerning the occurrence of the event A (due to the presence of focal events overlapping A and A^C).

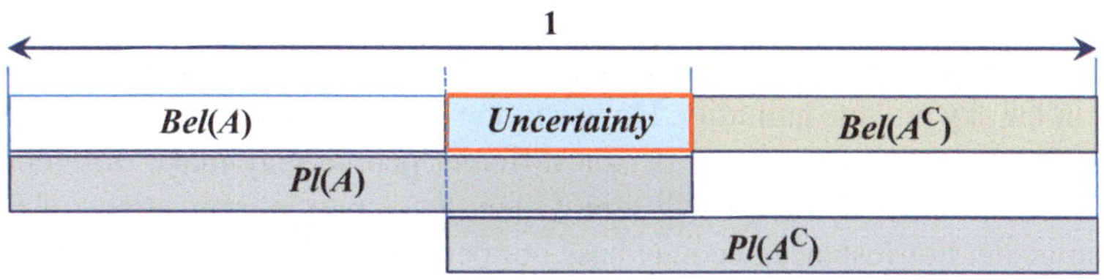

Fig. 5.1 Relation of the belief measure and its dual plausibility measure

Table 5.1 Conversion formulas in the DST

	M	Bel	Pl
$m(A) =$	$m(A)$	$\sum_{B \subseteq A} (-1)^{Card(A-B)} Bel(B)$	$\sum_{B \subseteq A} (-1)^{Card(A-B)} [1 - Pl(B^C)]$
$Bel(A) =$	$\sum_{B \subseteq A} m(B)$	$Bel(A)$	$1 - Pl(A^C)$
$Pl(A) =$	$\sum_{B \cap A \neq \Phi} m(B)$	$1 - Bel(A^C)$	$Pl(A)$

Moreover,

- if $A,B \in \mathcal{P}(\Omega)$ such that $A \subseteq B$, it is not required that $m(A) \leq m(B)$. In other words, the basic probability assignment m is not necessarily a monotone measure.
- If $A,B \in \mathcal{P}(\Omega)$ such that $A \subseteq B$, it is straightforward to show that $Bel(A) \leq Bel(B)$ and $Pl(A) \leq Pl(B)$. This means that both belief and plausibility functions defined in DST are monotone measures.

The knowledge of each of the three set functions m, Bel, and Pl is sufficient to determine the other two (Table 5.1 resumes the conversion formulas between these three functions).

Therefore, and as this has been shown, the DST can be considered as a "branch" of the monotone measures theory concerned by the particular class of belief monotone measures.

Probability Theory

Probability theory is certainly the most widely studied mathematical uncertainty modeling approach. The basic idea used by probability theory for uncertainty modeling consists on normalizing the global belief to one and assigning to each individual alternative, "x_n", $n = 1, \ldots, N$, a number in the unit interval $P\{x_n\} \in [0, 1]$ (called the probability density of the alternative x_n). As the global belief is normalized to one, then the probability densities must sum to "1", i.e., $\sum_{n=1}^{N} P\{x_n\} = 1$.

There are two common interpretations of the word "probability density." One is in terms of relative frequency. This intuitive interpretation means that the experience producing the uncertainty is repeated a very large number of times and that $P\{x_n\}$ measures/estimates/the chance of occurrence of the alternative x_n. A second interpretation of probability density is that it is a quantification of our degree of subjective personal belief that the alternative x_n will occur. To get a sense of what we mean by this, it may be helpful to consider the weather predicting example. When we think about the chances that it will rain today, we consider things like whether there are clouds in the sky and the humidity. However, the beliefs that we form based on these factors may vary from person to person different people may make different estimates of the probability that it will rain. Often, these two interpretations of probability coincide. For instance, we may base our personal beliefs about the chance that it will rain on an assessment of the relative frequency of rain on days with conditions like today.

Probability theory proposes a set function that quantifies the trust/belief/chance/ that the true alternative hits each crisp subset $A \subseteq \Omega$. This set function, called a *probability measure*, Pr, is defined as a mapping $\mathcal{P}(\Omega) \to [0, 1]$ that assigns a number Pr(A) to each subset $A \subseteq \Omega$ and satisfies the Kolmogorov axioms:

- $\Pr(\Omega) = 1$, and $\Pr(\Phi) = 0$;
- If $(A \cap B = \Phi)$ then $\Pr(A \cup B) = \Pr(A) + \Pr(B)$.

The second Kolmogorov axiom, or requirement, is generally referred to as the *additivity axiom* of probability measures. From this fundamental definition, it can be clearly seen that a probability measure is a special type of belief monotone measures, since the inequality $g(A \cup B) \geq g(A) + g(B)$ (in the definition of the belief monotone measure) is replaced with the stronger equality axiom. Moreover, the probability measure, considered as a belief monotone measure, satisfies the following property:

$$\Pr(A) = Bel(A) = Pl(A) = \sum\nolimits_{x \in A} m(x) = \sum\nolimits_{x \in A} \Pr\{x\} \text{ for all } A \subseteq \Omega.$$

This means that under the monotone measures theory, probability theory can be considered as a special "branch" of DST. It is worthwhile to stress the fact that probability theory is certainly the most powerful mathematical framework for uncertainty processing. Nevertheless, probability density values must be assigned to all alternatives even if no information is available (in this case, equal amount of probability values are assigned to all such alternatives). Moreover, probabilistic methods require *prior* probabilities which are very hard to find out.

Possibility Theory

Probability theory is an ideal mathematical framework for uncertainty modeling in situations where alternatives' frequencies are known or when evidence is based on outcomes of a sufficiently long series of independent random experiments. Possibility theory is another mathematical theory offering a new theoretical framework to model and to reason about uncertainty when formalizing incomplete knowledge expressed in terms of fuzzy (vague) propositions. The fundamental modeling concept of a possibility distribution is based on expressing the extent to which each individual alternative can occur. If $\Omega = \{x_1, \ldots, x_N\}$ denote a finite set of mutually exclusive alternatives, where the unique occurring alternative, i.e., the true alternative, is unknown, then $\pi(x_n) = 0$ (resp. $= 1$) means that the alternative is rejected as impossible (resp. is totally possible); and $0 < \pi(x_n) < 1$ means a partial possibility degree for the occurrence of the alternative x_n, $n = 1, 2, \ldots, N$. The only restriction of this distribution function is its normalization, requiring at least one alternative to have a possibility degree equal to 1.

From the primitive concept of the possibility distribution, two confidence set functions are induced. Namely a *possibility measure* Π such that:

$$\Pi(A) = max_{x \in A} \pi(x)$$

measuring the degree of consistency of the proposition that the true alternative belongs to $A \subseteq \Omega$ and a *necessity measure* N such that:

$$N(A) = min_{x \in A} [1 - \pi(x)]$$

Between these two measures, there is a duality relation (equivalent to that existing in evidence theory between plausibility and belief measures):

$$N(A) = 1 - \Pi\left(A^{C}\right) \text{ for all } A \subseteq \Omega.$$

Moreover, possibility and necessity degrees satisfy the following axioms:

$$\Pi(A \cup B) = max \{\Pi(A), \Pi(B)\}$$
$$N(A \cap B) = min \{N(A), N(B)\}$$

It is worthwhile to notice that a possibility measure $\Pi : \mathcal{P}(\Omega) \to [0, 1]$ is a monotone measure (it is straightforward to show that for all $A, B \subseteq \Omega$ if $A \subseteq B$ then $\Pi(A) \leq \Pi(B)$).

Moreover, as $\Pi(A \cup B) = max \{\Pi(A), \Pi(B)\}$ and the following property is fulfilled:

$$\Pi(A \cap B) \leq \Pi(A) + \Pi(B) - \Pi(A \cup B)$$

This means that the possibility measure is a plausibility monotone measure and that its dual necessity measure, N, is a belief monotone measure. Moreover, as determined by Klir [5] in the framework of evidence theory, that when all focal elements (i.e., events $A \subseteq \Omega$ for which $m(A) \neq 0$) are nested or consonant then:

- The derived belief measure becomes a necessity measure, i.e., $Bel(A \cap B) = min \{Bel(A), Bel(B)\}$.
- The derived plausibility measure becomes a possibility measure, i.e., $Pl(A \cup B) = max \{Pl(A), Pl(B)\}$.

A pictorial description of evidence, probability, and possibility theories based on monotone measures theory is depicted in Fig. 5.2. From this figure, we can clearly see that the five set-function measures are all monotone measures and that probability measures are also belief and plausibility at the same time.

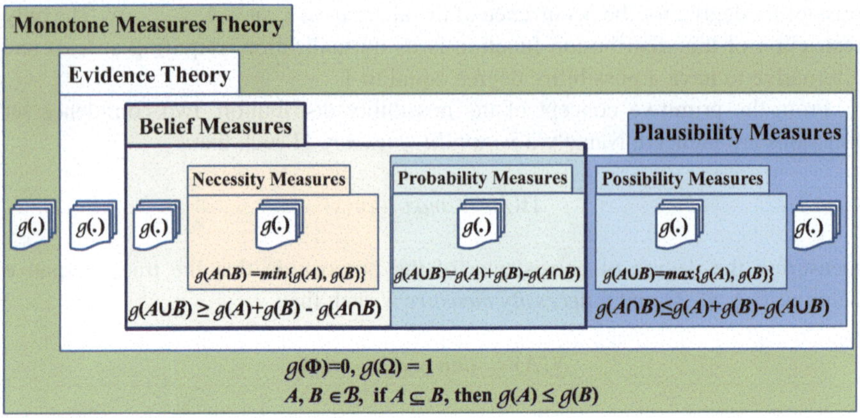

Fig. 5.2 Pictorial description of uncertainty theories

In the next sections, we will review the philosophy of several transformations proposed in the literature. The presentation is restricted to evidence-possibility and probability-possibility transformations.

5.4 Evidence-Possibility Transformations

As possibility, Dempster-Shafer and probability theories are conceived to operate on the same context of uncertainty; it is legitimate to study the transformation "bridges" and different links between these uncertainty theories. In this section, bridges and transformations between DST and possibility theory are presented. In fact, possibility theory may be viewed as a "special branch" of DST when all focal elements are nested, also said consonant as illustrated in Fig. 5.3. In this case, the derived belief, *Bel*, (resp. plausibility, *Pl*) measure in the framework of DST becomes the necessity, *N*, (resp. possibility, Π) measure.

In this case, the following properties hold for all $A, B \subseteq \Omega$:

- $\Pi(A \cup B) = max\ \{\Pi(A), \Pi(B)\}$;
- $\Pi(A \cup A^{C}) = 1$;
- $\Pi(A) = 1 - \Pi(A^{C})$;
- $\pi(x) = \Pi(\{x\}) = \sum_{A_n (x \in A_n)} m(A_n) = Pl(\{x\})$, for all $x \in \Omega$;
- For every $x \in A_1$ we have $\pi(x) = 1$;
- For every $x \in G_n = A_n - A_{n-1}$, we have $\pi(x) = \Pi(G_n)$. This means that all alternatives from G_n have the same value of possibility;
- If $x_n \in G_k$ and $x_m \in G_h$ (with $1 \leq k < h \leq N$), then $\pi(x_n) > \pi(x_m) > 0$;
- $m(A_n) = \Pi(G_n) - \Pi(G_{n+1}), n = 1, 2, \ldots, N - 1$.

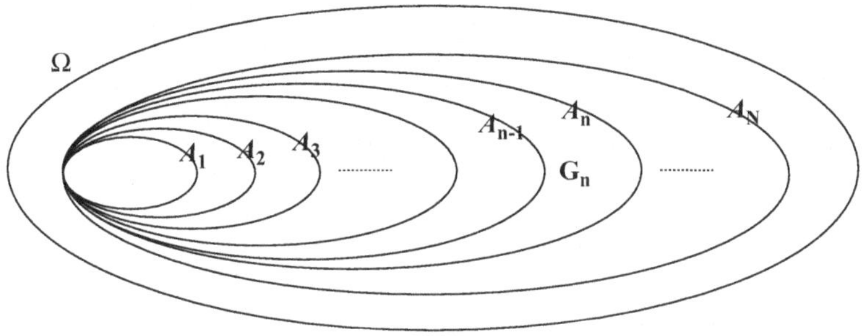

Fig. 5.3 Nested body of evidence ($A_1 \subseteq A_2 \subseteq \ldots \subseteq A_N$)

5.4.1 *Transforming a* b.p.a *into a Possibility Distribution*

Let $m(.)$ be a mass function on a finite set of alternatives Ω. The possibility distribution $\pi(.)$ induced by m, called *contour function*, can be obtained as the plausibility of different alternatives:

$$\pi(x) = Pl(\{x\}) = \sum_{x \in B} m(B) \text{ for all } x \in \Omega.$$

This transformation will be referred as the basic probability assignment to possibility distribution transformation ($b.p.a \to \pi$, for short). It is straightforward to check that $\pi(.)$ takes its values in the unit interval [0, 1]. Moreover, it is worthwhile to notice that:

- $\pi(x) = 1$ if and only if the alternative x belongs to all focal elements of the mass function m;
- $\pi(x) = 0$ if and only if the alternative x does not belong to any focal element of the mass function m;
- $\pi(.)$ is a normal possibility distribution if and only if all the focal elements have a common non empty intersection; otherwise, $\pi(.)$ is a subnormal possibility distribution.

Example Consider the alternatives set $\Omega = \{x_1, x_2, x_3, x_4, x_5\}$, and let us analyze the transformation of the following mass functions defined on the power set $\mathcal{P}(\Omega)$:
$m_1(\{x_1, x_2\}) = 0.4$, $m_1(\{x_1, x_3, x_4\}) = 0.2$, $m_1(\{x_1, x_2, x_5\}) = 0.3$, $m_1(\Omega) = 0.1$;
$m_2(\{x_1\}) = 0.2$, $m_2(\{x_2\}) = 0.1$, $m_2(\{x_3\}) = 0.1$, $m_2(\{x_4\}) = 0.2$, $m_2(\{x_5\}) = 0.3$;
$m_3(\{x_1, x_2\}) = 1.0$;
$m_4(\{x_1, x_2\}) = 0.4$, $m_4(\{x_3, x_4, x_5\}) = 0.6$;
$m_5(\Omega) = 1.0$;

Using the above mentioned ($b.p.a \to \pi$) transformation, we obtain:

- $\boldsymbol{m_1(.)}$:
  - $\pi_{m_1}(x_1) = m_1(\{x_1, x_2\}) + m_1(\{x_1, x_3, x_4\}) + m_1(\{x_1, x_2, x_5\}) + m_1(\Omega) = 1.0$,
  - $\pi_{m_1}(x_2) = m_1(\{x_1, x_2\}) + m_1(\{x_1, x_2, x_5\}) + m_1(\Omega) = 0.8$,

- $\pi_{m_1}(x_3) = m_1(\{x_1, x_3, x_4\}) + m_1(\Omega) = 0.3,$
- $\pi_{m_1}(x_4) = m_1(\{x_1, x_3, x_4\}) + m_1(\Omega) = 0.3,$
- $\pi_{m_1}(x_5) = m_1(\{x_1, x_2, x_5\}) + m_1(\Omega) = 0.4.$

Notice that $\pi_{m_1}(.)$ is a normal possibility distribution since $\{x_1\}$ belongs to all focal elements.

- $\boldsymbol{m_2}$**(.)**: (Bayesian *b.p.a*).
  - $\pi_{m_2}(x_1) = m_2(\{x_1\}) = 0.2,$
  - $\pi_{m_2}(x_2) = m_2(\{x_2\}) = 0.1,$
  - $\pi_{m_2}(x_3) = m_2(\{x_3\}) = 0.1,$
  - $\pi_{m_2}(x_4) = m_2(\{x_4\}) = 0.2,$
  - $\pi_{m_2}(x_5) = m_2(\{x_5\}) = 0.3.$

- $\boldsymbol{m_3}$**(.)**: (encodes an imprecise information, also called partial ignorance),
  - $\pi_{m_3}(x_1) = m_3(\{x_1, x_2\}) = 1.0,$
  - $\pi_{m_3}(x_2) = m_3(\{x_1, x_2\}) = 1.0.$

- $\boldsymbol{m_4}$**(.)**:
  - $\pi_{m_4}(x_1) = m_4(\{x_1, x_2\}) = 0.4,$
  - $\pi_{m_4}(x_2) = m_4(\{x_1, x_2\}) = 0.4,$
  - $\pi_{m_4}(x_3) = m_4(\{x_3, x_4, x_5\}) = 0.6,$
  - $\pi_{m_4}(x_4) = m_4(\{x_3, x_4, x_5\}) = 0.6,$
  - $\pi_{m_4}(x_5) = m_4(\{x_3, x_4, x_5\}) = 0.6.$

- $\boldsymbol{m_5}$**(.)**: (encodes the total ignorance *p.b.a*).
  - $\pi_{m_5}(x_1) = m_5(\Omega) = 1.0,$
  - $\pi_{m_5}(x_2) = m_5(\Omega) = 1.0,$
  - $\pi_{m_5}(x_3) = m_5(\Omega) = 1.0,$
  - $\pi_{m_5}(x_4) = m_5(\Omega) = 1.0,$
  - $\pi_{m_5}(x_5) = m_5(\Omega) = 1.0.$

5.4.2 *Transforming a Possibility Distribution into a* b.p.a

Let $\Omega = \{x_1, x_2, x_3, \ldots, x_N\}$ denote a finite set of alternatives, and $\pi : \Omega \rightarrow [0, 1]$ a normal possibility distribution defined on Ω. Let us also assume that within the N possibility values $\pi(x_n)$, $n = 1, 2, \ldots, N$, we have K distinct possibility values π_k, $k = 1, 2, \ldots, K$ such that $1 = \pi_1 \geq \pi_2 \geq \cdots \geq \pi_K$. $B_1, B_2, \ldots, B_K$ and B_{K+1} are the disjoint subsets of Ω such that B_K, $k = 1, 2, \ldots, K$, contains all alternatives x_n for which $\pi(x_n) = \pi_k$ and $\pi(x_n) = 0$ when $x_n \in B_{K+1}$ (Fig. 5.4).

The following properties hold:

1. Let $A_n = \bigcup_{k=1}^{n} B_k$ for $n = 1, 2, \ldots, K$, then subsets $A_1, A_2, \ldots, A_K$ are nested;
2. Let $m(A_n) = \pi_n - \pi_{k+1}$ for $n = 1, 2, \ldots, K$. Then, $m(.)$ is a basic belief assignment, *bpa*, (or a mass function) on focal elements A_n;

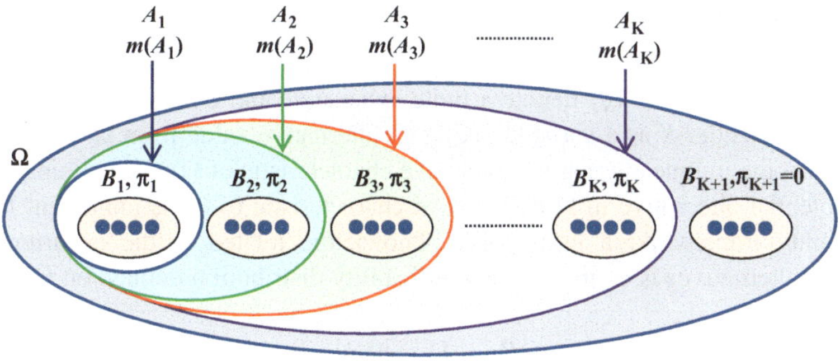

Fig. 5.4 Possibility theory as a branch of DST, case when focal elements are nested

3. Let *Bel*(.) be the belief function corresponding to *m*(.) defined above, then *Bel*(.) is a consonant function.

Example Let π denote the following possibility distribution defined on $\Omega = \{x_1, x_2, x_3, x_4\}$:

$\pi(x_1) = 0.7$, $\pi(x_2) = 1.0$, $\pi(x_3) = 0.8$, $\pi(x_4) = 0.7$. The disjoint subsets for π are $B_1 = \{x_2\}$, $B_2 = \{x_3\}$, $B_3 = \{x_1, x_4\}$, and the corresponding focal elements as well as the *bpa m*(.) are:

$A_1 = B_1 = \{x_2\} : m(A_1) = 0.2;$
$A_2 = B_1 \cup B_2 = \{x_2, x_3\} : m(A_2) = 0.1;$
$A_3 = B_1 \cup B_2 \cup B_3 = \{x_1, x_2, x_3, x_4\} : m(A_3) = 0.7.$

5.5 Probability-Possibility Transformations

Probability theory and possibility theory are suitable for modeling different types of knowledge when describing uncertainty. If enough information and accurate predictive models are available for dealing with uncertainty, then the use of probability theory in certainly advantageous. On the other hand, if the available knowledge is "weak" or using approximate predictive models, the use of possibility theory sounds more convenient. In situations where information regarding an observed phenomenon is given in both probabilistic and possibilistic terms, the two descriptions should be, in some sense, consistent. In this case, transformation bridges between the two uncertainty framework models are of high interest. In fact, an important link is made between the two theories when we view necessity and possibility measures as lower and upper probabilities. This can be done because of the nested structure of evidence. This allows to bridge the interpretation of probability theory to the interval-valued probabilities of possibilistic type.

In this section, different consistency concepts between probability and possibility measures are, first, reviewed. Major bridges allowing the transformation of a probability measure into a possibility measure, and conversely, are then presented.

5.5.1 Probability-Possibility Consistency Concepts

Let $\Omega = \{x_1, x_2, x_3, \ldots, x_N\}$ denote a finite exhaustive and exclusive set of alternatives, and consider X as a variable taking an alternative value from Ω. Uncertainty means that the unique alternative (exclusive characteristic of Ω) that certainly exists as a potential alternative in Ω (exhaustive characteristic of Ω) is unknown. In this uncertainty framework, a probabilistic knowledge related to the occurrence of different alternatives is expressed as a probability distribution defined on Ω:

$$\begin{aligned} P: \quad & \Omega \rightarrow [0,1] \\ & x_n \rightarrow P(x_n) \end{aligned}$$

and the "couple" (X, P) is called a random variable.

In the same uncertainty framework, possibility theory offers another knowledge representation form expressed as a normal possibility distribution defined on Ω:

$$\begin{aligned} \pi: \quad & \Omega \rightarrow [0,1] \\ & x_n \rightarrow \pi(x_n) \end{aligned}$$

and the "couple" (X, π) is called a possibilistic variable.

The possibilistic representation is weaker that the probabilistic one for the main reason that it leads to a "quasi-qualitative" model of uncertainty representation that is to some extent less expressive than probability but also less demanding in terms of information constraints.

Two interesting questions arise at this level:

(a) *How can we study the link in terms of coherence between* the two knowledge representation forms (X, P) and (X, π)?
(b) How can we transform one knowledge representation form into the other one?

It is obvious that when transforming (X, P) to (X, π), some information is lost since we go from point value probabilities to interval values ones. Also, the converse transformation from (X, π) to (X, P) is appealed to add information to the possibilistic incomplete knowledge.

In this section, we will focus on the coherence issue of the two knowledge representation forms (X, P) to (X, π), both describing the same variable X. L. Zadeh [8] has illustrated this coherence question (called possibility-probability consistency) through a simple example. In fact, consider the statement: "Hans ate x eggs for breakfast," with x taking values in $\Omega = \{1, 2, 3, \ldots, 8\}$. In this example, $\pi(x)$ can be interpreted as the degree of ease with which Hans can eat x eggs, whereas $P\{x\}$ is the probability of Hans eating x eggs for breakfast. Table 5.2 shows realistic possibility and probability distributions.

What we can notice is that a high possibility does not imply a high degree of probability, nor does a low degree of probability imply a low degree of possibility.

Table 5.2 Realistic possibility and probability distributions associated with X

x	1	2	3	4	5	6	7	8
$\pi(x)$	1	1	1	1	0.8	0.6	0.4	0.2
$P(x)$	0.1	0.8	0.1	0	0	0	0	0

5.5.1.1 Probability-Possibility Consistency Principle

The weakest consistency principle [9] is based on the fact that an event that is probable to some degree must be possible at least to the same degree. In other words, what is probable must be possible first. This consistency principle can be formally expressed by the following inequality:

$$P(x_n) \leq \pi(x_n), \quad \forall x_n \in \Omega$$

or in terms of events:

$$\Pr(A) \leq \Pi(A) \quad \forall A \subseteq \Omega$$

where Pr (resp. Π) is the probability (resp. possibility) measure associated with P (resp. π). In such case, the possibility distribution π is said to be *dominating the probability distribution P*. It is worthwhile to notice that the strongest consistency condition would require that any event with nonzero probability must be fully possible (i.e., If $P(x_n) > 0 \Rightarrow \pi(x_n) = 1$, $\Rightarrow$ stands for implication). Moreover, if the two variables (X, P) and (X, π) satisfy the probability-possibility consistency principle, then $N(A) \leq \Pr(A)$ holds for all $A \subseteq \Omega$ (where $N(.)$ denotes the necessity measure associated with Π). This property can be easily proved by recalling that $N(A) = 1 - \Pi(A^C)$. Therefore, if the two variables (X, P) and (X, π) satisfy the Dubois-Prade's consistency principle, then:

$$N(A) \leq \Pr(A) \leq \Pi(A) \quad \forall A \subseteq \Omega$$

In this case, the probability measure, Pr, and the possibility measure, Π, are said to be consistent. Thus, and according to [10], given the interpretation of possibility and necessity measures as upper and lower probability measures, every normal possibility distribution π induces a family $\mathrm{IPr}(\pi)$ of probability measures satisfying probability-possibility consistency principle:

$$\mathrm{IPr}(\Pi) = \{\Pr : N(A) \leq \Pr\{A\} \leq \Pi(A), \text{for all } A \subseteq \Omega\}$$

where Π and N are the possibility and necessity measures associated with π. Since there is no one-to-one relation between possibility and probability, a transformation of a possibility distribution π into a probability measure Pr can therefore only ensure that Pr is a member of $\mathrm{IPr}(\Pi)$ and Pr is selected (among the members of $\mathrm{IPr}(\Pi)$) according to some principle (for instance, preference preservation, information content maximization, etc.).

5.5.1.2 Least Commitment Principle

The principle of least commitment between a probability measure and a possibility measure means that both measures contain the same amount of uncertainty (i.e., no added informational content). A probability/possibility, or conversely, transformation is said to satisfy the least commitment principle (also called *insufficient reason principle*) if the vacuous possibility distribution (i.e., all alternatives are equally possible with full unit possibility) is transformed into a uniformly distributed probability distribution (and conversely).

5.5.1.3 Zadeh's Consistency Principle

The *possibility/probability consistency principle* established by Zadeh [8] represents an approximate formalization of the heuristic observation that a lessening of the possibility of an event tends to lessen its probability—but not vice-versa. In fact, Zadeh defined the probability-possibility consistency principle such as "a high degree of possibility does not imply a high degree of probability, nor does a low degree of probability imply a low degree of possibility." Formally, Zadeh's consistency principle can be expressed by:

$$\pi(x_n) = 0 \Rightarrow P(x_n) = 0$$

(i.e., an alternative must be possible *prior* to being probable) and:

$$\pi(x_m) \leq \pi(x_n) \Rightarrow P(x_m) \leq P(x_n)$$

Moreover, Zadeh defined the *degree of consistency* of the probability distribution $P(.)$ with the possibility distribution $\pi(.)$ by:

$$Cons(\pi, P) = \sum_{n=1}^{N} \pi(x_n).P(x_n)$$

This definition does not express a formal relationship between $P(.)$ and $\pi(.)$. Some characteristics of Zadeh's degree of consistency are given as follows:

1. $0 \leq Cons(\pi, P) \leq 1$.
2. $Cons(\pi, P) = 0$ if and only if Support(π) $\cap$ Support(Pr) $= \Phi$ (i.e., there is no alternative $x \in \Omega$ for which: $\pi(x) \neq 0$ and $P(x) \neq 0$ at the same time).
3. $Cons(\pi, P) = 1$ if and only if $\pi(x) = 1$ for all $x \in \Omega$ for which $P(x) \neq 0$ (i.e., all alternatives for which $P(x) \neq 0$ are totally possible: $\pi(x) = 1$).
4. If π is the total ignorance possibility distribution (i.e., $\pi(x) = 1$ for all $x \in \Omega$), then $Cons(\pi, P) = 1$.
5. If $\pi = 0$ (i.e., $\pi(x) = 0$ for all $x \in \Omega$), then $Cons(\pi, P) = 0$. This means that the impossible event has no consistency with any probability distribution.

6. If P is the total ignorance probability distribution (i.e., $P(x) = 1/N$ for all $x \in \Omega$), then $Cons(\pi, P) =$ Mean possibility value $= \frac{1}{N} \sum_{n=1}^{N} \pi(x_n)$.

5.5.1.4 Preference Preservation Principle

An important informative facet of a possibility distribution is related to alternatives' preference ordering. In fact, for given a possibility distribution π and two alternatives x_n and x_m from Ω such that $\pi(x_n) > \pi(x_m)$, then the alternative x_n is said to be preferred to the alternative x_m by the possibility distribution π. The *preference preservation principle* (also called *order preservation principle*) extends the Zadeh's consistency principle in order to have $\pi(x_m) \leq \pi(x_n) \Leftrightarrow P(x_m) \leq P(x_n)$ for all x_n, $x_m \in \Omega$.

5.5.2 *Probability-Possibility Transformation Methods*

The problem of converting possibility distributions into probability distributions, and conversely, is of high importance in any problem where heterogeneous forms of uncertainty (probabilistic, imprecision, ambiguity, etc.) must be dealt with. This transformation raises philosophically interesting debate since possibility distributions are considered as less informative and weaker than probability distributions (i.e., possibility degrees can be viewed as upper bounds on probability degrees). In fact, turning a probability distribution into a possibility distribution may be useful in the presence of other weak sources of information, when computing with possibilities is simpler than computing with probabilities or when the probability distribution is estimated from few, or unreliable, training data. Moreover, leading this transformation (in terms of knowledge representation) means that "some" information is lost because we go from point-valued probabilities to interval-valued ones (i.e., this transformation leads, for all $A \subseteq \Omega$, to find a bracketing of $\Pr\{A\}$ in terms of an interval $[N(A), \Pi(A)]$). It is worthwhile to recall that this bracketing is never tight since it takes one of the two following forms: $[N(A), 1]$ and $[0, \Pi(A)]$.

Opposite transformations (turning a possibility distribution into a probability distribution) are assumed to "add" information to the possibilistic incomplete knowledge. Nevertheless, they are of great interest in the scope of decision-making. Different possibility-probability transformations have been suggested in the literature.

5.5.2.1 Ratio Scale Transformation

This probability-possibility ratio scale transformation method (also called Klir, or normalized transformation method) is based on Shackle's interpretation of the possibility measure of an event [11] as being the absence of surprise when this event occurs. An event which often occurs is not very surprising and, thus, seems very possible that it happens. On the contrary, events which are not very possible do not often occur and are considered as surprising when they occur. This interpretation leads to suppose that a rarely occurring event must be less possible than events which often occur. Therefore, and using Shackle's possibility interpretation, it is quite simple to deduce a possibility distribution by normalizing a given probability distribution. Let $\Omega = \{x_1, x_2, x_3, \ldots, x_N\}$ denote a finite exhaustive and exclusive set of alternatives, and consider the probabilistic variable (X, P) defined on Ω. The possibility distribution obtained by a ratio scale transformation (denoted by $\Pr \to \pi$) is given by:

$$\pi(x_n) = P(x_n)/[max_{x_m \in A} P(x_m)] \text{ for all } x_n \in \Omega$$

Thus, most frequent alternatives receive a degree of possibility equal to 1.

It is worthwhile to notice that this transformation method satisfies the preference preservation principle and the associated probability and possibility measures are consistent (i.e., satisfy Dubois-Prade's consistency principle). Using the same reasoning approach, if the possibilistic variable (X, π) is known, then the possibility-probability transformation (denoted by $\pi \to \Pr$) is given by:

$$P(x_n) = \pi(x_n)/\left[\sum\nolimits_{x_m \in A} \pi(x_m)\right] \text{ for all } x_n \in \Omega$$

5.5.2.2 Maximal Specificity ($\Pr \to \pi$) Transformation

Let $\Omega = \{x_1, x_2, x_3, \ldots, x_N\}$ denote a finite exhaustive and exclusive set of alternatives, and consider a probabilistic variable (X, P) and a normal possibilistic variable (X, π) both defined on Ω. Assuming known a probabilistic variable (X, P), Dubois and Prade [9] have proposed to determine the maximally specific possibility distribution, (X, π), i.e., maximal specificity ($\Pr \to \pi$) transformation. Without loss of generality, assume that the alternatives x_1, x_2, x_3, …, x_N are ordered as shown in Fig. 5.5.

Notice that this ordering is related to possibility values and does not concern probability values. *M*. Delgado [12] has shown that Zadeh's consistency principle between (X, P) and (X, π), i.e., $\pi(x_n) = 0 \Rightarrow P(x_n) = 0$ and $\pi(x_m) \le \pi(x_n) \Rightarrow P(x_m) \le P(x_n)$, is satisfied if and only if the following condition holds:

$$\pi(x_n) \ge \sum_{m=n}^{N} P(x_m)$$

Let us now go back to the ($\Pr \to \pi$) transformation. Assume that we only have a probabilistic variable (X, P) defined on Ω and that the alternatives x_1, x_2, x_3, …, x_N

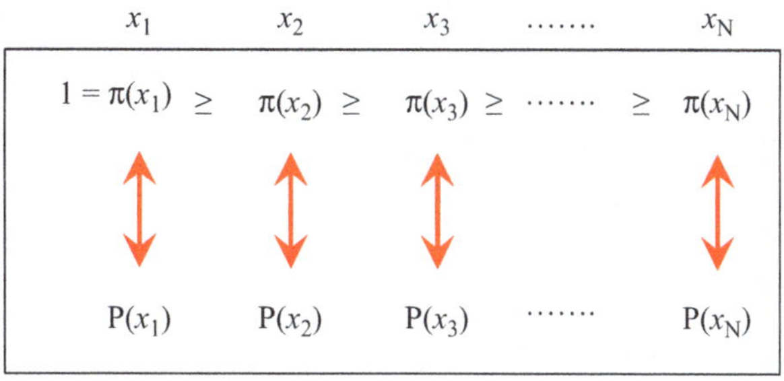

Fig. 5.5 Maximal specificity ($Pr \rightarrow \pi$) transformation

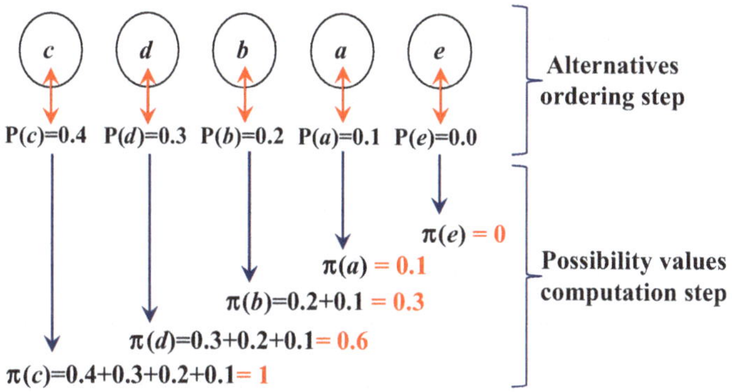

Fig. 5.6 Maximal specificity based transformation application

are ordered so that $P(x_1) \geq P(x_2) \geq P(x_3) \geq \cdots \geq P(x_N)$ and that we are looking to transform the probability distribution P into a possibility distribution π. Then, using Delgado's result it is straightforward to show that the maximally specific possibility distribution satisfying both probability/possibility consistency and preference preservation principles is given by [9]:

$$\pi(x_n) = \sum_{m=n}^{N} P(x_m)$$

Example Let us consider the following probability distribution defined on $\Omega = \{a, b, c, d, e\}$: $P(a) = 0.1$, $P(b) = 0.2$, $P(c) = 0.4$, $P(d) = 0.3$, and $P(e) = 0$.

The first step for the application of the maximal specificity transformation method consists on alternatives ordering, Fig. 5.6. Once the alternatives are ranked, the second step consists on the direct computation of the possibility degrees. Finally, the resulting possibility distribution is given by: $\pi(a) = 0.1$, $\pi(b) = 0.3$, $\pi(c) = 1$, $\pi(d) = 0.6$, and $\pi(e) = 0$.

It is worthwhile to notice that a major weakness of this transformation method is due to the fact that two alternatives, x_n and x_m, having the same occurrence probability, $P(x_n) = P(x_m)$, will have two different possibility values, $\pi(x_n) \neq \pi(x_m)$.

For instance, in the previous example, if the probability distribution is given by $P(a) = 0.1$, $P(b) = 0.2$, $P(c) = 0.5$, $P(d) = 0.2$, and $P(e) = 0$, then two distinct resulting maximal specificity possibility distribution are given by $\pi(a) = 0.1$, $\pi(b) = 0.5$, $\pi(c) = 1$, $\pi(d) = 0.3$, and $\pi(e) = 0$ (if the alternatives are ordered as $(P(c) \geq P(b) \geq P(d) \geq P(a) \geq P(e))$ and $\pi(a) = 0.1$, $\pi(b) = 0.3$, $\pi(c) = 1$, $\pi(d) = 0.5$, and $\pi(e) = 0$ (if the alternatives are ordered as $(P(c) \geq P(d) \geq P(b) \geq P(a) \geq P(e))$.

Finally, this ($\Pr \rightarrow \pi$) transformation method can be used to transform a possibilistic variable (X, π) into a probabilistic variable (X, P) for which the possibility distribution is the most specific transformation. In fact, assume that the alternatives $x_1, x_2, x_3, \ldots, x_N$ are ordered so that $1 = \pi(x_1) \geq \pi(x_2) \geq \pi(x_3) \geq \cdots \geq \pi(x_N)$, then we can write the following N equations:

$$\pi(x_1) = p_1 + p_2 + p_3 + \cdots + p_N = 1$$
$$\pi(x_2) = p_2 + p_3 + \cdots + p_N$$
$$\pi(x_3) = p_3 + \cdots + p_N$$
$$\pi(x_{N-1}) = p_{N-1} + p_N$$
$$\pi(x_N) = p_N$$

where $p_n = P(x_n)$, $n = 1, 2, \ldots, N$. Therefore, as all the possibility values $\pi(x_n)$ are assumed known, the reverse transformation is simply given by the iterative equation:

$$p_N = \pi(x_N)$$
$$p_n = \pi(x_n) - \pi(x_{n+1}), n = N-1, N-2, \ldots, 1$$

Example Let us consider the following possibility distribution defined on $\Omega = \{a, b, c, d, e\}$: $\pi(a) = 1$, $\pi(b) = 0.8$, $\pi(c) = 0.4$, $\pi(d) = 0.3$, and $\pi(e) = 0.1$. The reverse transformation is obtained by resolving the following equations:

$$\pi(a) = p_1 + p_2 + p_3 + p_4 + p_5 = 1$$
$$\pi(b) = p_2 + p_3 + p_4 + p_5 = 0.8$$
$$\pi(c) = p_3 + p_4 + p_5 = 0.4$$
$$\pi(d) = p_4 + p_5 = 0.3$$
$$\pi(e) = p_5 = 0.1$$

(where $p_1 = P(a)$, $p_2 = P(b)$, $p_3 = P(c)$, $p_4 = P(d)$, and $p_5 = P(e)$). The resolution of these equations leads to $P(a) = 0.2$, $P(b) = 0.4$, $P(c) = 0.1$, $P(d) = 0.2$, and $P(e) = 0.1$. It is worthwhile to notice that this reverse possibility-probability transformation (i.e., $\pi \rightarrow \Pr$) does not satisfy the preference preservation principle!

5.5.2.3 Pignistic ($\pi \rightarrow$ Pr) Transformation

Possibility theory offers a quasi-qualitative representation model of uncertainty that is to some extent less expressive and, thus, weaker than the probabilistic knowledge representation. Therefore, transforming a possibility distribution into a probability distribution is appealed to add information to the possibilistic incomplete knowledge. For doing so, the ($\pi \rightarrow$ Pr) transformations are generally guided by the principle of insufficient reason, also called *least commitment principle*, which aims at finding the probability distribution that contains as much uncertainty as possible while retaining the main features of the possibility distribution [9, 13]. The pignistic ($\pi \rightarrow$ Pr) transformation proposed by D. Dubois and H. Prade is based on transforming, first, the possibility distribution π into a basic probability assignment function $m(.)$ defined on the corresponding nested body of evidence and, then, transforming the consonant function $m(.)$ into a probability distribution.

Let $\Omega = \{x_1, x_2, \ldots, x_N\}$ denote a finite set of alternatives, and $\pi : \Omega \rightarrow [0, 1]$ a normal possibility distribution defined on Ω. Let us also assume that the N possibility values $\pi(x_n)$, $n = 1, 2, \ldots, N$, are arranged in a decreasing order such that $1 = \pi(x_1) \geq \pi(x_2) \geq \cdots \geq \pi(x_N)$ (Fig. 5.7).

Let A_n, $n = 1, 2, \ldots, N$, denote the resulting nested focal element ($A_1 \subseteq A_2 \subseteq \ldots \subseteq A_N$), with $|A_n| = Card(A_n) = n$. As previously shown, the obtained *b.p.a* is simply given by: $m(A_n) = \pi(x_n) - \pi(x_{n+1})$, $n = 1, 2, \ldots, N$ with the convention that $\pi(x_{N+1}) = 0$. Notice that a given alternative x_n belongs to the successive nested subsets $A_n, A_{n+1}, \ldots, A_N$. Thus, the pignistic transformation of the *b.p.a* function $m(.)$ into a probability distribution is simply given by:

$$P(x_n) = \sum_{m=n}^{N} \frac{m(A_m)}{Card(A_m)}, n = 1, 2, \ldots, N.$$

Therefore, the pignistic ($\pi \rightarrow$ Pr) transformation is given by:

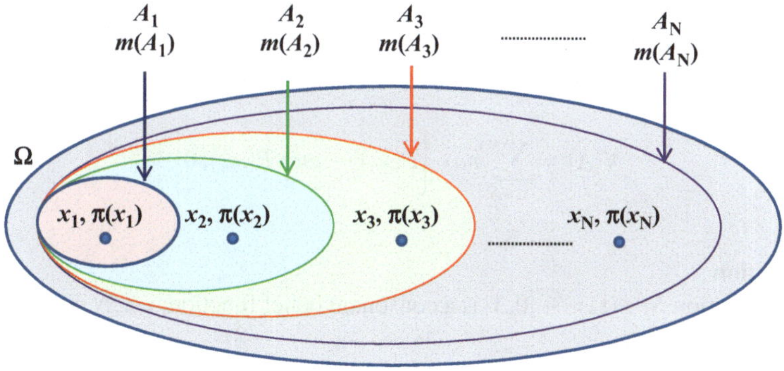

Fig. 5.7 Possibility distribution transforming into a *b.p.a*

$$P(x_n) = \sum_{m=n}^{N} \frac{\pi(x_m) - \pi(x_{m+1})}{m}, \ n = 1, 2, \ldots, N$$

It can be easily verified that the pignistic ($\pi \rightarrow$ Pr) transformation yields the probability/possibility consistency and preference ordering principles. Moreover, this transformation fulfils the least commitment principle. In fact, if the considered possibility distribution is the vacuous possibility distribution (i.e., $\pi(x_n) = 1$, $n = 1, 2, \ldots, N$), then the resulting pignistic ($\pi \rightarrow$ Pr) transformation is the uniform probability distribution (i.e., $P(x_n) = 1/N$, $n = 1, 2, \ldots, N$).

Finally, Dubois et al. [9] have shown that the resulting pignistic probability is also the gravity center of the set $\text{IPr}(\Pi) = \{ \text{Pr} : N(A) \leq \text{Pr}\{A\} \leq \Pi(A) \text{ for all } A \subseteq \Omega\}$ of probability distributions dominated by Π.

5.5.2.4 Dubois-Prade's Symmetric Transformation

Dubois and Prade [9] proposed another (Pr $\rightarrow \pi$) transformation approach based upon the concept of occurrence necessity of an event. In fact, considering a biased coin so that heads are the most frequent outcomes (i.e., $1 \geq P\{x_1 = \text{Head}\} = p_1 \geq ½ \geq p_2 = P\{x_2 = \text{Tail}\} \geq 0$), then a natural measurement of the necessity for getting heads (i.e., $N(\{x_1 = \text{Head}\})$) can be expressed as the excess of probability in favor of heads (i.e., $N(\{x_1 = \text{Head}\}) = p_1 - p_2$). The necessity of getting tails can be "considered" as null (i.e., $N(\{x_2 = \text{Tails}\}) = 0$) since $p_1 \geq p_2$. Using this necessity definition leads to consider the following corresponding possibility degrees: $\pi(x_1) = \Pi(\{x_1\}) = 1 - N(\{x_2\}) = 1$ and $\pi(x_2) = \Pi(\{x_2\}) = 1 - N(\{x_1\}) = 1 - p_1 + p_2 = 2p_2$. Notice that this necessity definition sounds realistic, since if $p_1 = p_2 = ½$, then $\pi(x_1) = \pi(x_2) = 1$ (i.e., both alternatives are fully possible) and if $p_1 = 1$, then $\pi(x_1) = 1$ and $\pi(x_2) = 0$ (i.e., having a Head is fully possible and having a Tail is fully impossible). The generalization of this basic idea into the general case of a finite exhaustive and exclusive set of alternatives, $\Omega = \{x_1, x_2, \ldots, x_N\}$ on which a probabilistic variable (X, P) is defined, is then conducted as follows. The degree of necessity of an event $A \subseteq \Omega$ is defined as the accumulation of the extra amount of probability of alternatives in A assigned to the most frequent alternative outside A (i.e., alternatives from A^C), Fig. 5.8:

$$N(A) = \sum_{x_n \in A} max \left[P(x_n) - \underset{x_m \in A^C}{max} P(x_m), 0 \right]$$

Proposition

The set function $N: \mathcal{P}(\Omega) \rightarrow [0, 1]$ is a consonant belief function, i.e., N is a necessity measure:

- $N(\Phi) = 0$; $N(\Omega) = 1$;
- $N(A \cap B) = min\{N(A), N(B)\}, \forall A, B \subseteq \Omega$.

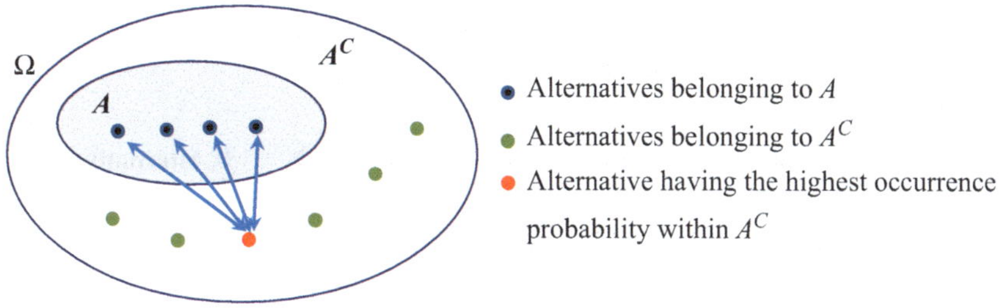

Fig. 5.8 Dubois-Prade probability-possibility transformation approach

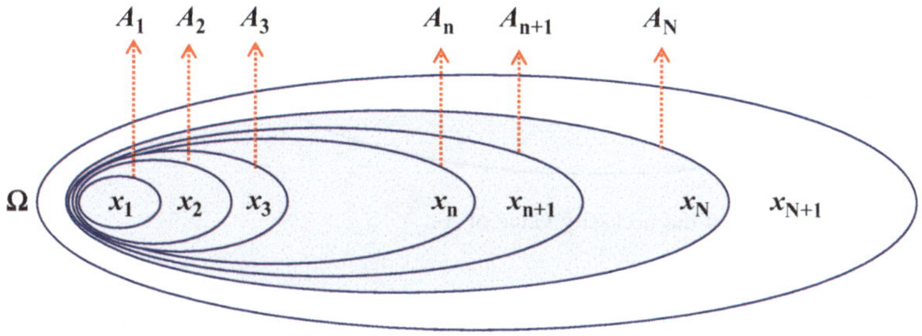

Fig. 5.9 Nested body of evidence

Proof Assume the x_n's are ranked so that $p_1 \geq p_2 \geq p_3 \geq \cdots \geq p_N$, where $p_n = P(x_n)$, $n = 1, 2, \ldots, N$, and let A_n denote the set $\Omega = \{x_1, x_2, \ldots, x_n\}$, $A_0 = \Phi$ and x_{n+1} such that $p_{n+1} = 0$ by convention, Fig.5.9.

Using the definition of the proposed set function N, we can easily show that:

$$N(A_n) = \sum_{x_n \in A_n} max\left[P(x_n) - \underset{x_m \in A_n^C}{max} \mathrm{P}(x_m), 0\right]$$
$$= \sum_{m=1,2,\ldots,n} [P(x_m) - P(x_{n+1})]$$

It is worthwhile to notice, at this level that:

- $N(A_n) = \sum_{m=1}^{n} [P(x_m) - P(x_{m+1})], \quad n = 1, 2, \ldots, N;$
- $N(A_1) \leq N(A_2) \leq N(A_3) \leq \cdots \leq N(A_N) = 1;$
- $N(A_n) = \left[\sum_{m=1,2,\ldots,n} \mathrm{P}(x_m)\right] - nP(x_{n+1}) \leq \Pr\{A_n\} = \left[\sum_{m=1,2,\ldots,n} P(x_m)\right]$

Moreover, we can easily show that for all $A \subseteq \Omega$, we have

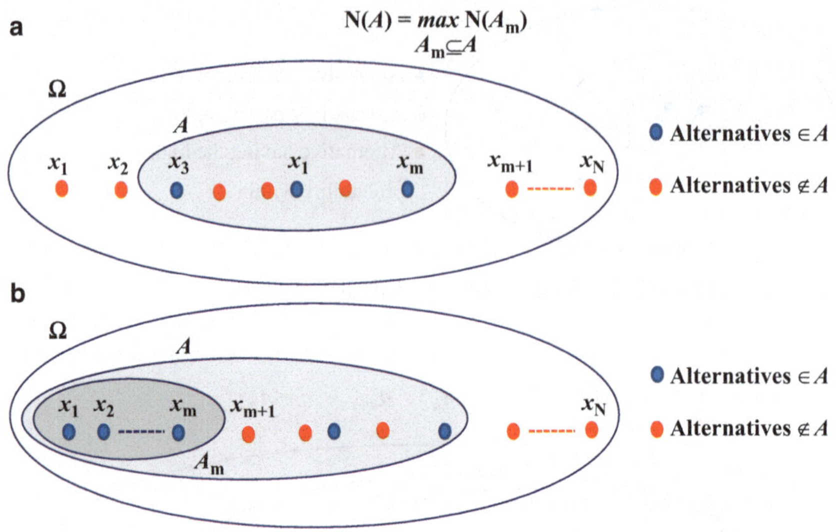

Fig. 5.10 Computation of the necessity value of A

$$N(A) = \max_{A_m \subseteq A} N(A_m).$$

In fact, if the most probable alternative (i.e., x_1) does not belong to A, then $N(A) = 1$, Fig. 5.10a. Assuming that A_m is the maximum set from the family of nested sets $\{A_n\}$ such that $A_m \subseteq A$, then it is straightforward to show that alternatives belonging to A/A_m will have no impact on the computation on $N(A)$ (since, $P(x_{m+1}) = \max_{x_n \in A_m^C} P(x_n)$ and x_{m+1} does not belong to A, Fig. 5.10b). Hence,

$$N(A) = \max_{A_m \subseteq A} N(A_m)$$

Therefore, to show that N is a necessity measure, let us consider two events A and $B \subseteq \Omega$, and let A_{m1}, A_{m2} and A_{m3} be the greatest sets, form the nested A_n family, respectively, contained in A, B, and $A \cap B$. Then, either $A_{m1} \subseteq A_{m2}$ or $A_{m2} \subseteq A_{m1}$. If $A_{m1} \subseteq A_{m2}$ (resp. $A_{m2} \subseteq A_{m1}$), then $A_{m1} \subseteq A \cap B$ and thus $A_{m1} = A_{m3}$ (resp. $A_{m2} \subseteq A \cap B$ and thus $A_{m2} = A_{m3}$). Therefore, $N(A \cap B) = N(A_{m1})$ (resp. $N(A_{m2})$). This means that $N(A \cap B) = \min\{N(A), N(B)\}$, $\forall A, B \subseteq \Omega$, i.e., N is a necessity measure. This approach means that $N(A)$ is, in fact, viewed as the degree of impossibility of the opposite event A^C. Based on the use of this necessity measure, the possibility degrees can be easily obtained from the probability values as follows:

$$\pi(x_n) = \Pi\{(x_n)\} = 1 - N(\Omega - \{x_n\}) = 1 - N(A_{n-1})$$
$$= 1 - \sum_{m=1}^{n-1} [P(x_m) - P(x_n)], n > 1$$

and $\pi(x_1) = 1$ (normalization). Using, $\sum_{m=1}^{N} P(x_m) = 1$, we get:

$$\pi(x_n) = n\,P(x_n) + \sum_{m=n+1}^{N} P(x_m) \quad \forall n \in \{2, \ldots, N\}$$

Without assuming that the $P(x_m)$'s are ordered, this probability-possibility transformation expression can be written as follows:

$$\pi(x_n) = \sum_{m=1}^{N} min\,[P(x_m), P(x_n)] \quad \forall n \in \{1, 2, \ldots, N\}$$

Here follows some important properties of this probability-possibility transformation:

- This transformation satisfies the preference preservation principle. In fact, from the transformation expression, we can easily see that for $n \in \{1, \ldots, N-1\} : \pi(x_n) - \pi(x_{n+1}) = n[P(x_n) - P(x_{n+1})]$, thus,
  - $\pi(x_n) = \pi(x_{n+1}) \Leftrightarrow P(x_n) = P(x_{n+1})$ and,
  - $\pi(x_n) > \pi(x_{n+1}) \Leftrightarrow P(x_n) > P(x_{n+1})$.
- If the probabilistic variable (X, P) is uniformly distributed (i.e., $P(x_n) = 1/N$, for all $n \in \{1, \ldots, N\}$), then the obtained possibilistic variable (X, π) is the full ignorance possibilistic variable (i.e., $\pi(x_n) = 1$, for all $n \in \{1, \ldots, N\}$);
- This transformation is bijective. In fact, assuming that we are given a possibilistic variable (X, π), then the associated possibilistic variable (X, P) (i.e., if (X, P) is transformed using the Dubois-Prade's transformation, then the possibilistic variable (X, π) will be obtained) can be obtained by the resolution of the linear system of $N - 1$ equations with $N - 1$ unknowns (recall that $\pi(x_1) = 1$) given by $\pi(x_n) - \pi(x_{n+1}) = n[P(x_n) - P(x_{n+1})]$. Therefore, we can easily verify that:

$$P(x_n) = \sum_{m=n}^{N} \frac{1}{m}\left[\pi(x_m) - \pi(x_{m+1})\right]$$

  with the convention $\pi(x_{N+1})$;
- This probability-possibility transformation satisfies probability/possibility consistency principle (i.e., the induced necessity and possibility measures constitute a bracketing to the considered probability measure: $N(A) \leq \Pr\{A\} \leq \Pi(A)$, for all $A \subseteq \Omega$). This property can be easily proved as follows:
  - $N(A) = N(A_n)$ (where A_n is the largest $A_n \subseteq A$)

$$N(A) \leq \Pr\{A_n\} \left(\text{since}, N(A_n) = \sum_{m=1}^{n} [P(x_m) - P(x_{m+1})] = \Pr\{A_n\} - a \text{ with } a \geq 0 \right)$$

$$\leq \Pr\{A_n\} + \Pr\{A - A_n\} = \Pr\{A\}$$

– $\Pi(A) = 1 - N(A^C) \geq 1 - \Pr\{A^C\} = \Pr\{A\}$.

References

1. Z. Wang, G. Klir, *Generalized Measure Theory* (Springer, New York, 2009)
2. G. Choquet, Theory of capacities. Ann. Inst. Fourier **5**, 131–295 (1954)
3. M. Sugeno, Theory of fuzzy integrals and its applications, Doctoral Thesis. Tokyo Institute of Technology, 1974
4. T. Murofushi, M. Sugeno, An interpretation of fuzzy measures and the Choquet integral as an integral with respect to a fuzzy measure. Fuzzy Sets Syst. **29**, 201–227 (1989)
5. G.J. Klir, B. Yuan, *Fuzzy Sets and Fuzzy Logic: Theory and Applications* (Prentice Hall, Upper Saddle River, 1995), p. 563
6. A.P. Dempster, Upper and lower probabilities induced by a multivalued mapping, in *Classic Works of the Dempster-Shafer Theory of Belief Functions*, (Springer, Berlin, 2008), pp. 57–72
7. G. Shafer, *A Mathematical Theory of Evidence*, vol 42 (Princeton University Press, Princeton, 1976)
8. L.A. Zadeh, Fuzzy sets as a basis for a theory of possibility. Fuzzy Sets Syst. **1**, 3–28 (1978)
9. D. Dubois, H. Prade, S. Sandri, On possibility/probability transformations, in *Fuzzy logic*, (Springer, Berlin, 1993), pp. 103–112
10. G. De Cooman, D. Aeyels, Supremum preserving upper probabilities. Inf. Sci. **118**, 173–212 (1999)
11. G.L.S. Shackle, *Decision Order and Time in Human Affairs* (Cambridge University Press, Cambridge, 2010)
12. M. Delgado, S. Moral, On the concept of possibility-probability consistency. Fuzzy Sets Syst. **21**, 311–318 (1987)
13. P. Smets, Constructing the Pignistic probability function in a context of uncertainty. UAI **89**, 29–40 (1989)

Chapter 6
Possibility Integral

6.1 Introduction

Decision-making is a very important engineering task that consists of choosing the most adequate decision/alternative among the ones available. For doing so, each alternative is evaluated with respect to several information sources (sensors, features, attributes, criterion, human experts, etc.). Several applications share the same "philosophy" from an engineering point of view: pattern recognition, risk evaluation, multi-criterion decision-making, situation assessment, classification, medical diagnosis, etc. In all these applications, the first task consists of computing, for each alternative (decision, class, assumption, etc.), a degree of support/satisfaction of this alternative by each information source. These support values are related to the discrimination capacity and the distortion (due to noise, nonlinearities, etc.) of different information sources. The second task, called *information aggregation*, consists of aggregating different support values in order to obtain a global support associated with each alternative. Finally, from a decision point of view, the objective consists of ranking different alternatives and, thus, finding the best one (i.e., the alternative with the highest support).

Adequateness ranking of different alternatives strongly suffers from informational dependency between different information sources. To face this difficulty, many practical applications consider each information source as being associated with an "importance" weight. This approach seems adequate to model the discrimination capacity of different sources but is not really suitable for modeling information sources dependency. Moreover, the importance weight of a subset of information sources is more or less similar to the simple addition of individual information sources forming the given subset. To model this interaction (or synergy) phenomenon, individual importance weights can be replaced by a nonadditive set function defined on the set of information sources.

B. Solaiman, É. Bossé, *Possibility Theory for the Design of Information Fusion Systems*, Information Fusion and Data Science,
https://doi.org/10.1007/978-3-030-32853-5_6

Table 6.1 An example of the students ranking problem

Student	Mathematics	Physics	Literature	Global evaluation (weighted mean)
A	18	16	10	15.25
B	10	12	18	12.75
C	14	15	15	14.62

This leads to consider the following students ranking: (1) A, (2) C, and (3) B

To illustrate the interacting information sources issue, let us consider the classical example of students ranking (as a multi-criterion decision-making problem). In this problem, students are evaluated according to their level in three subjects (i.e., information sources): mathematics, physics, and literature. Higher importance weights are attributed to mathematics and physics, and these two subjects are considered as equally important. This leads, for instance, to assign the following importance weights: three for mathematics, three for physics, and two for literature. Let us consider three different students (i.e., alternatives) and compute the average evaluation of these students using a simple weighted mean (i.e., global support) of their marks (i.e., support values of different students according to the used information sources). The obtained results are shown in Table 6.1.

If the school wants to favor students without weak points, then this ranking is not satisfactory (student A has a severe weakness in literature but still first ranked student, whereas student C has no weak marks but ranked second). This is due to the fact that both subjects (mathematics and physics) are strongly dependent/correlated/redundant and have important weights (it often happens that students interested in science and strong in mathematics are also strong in physics and vice versa). Therefore, this support values aggregation approach leads to overestimate (resp. underestimate) students who are strong (resp. weak) in scientific subjects. Moreover, imagine some more sophisticated rules used for decision-maker's preferences (such as students strong in scientific as well as literature are rather uncommon and thus must be favored; or students' particularity strong in mathematics and literature should be the best favored, etc.). Modeling such dependency relations cannot be easily realized using the simple importance weights associated with each information source. Therefore, modeling information sources interactions as well as the way to integrate the dependency relations into the "supports" aggregation constitutes a crucial issue in nearly all decision-making problems.

The following section is devoted to give a brief introduction to aggregation functions. In Sect. 6.3, basic definitions and examples related to monotone measures (allowing to model the interaction relations of different subsets of information sources) are detailed. The concept of fuzzy integrals allowing to combine in a nonlinear manner the two pieces of knowledge, i.e., the monotone measure as well as the support values, is also detailed. The two most applied fuzzy integrals (i.e., discrete Choquet and discrete Sugeno integrals) are then detailed, and their respective properties are given (Sects. 6.4 and 6.5). The use of possibility measures, as a special type of monotone measures, is then discussed using both discrete Choquet and Sugeno integrals (Sect. 6.6). The result of this use is generally referred to as possibility integrals. Finally, the application of possibility integrals in pattern recognition and decision-making applications is presented in Sect. 6.7.

6.2 Aggregation Functions

Aggregation refers to the process of combining numerical values into a single one. In multisource information fusion and multi-criteria decision-making, aggregation functions play an important role where the values to be aggregated are typically preference/satisfaction of decision support degrees and thus generally considered as belonging to the unit interval [0,1]. Let $\Omega = \{S_1, S_2, \ldots, S_n, \ldots, S_N\}$ denote a finite set of N information sources (or relevant criteria) of interest in a decision-making problem, and let $x_n \in [0, 1]$ the support degree to which the information source/criterion S_n is satisfied. Precisely, the function,

$$h : \Omega \rightarrow [0, 1]$$
$$S_n \rightarrow x_n = h(S_n)$$

is called the *utility*, *evidence*, or *support function*. The meaning of an "h" value is interpreted as follows: $h(S_n) = 1$ means full satisfaction (i.e., a total support); $h(S_n) = 0$ means full non-satisfaction (i.e., a total reject). A value $0 < h(S_n) < 1$ indicates a partial degree of satisfaction that the solution (i.e., decision, alternative) meets the requirement of S_n without considering the degree of importance, or trust, of S_n to the desired solution. The vector $X = [x_1 \ldots x_n \ldots x_N] \in [0, 1]^N$ is called the information sources *support vector* or *profile* (Fig. 6.1).

The aggregation process consists of combining the N-tuple support vector components in order to produce a global support, or satisfaction, value. From a mathematical point of view, here follows a formal definition of an aggregation function.

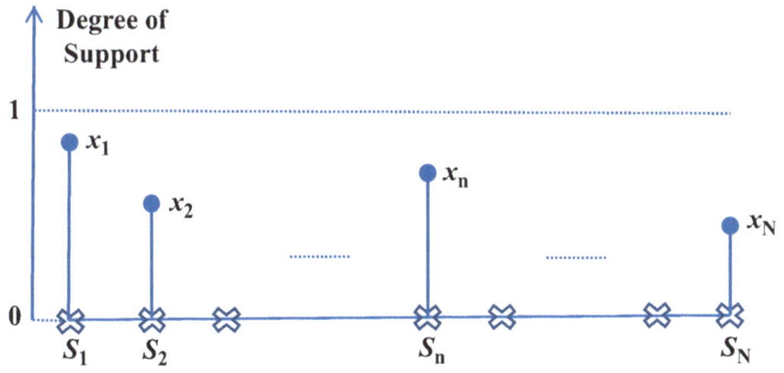

Fig. 6.1 The concept of a support vector

Definition: ***Aggregation function***
A mapping $F : [0, 1]^N \rightarrow [0, 1]$ is called an *aggregation function* if it satisfies the three following conditions:

1. *Upper boundary condition:* $F(1, 1, \ldots, 1) = 1$.
2. *Lower boundary condition:* $F(0, 0, \ldots, 0) = 0$.
3. *Non-decreasing monotonicity condition:* if $x_n \geq y_n$ for all $n = 1, 2, \ldots, N$, then:

$$F(x_1, x_2, \ldots, x_N) \geq F(y_1, y_2, \ldots, y_N).$$

Three major types of aggregation functions can be identified:

- *Conjunctive*: An aggregation function F is called a *conjunctive-type* aggregation function if:

$$F(x_1, x_2, \ldots, x_N) \leq \min_{n=1}^{N} x_n \quad \text{for all } X = [x_1 \ldots x_n \ldots x_N] \in [0, 1]^N.$$

- *Disjunctive*: An aggregation function F is called a *disjunctive-type* aggregation function if:

$$F(x_1, x_2, \ldots, x_N) \geq \max_{n=1}^{N} x_n \quad \text{for all } X = [x_1 \ldots x_n \ldots x_N] \in [0, 1]^N.$$

- *Mean*: An aggregation function F is called a *mean-type* aggregation function if:

$$\min_{n=1}^{N} x_n \leq F(x_1, x_2, \ldots, x_N) \leq \max_{n=1}^{N} x_n \quad \text{for all}\, X = [x_1 \ldots x_n \ldots x_N] \in [0, 1]^N.$$

It is worthwhile to notice that all mean-type aggregation functions are *idempotent*, i.e., if all $x_n = x$ then, $F(x_1, x_2, \ldots, x_N) = x$.

Some examples of aggregation functions used in the literature are given as follows:

- The *minimum* aggregation function: $F(x_1, x_2, \ldots, x_N) = \min_{n=1}^{N} x_n$;
- The *maximum* aggregation function: $F(x_1, x_2, \ldots, x_N) = \max_{n=1}^{N} x_n$;
- The *arithmetic mean* aggregation function: $F(x_1, x_2, \ldots, x_N) = \sum_{n=1}^{N} x_n / N$;
- The *geometric mean* aggregation function: $F(x_1, x_2, \ldots, x_N) = \left(\prod_{n=1}^{N} x_n \right)^{1/N}$;
- The *harmonic mean* aggregation function: $F(x_1, x_2, \ldots, x_N) = \sum_{n=1}^{N} \frac{1}{x_n} / N$;

- *Ordered weighted average* (OWA) aggregation function. This aggregation function is a mean-type function introduced by Yager [1]. Let $W = [\omega_1, \omega_2, \ldots, \omega_N]$ be a weighting vector such that: $\omega_n \in [0, 1]$, for all $n = 1, \ldots, N$ such that:

$$\sum_{n=1}^{N} \omega_n = 1, \text{and assume} X \downarrow = \left[x_{\text{id}(1)} \ldots x_{\text{id}(n)} \ldots x_{\text{id}(N)}\right]$$

the re-ordered support vector (i.e., $x_{\text{id}(n)}$ is the nth largest support value in the collection of N values $x_1 \ldots x_n \ldots x_N$). The OWA aggregation function is given by:

$$F(x_1, x_2, \ldots, x_N) = \sum_{n=1}^{N} \omega_n . x_{\text{id}(n)}.$$

Notice that ω_1 affects the most important support value; ω_2 affects the second most important support value, etc. It is worthwhile also to notice that the weighting vector $W = [1, 0, \ldots, 0]$ (resp. $W = [0, 0, \ldots, 1]$) leads to the *maximum* (resp. *minimum*) aggregation function, and $W = [1/N, 1/N, \ldots, 1/N]$ leads to the arithmetic *mean* aggregation function.

6.3 Monotone Measures and Fuzzy Integrals

Within aggregation functions, fuzzy integrals are known to be one of the most powerful and flexible functions allowing the aggregation of information under different assumptions on the independence (among information sources), or criteria in Multiple Criteria Decision Making (MCDM). The main idea of fuzzy integrals is to combine, in a nonlinear manner, two pieces of distinguished information, namely, the support values (delivered by different information sources, or criteria) according to a monotone measure defined on the subsets of information sources. This monotone measure represents the "background knowledge" on the information sources expressed as a set function measuring the importance/relevance of different subsets of information sources.

In this section, we focus on computing a global degree of support of the N information sources from $\Omega = \{S_1, S_2, \ldots, S_n, \ldots, S_N\}$ by taking into account the "weights of importance" of different information sources, or criteria, and this, for a "single alternative, or decision." In other words, the problem of ranking different alternatives and the decision-making process are not considered.

The major drawback of aggregation functions (given in the previous section) is that information sources are assumed independent so that the interaction, or correlation, between them is not considered. Therefore, and in order to offer a flexible representation of the complex interaction phenomena between different information sources, or criteria, the substitution of the weighting vector $W = [\omega_1, \omega_2, \ldots, \omega_N]$ by a nonadditive set function defined on 2^{Ω} (i.e., the power set of Ω) through the use of monotone measures is proposed. The concept of monotone measures has been introduced by Sugeno [2] as a solution to the rigidness problem of classical measures, i.e., additivity characteristic. In fact, in monotone measures, additivity is substituted by monotonicity, which is a weaker condition allowing, thus, to reach a higher level of flexibility. After recalling the basic definitions and concepts related to monotone measures, the application of fuzzy integrals is then detailed in the following sections.

6.3.1 Monotone Measures Definition

Monotone measures are used to represent the interactions between information sources, namely, the aspects of independency, complementarity, and redundancy. Consider an exhaustive and finite set of mutually exclusive information sources $\Omega = \{S_1, \ldots, S_n, \ldots, S_N\}$. The set function $g : 2^{\Omega} \to [0, 1]$ is called a *monotone measure* [3] (also called a *Choquet capacity*) if it satisfies the following axioms:

1. *Boundary conditions*: $g(\Phi) = 0$ and $g(\Omega) = 1$ (i.e., normality).
2. *Non-decreasing monotonicity*: $A, B \in 2^{\Omega}$, if $A \subseteq B$ then $g(A) \leq g(B)$.

The value $g(A)$ is assumed to reflect the weight, worth, importance, reliability, relevance, and satisfaction (or other similar concepts) of each subset $A \subseteq \Omega$ of information sources. Therefore, $g(\Omega) = 1$ means that the whole set of information sources is considered as being totally satisfactory (or, having the total importance).

It is worthwhile to notice that monotone measures could be considered as an extension of the concept of probability measures, replacing the additivity property by the weaker condition of monotonicity. One of the reasons behind the name of "fuzzy measure" (that is used to be called[1] monotone measure) is that g may be considered as defining a fuzzy set on 2^{Ω} where each singleton, i.e., a subset A from Ω, has a membership value $g(A)$ reflecting the "importance" of A for the decision process.

Several monotone measures cases can be encountered. Let A and B denote two subsets from Ω such that $A \cap B = \Phi$.

A monotone measure g is said to be:

- *Additive*: If $g(A \cup B) = g(A) + g(B)$. In this case the monotone measure reflects the fact that there is no interaction between the two subsets A and B.
- *Superadditive*: If $g(A \cup B) \geq g(A) + g(B)$. This type of monotone measures reflects the fact that there is a relevance/importance reinforcement by the adjunction of the two subsets A and B.

[1]See footnote of Chap. 5.

- *Subadditive*: If $g(A \cup B) \leq g(A) + g(B)$. In this case, the monotone measure reflects the fact that there is a relevance/importance redundancy by the adjunction of the two subsets A and B.

Theorem *Fundamental theorem on the aggregation of monotone measures*

Let us consider a finite set Ω, and let $g_1, g_2, \ldots, g_N$ be a collection of N monotone measures defined on 2^Ω.

If $F : [0, 1]^N \to [0, 1]$ is an aggregation function, then the set function $g : 2^\Omega \to [0, 1]$ defined by $g(A) = F(g_1(A), g_2(A), \ldots, g_N(A))$ for all $A \subseteq \Omega$ is a monotone measure.

The demonstration of this theorem is straightforward. In fact:

- For $A = \Phi$ we have $g(\Phi) = F(g_1(\Phi), g_2(\Phi), \ldots, g_N(\Phi)) = F(0, 0, \ldots, 0) = 0$.
- For $A = \Omega$ we have $g(\Omega) = F(g_1(\Omega), g_2(\Omega), \ldots, g_N(\Omega)) = F(1, 1, \ldots, 1) = 1$.
- If $A, B \in 2^\Omega$ such that $A \subseteq B$, then: $g(A) = F(g_1(A), g_2(A), \ldots, g_N(A)) \leq F(g_1(B), g_2(B), \ldots, g_N(B))$, since $g_n(A) \leq g_n(B)$ for $n = 1, 2, \ldots, N$ (due to the fact that F is an aggregation function).

This theorem provides a very general approach to construct monotone measures based on other existing monotone measures. For instance, if $g_1, g_2, \ldots, g_N$ is a collection of N monotone measures defined on 2^Ω, then:

- $g : 2^\Omega \to [0, 1]$ defined by $g(A) = g_1(A).g_2(A)\ldots g_N(A)$ for all $A \subseteq \Omega$, is a monotone measure (using the simple product aggregation function, i.e., $F(x_1, x_2, \ldots, x_N) = x_1.\, x_2 \ldots x_N$).
- $g : 2^\Omega \to [0, 1]$ defined by $g(A) = max\,(g_1(A), g_2(A), \ldots, g_N(A))$ for all $A \subseteq \Omega$, is a monotone measure (using the *maximum* aggregation function (i.e., $F(x_1, x_2, \ldots, x_N) = max\,(x_1.\, x_2 \ldots x_N)$).
- $g : 2^\Omega \to [0, 1]$ defined by $g(A) = min\,(g_1(A), g_2(A), \ldots, g_N(A))$ for all $A \subseteq \Omega$, is a monotone measure (using the *minimum* aggregation function (i.e., $F(x_1, x_2, \ldots, x_N) = min\,(x_1.\, x_2 \ldots x_N)$).
- If $\omega_n \in [0, 1]$, $n = 1, 2, \ldots, N$ is a collection of coefficients such that $\sum_{n=1}^{N} \omega_n = 1$,

  then $g : 2^\Omega \to [0, 1]$ defined as the linear combination $g(A) = \sum_{n=1}^{N} \omega_n g_n(A)$ for all $A \subseteq \Omega$, is a monotone measure.

Remark

The definition of monotone measures can be extended to the qualitative case. In fact, a monotone measure $g : 2^\Omega \to L$ is said to be *qualitative* (or q-monotone measure, for short) if its range is a finite totally ordered set (i.e., $L = \{0 = \alpha_0 < \alpha_1 < \cdots < \alpha_K = 1\}$). This means that classical numerical operations (addition, subtraction, multiplication, etc.) are not valid anymore, and only smaller, higher, *minimum*, and *maximum* operations can be operated on this range set.

6.3.2 Special Monotone Measures

Major examples of fuzzy measures include the Sugeno λ-measure, cardinality-based measures, prioritization measures, probability measures, and possibility measures.

6.3.2.1 The Sugeno λ-measure

Let us consider $\Omega = \{S_1, S_2, \ldots, S_n, \ldots, S_N\}$. For $A, B \in 2^\Omega$ such that $A \cap B = \Phi$, the Sugeno λ-measure for the union set $A \cup B$ is defined as follows:

$$g_\lambda(A \cup B) = g_\lambda(A) + g_\lambda(B) + \lambda . g_\lambda(A) . g_\lambda(B)$$

for some $\lambda > -1$.

The Sugeno λ-measure is built in an incremental way from a set of the so-called densities (i.e., measures "predefined" on singleton information sources $g_\lambda(\{S_n\}), n = 1, 2, \ldots, N$). In fact, given the fuzzy densities $g_\lambda(\{S_n\}, n = 1, 2, \ldots, N$, the Sugeno λ-measure of any subset $A \subseteq \Omega$ is given by:

$$g_\lambda(A) = \frac{1}{\lambda}\left\{\left[\prod_{S_n \in A}(1 + \lambda g_\lambda(\{S_n\}))\right] - 1\right\} \quad \text{for } \lambda \neq 0, \text{and}$$

$$g_\lambda(A) = \sum_{S_n \in A} g_\lambda(\{S_n\}) \quad \text{for } \lambda = 0$$

The parameter λ describes the degree of additivity the monotone measure holds. Three cases are considered:

- *Positive interaction*: if $\lambda > 0$, then $g_\lambda(A \cup B) > g_\lambda(A) + g_\lambda(B)$. This corresponds to a super-additivity case (i.e., Sugeno's measure is a belief measure).
- *Absence of interaction*: if $\lambda = 0$, then $g_\lambda(A \cup B) = g_\lambda(A) + g_\lambda(B)$. This corresponds to the additive case (i.e., Sugeno's measure is a probability measure).
- *Negative interaction*: if $-1 < \lambda < 0$, then $g_\lambda(A \cup B) < g_\lambda(A) + g_\lambda(B)$. This corresponds to the sub-additive case (i.e., Sugeno's measure is a plausibility measure).

Moreover, if A^C denotes the complement of A, then $g_\lambda(A^C) = [1 - g_\lambda(A)]/[1 + \lambda g_\lambda(A)]$. As $g_\lambda(\Omega) = 1$, Sugeno showed that λ is determined, in a second step, by solving the polynomial equation:

$$\lambda + 1 = \prod_{n=1}^{N} [1 + \lambda . g_\lambda(\{S_n\})]$$

where there exists one real solution such that $\lambda > -1$. To illustrate the role of the parameter λ for the description of the degree of additivity of the Sugeno λ-measure

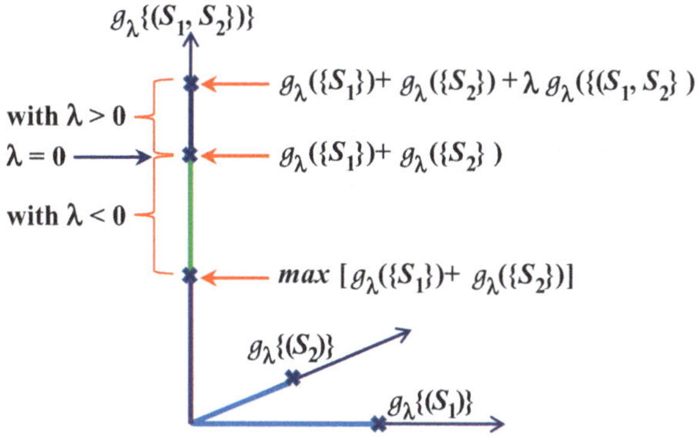

Fig. 6.2 Information sources interaction depending on the variation of λ

[4], consider a set of two information sources $\Omega = \{S_1, S_2\}$, and let $g_\lambda(\{S_1\})$ and $g_\lambda(\{S_2\})$ denote the two Sugeno λ-measure densities; see Fig. 6.2. The decision-maker may consider that it is sufficient that one of the two sources S_1 and S_2 shows a full support. In this case S_1 and S_2 act disjunctively (negative interaction). The union of the two sources has no additional benefit, and the importance of the pair $g_\lambda(\{S_1, S_2\})$ is almost the same as the importance of the most supporting support. In this case, the two sources are said to be *redundant* ($\lambda < 0$). If the decision-maker considers that the importance of a single source for the decision is almost zero while the simultaneous support of the pair of sources is highly important (i.e., conjunctive behavior), then the sources are said to be *complementary* ($\lambda < 0$). Finally, if the decision-maker thinks that the importance of the pair of sources is approximately the sum of the individual importance of each source, then both sources are considered as acting independently, and there is *no interaction* between them ($\lambda = 0$).

Example Let $\Omega = \{S_1, S_2, S_3\}$ be a finite set of three information sources, and assume the following density values: $g_\lambda(\{S_1\}) = 0.4, g_\lambda(\{S_2\}) = 0.3$ and $g_\lambda(\{S_3\}) = 0.2$.

In order to determine the Sugeno λ-measure, the first step consists of determining the value of λ by solving the polynomial equation $\lambda + 1 = (0.4\lambda + 1)(0.3\lambda + 1)(0.2\lambda + 1)$ (i.e., $0.024\lambda^3 + 0.26\lambda^2 - 0.1\lambda = 0$). The three roots resulting of the resolution of this equation are $\lambda_1 = 0$, $\lambda_2 = -11.87$, and $\lambda_3 = 0.3719$. As $\lambda_2 \notin]0, \infty[$, only λ_1 and λ_2 should be retained. $\lambda_1 = 0$ leads to an additive Sugeno λ-measure, whereas $\lambda_3 = 0.3719$ leads to the following superadditive Sugeno λ-measure: $g_\lambda(\{S_1\}) = 0.4, g_\lambda(\{S_2\}) = 0.3$, and $g_\lambda(\{S_3\}) = 0.2$; $g_\lambda(\{S_1, S_2\}) = 0.7446$, $g_\lambda(\{S_1, S_3\}) = 0.6298$, $g_\lambda(\{S_2, S_3\}) = 0.5223$, and $g_\lambda(\{S_1, S_2, S_3\}) = 1$.

6.3.2.2 Cardinality-Based Monotone Measures

Let us consider a finite set of N information sources $\Omega = \{S_1, S_2, \ldots, S_n, \ldots, S_N\}$. The set function $g : 2^\Omega \to [0, 1]$ is called a *cardinality-based monotone measure* if:

- g is a monotone measure.

- $g(A) = f(|A|)$ for all subsets $A \subseteq \Omega$ where $|A|$ denotes the cardinality of A (i.e., number of elements belonging to A). This means that $g(A)$ depends upon the cardinality of A.

Cardinality-based monotone measures are also called, in the literature, *symmetric fuzzy measures* and are defined as verifying the following property: for any A, $B \subseteq \Omega$ for which $|A| = |B|$, then $g(A) = g(B)$. Here follows some examples of cardinality-based monotone measures:

$$\begin{aligned} g_1 &: 2^{\Omega} \to [0,1] \\ A &\to g_1(A) = 1 \quad \text{if } |A| \neq 0 \text{ (i.e., } A \text{ contains at least one element);} \\ &\qquad\;\;\, = 0 \quad \text{if } A = \Phi. \end{aligned}$$

$$\begin{aligned} g_2 &: 2^{\Omega} \to [0,1] \\ A &\to g_2(A) = 0 \quad \text{if } |A| < N \text{ (i.e., } A \subset \Omega); \\ &\qquad\;\;\, = 1 \quad \text{if } A = \Omega. \end{aligned}$$

$$\begin{aligned} g_3 &: 2^{\Omega} \to [0,1] \\ A &\to g_3(A) = 0 \qquad \text{if } A = \Phi; \\ &\qquad\;\;\, = |A|/N \quad \text{if } |A| \leq N. \end{aligned}$$

g_3 is also an additive monotone measure (i.e., $g_3(A \cup B) = g_3(A) + g_3(B)$). It is generally called the *uniform capacity*.

$$\begin{aligned} g_4 &: 2^{\Omega} \to [0,1] \\ A &\to g_4(A) = 0 \qquad \text{if } A = \Phi; \\ &\qquad\;\;\, = (|A|/N)^{q} \quad \text{if } |A| \leq N. \end{aligned}$$

with $q > 0$. g_4 is named the *exponential cardinality capacity*.

6.3.2.3 Prioritization Monotone Measure

Consider an exhaustive and finite set of mutually exclusive information sources $\Omega = \{S_1, S_2, \ldots, S_n, \ldots, S_N\}$, and let us define an order relation (>) on Ω (allowing to these information sources to have different priority levels for the decision-making process) such that $S_1 > S_2 > \cdots > S_n > \cdots > S_N$ ($S_m > S_n$ means that the information source S_m has a higher priority over the information source S_n).

It is important to stress the fact that this ordering is independent of the information sources ranking (as a function of the support values delivered by these information sources with respect to the decision to be made).

The basic idea of prioritization is that lack of satisfaction to higher priority information sources is not easily compensated by satisfaction to lesser priority information sources. In other words, we are not willing to let high priority information sources to compensate poor priority ones.

Let us denote $B_n = \{S_{\mathrm{Pr}(1)}, S_{\mathrm{Pr}(q)}, \ldots, S_{\mathrm{Pr}(n)}\}$, $n = 1, 2, \ldots, N$ the subset of "n" higher priority information sources ($S_{\mathrm{Pr}(q)}$ stands for the qth highest priority information source), and assume that each B_n is associated with the value $v_n = n/N$. Each subset of information sources $B \subseteq \Omega$ is associated with N values $G_n(B)$ such that $G_n(B) = 1$ if $B_n \subseteq B$ and $G_n(B) = 0$ if $B_n \not\subset B$. For instance, $G_1(B) = 1$ if and only if $\{S_{\mathrm{Pr}(1)}\} \subseteq B$ (i.e., when the highest priority information source belongs to B), $G_2(B) = 1$ if and only if $\{S_{\mathrm{Pr}(1)}, S_{\mathrm{Pr}(2)}\} \subseteq B$ (i.e., when the two highest priority information sources belong to B), etc. In other words, $G_n(B)$ indicates if, or if not, the "n" highest priority information sources are contained in the subset B. Using the values v_n and $G_n(B)$, the following prioritization monotone measure can be defined:

$$g_{\mathrm{P}} : 2^{\Omega} \rightarrow [0, 1]$$
$$B \rightarrow g_{\mathrm{P}}(B) = \max_{n=1}^{N} \{v_n . G_n(B)\}$$

This leads to $g_P(B) = n_0/N$ where B_{n_0} is the largest set B_n contained in B. It is straightforward to show that g_P is a monotone measure (i.e., $g_P(\Phi) = 0$, $g_P(\Omega) = 0$, and if $A \subseteq B$, then $g_P(A) \leq g_P(B)$).

Here follow the values of this monotone measure for some subsets of Ω:

- Single singleton subsets: $g_P(\{S_{\mathrm{Pr}(1)}\}) = 1$, $g_P(\{S_{\mathrm{Pr}(n)}\}) = 0$ for all $n \neq 1$ (i.e., only the singleton subset containing the highest priority information source $S_{\mathrm{Pr}(1)}$ has a nonzero measure).
- Subsets containing two singletons:
  - $g_P(\{S_{\mathrm{Pr}(1)}, S_{\mathrm{Pr}(2)}\}) = 2/N$;
  - $g_P(\{S_{\mathrm{Pr}(1)}, S_{\mathrm{Pr}(m)}\}) = 1/N$ (for all $m \neq 2$);
  - $g_P(\{S_{\mathrm{Pr}(n)}, S_{\mathrm{Pr}(k)}\}) = 0$ (if neither "n" nor "m" is 1).
- Any subset $B \subseteq \Omega$ such that $S_{\mathrm{Pr}(1)} \notin B$ will have a zero measure value.

6.4 Discrete Choquet Integral

Within aggregation operators, fuzzy integrals are known to be of the most powerful and flexible functions allowing to aggregate support information, according to a monotone measure (under different assumptions related to the independence of the information sources and representing the background knowledge related to the importance, or relevance of the information sources).

Let $\Omega = \{S_1, \ldots, S_n, \ldots, S_N\}$ denote a finite set of N information sources of interest in a decision-making problem, and assume $h(S_n) = x_n \in [0, 1]$ the support degree of that an alternative/decision meets the requirement of the information

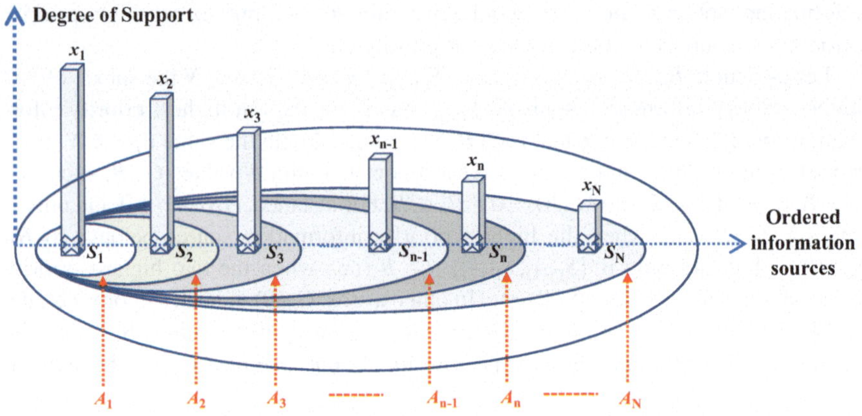

Fig. 6.3 Knowledge structuring in the framework of the discrete Choquet integral

source S_n. h is called the *evidence* or the *support function*. Moreover, let us consider a monotone measure $g : 2^{\Omega} \rightarrow [0, 1]$ assumed to reflect the weight, worth, importance, reliability, relevance, and satisfaction (or other similar concepts) of each subset $A \subseteq \Omega$ of information sources.

Without loss of generality, we assume different sources as being arranged such that their support values are decreasing, i.e., $h(S_1) = x_1 \geq \cdots \geq x_n \geq \cdots \geq x_N = h(S_N)$ (Fig. 6.3), where S_1 (resp. S_N) is assumed to show the greatest (resp. lowest) support value.

One of the most known fuzzy integrals, called *discrete Choquet integral*, was introduced by G. Choquet in 1954 [5].

The discrete Choquet integral (also named *monotone expectation*) focuses the attention on the nested subsets $A_1 \subseteq A_2 \subseteq \ldots \subseteq A_N = \Omega$ of the ordered information sources (ordering is with respect to support values) (Fig. 6.3):

$A_1 = \{S_1\}$: subset formed by the most satisfied information source
$A_2 = \{S_1, S_2\}$: subset formed by the two most satisfied information sources
$A_n = \{S_1, S_2, \ldots, S_n\}$, $n = 1, \ldots, N$: the subset of the "n" most satisfied information sources

Therefore, the discrete Choquet integral "restricts the attention" on the N nested subsets $A_1 \subseteq A_2 \subseteq \ldots \subseteq A_N = \Omega$ instead of considering the 2^{N-1} subsets of the power set 2^{Ω} (the set of all subsets of information sources). Moreover, the discrete Choquet integral characterizes each subset $A_n = \{S_1, S_2, \ldots, S_n\}$, $n = 1, 2, \ldots, N$, by two values (Fig. 6.4):

- The least supporting value of information sources contained in A_n, i.e., $h(S_n)$
- The additional worth/importance of S_n in supporting the global decision, i.e., $g(A_n) - g(A_{n+1}) \geq 0$ (due to the monotonicity property of the fuzzy measure $g(.)$)

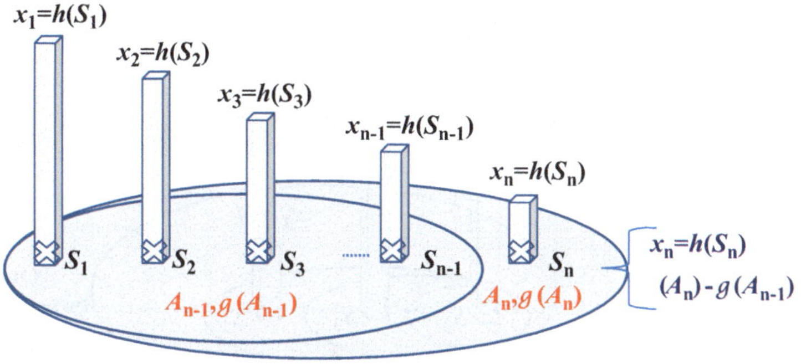

Fig. 6.4 Subset A_n characterization by the discrete Choquet integral

Given the fact that the additional worth values of $S_n : \omega_n = g(A_n) - g(A_{n-1})$, $n = 1, 2, \ldots, N$, range in the unit interval [0,1] and that $\sum_{n=1}^{N} \omega_n = 1$, the main idea behind the discrete Choquet integral consists in considering the coefficients ω_n as weighting coefficients of the linear combination aggregation function of the information sources support values.

The discrete Choquet integral of h with respect to g is, thus, defined as follows:

$$\int_{C} h\,\mathrm{d}g = \sum_{n=1}^{N} h(S_n)\,[g(A_n) - g(A_{n-1})]$$
$$= h(S_1)g(A_1)$$
$$+ h(S_2)[g(A_2) - g(A_1)]$$
$$+ h(S_3)[g(A_3) - g(A_2)]$$
$$+ \ldots$$
$$+ h(S_N)[g(A_N) - g(A_{N-1})]$$

($g(A_0) = 0$ by convention).

Example Let $\Omega = \{S_1, S_2, S_3\}$ be a finite set of three information sources, and let g_λ denote the Sugeno λ-measure previously determined in this chapter:

$$g_\lambda(\{S_1\}) = 0.4, g_\lambda(\{S_2\}) = 0.3, g_\lambda(\{S_3\}) = 0.2;$$
$$g_\lambda(\{S_1, S_2\}) = 0.7446, g_\lambda(\{S_1, S_3\}) = 0.6298, g_\lambda(\{S_2, S_3\})$$
$$= 0.5223 \text{ and } g_\lambda(\{S_1, S_2, S_3\}) = 1.$$

Assuming the support vector $X = [x_1 = h(S_1) = 0.4, x_2 = h(S_2) = 1,\ x_3 = h(S_3) = 0.9]$ (information sources ordering and the corresponding nested subsets are illustrated in Fig. 6.5), then the discrete Choquet integral of h with respect to g_λ is given by:

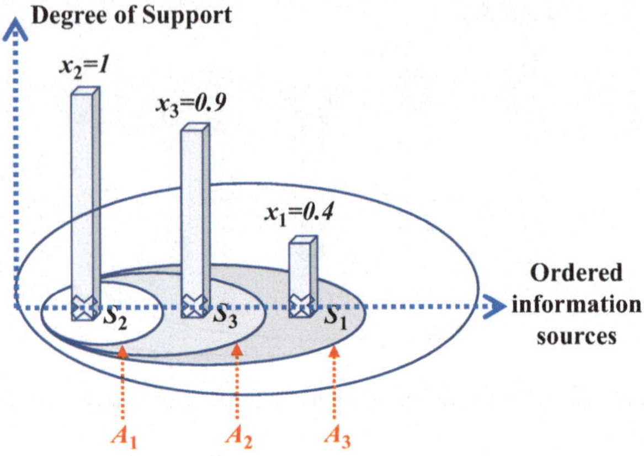

Fig. 6.5 Knowledge structuring for the discrete Choquet integral with respect to g_λ

$$\begin{aligned}\int_{C} h\,\mathrm{d}g_\lambda &= h(S_2)g_\lambda(\{S_2\}) + h(S_3)\left[g_\lambda(\{S_2,S_3\}) - g_\lambda(\{S_2\})\right] \\ &\quad + h(S_1)\left[g_\lambda(\{S_1,S_2,S_3\}) - g_\lambda(\{S_2,S_3\})\right] \\ &= 1\times 0.3 + 0.9\times[0.5223 - 0.3] + 0.4\times[1 - 0.5223] \\ &= 0.69115.\end{aligned}$$

6.4.1 Important Properties of the Discrete Choquet Integral

The discrete Choquet integral with respect to the monotone measure g possesses some interesting properties including:

- The discrete Choquet integral with respect to the monotone measure g is a mean-type aggregation operator. That is, for any monotone measure $g(.)$ we have:
  - If $h(S_n) = 0$ for all $n = 1, 2, \ldots, N$, then $\int_C h\,\mathrm{d}g = 0$.
  - If $h(S_n) = 1$ for all $n = 1, 2, \ldots, N$, then $\int_C h\,\mathrm{d}g = 1 = g(\Omega)$.
  - If $m = \min_{n=1}^{N} h(S_n)$ and $M = \max_{n=1}^{N} h(S_n)$, then:

$$m \le \int_C h\,\mathrm{d}g \le M$$

(i.e., the overall evaluation of the discrete Choquet integral with respect to the monotone measure g is comprised between the minimum and the maximum of all support values).

– Idempotent property: If $h(S_n) = \alpha$ for all $n = 1, 2, \ldots, N$, then $\int_C h\,\mathrm{d}g = \alpha$.

- Monotonicity non-decreasing with respect to the support function: If $h_1(S_n) \le h_2(S_n)$ for all $S_n \in \Omega$, then:

$$\int_C h_1 \mathrm{d}g \le \int_C h_2 \mathrm{d}g$$

(i.e., an increase of the support cannot lead to a decrease of the overall evaluation of the discrete Choquet integral).

- Monotonicity non-decreasing with respect to the monotone measure: If $g_1(A) \le g_2(A)$ for all $A \in 2^\Omega$, then:

$$\int_C h\,\mathrm{d}g_1 \le \int_C h\,\mathrm{d}g_2.$$

- For any nonnegative constant α:
  – $\int_C (\alpha.h)\,\mathrm{d}g = \alpha \int_C h\,\mathrm{d}g$ (i.e., compatibility with a ratio scale);
  – $\int_C (\alpha + h)\,\mathrm{d}g = \alpha + \int_C h\,\mathrm{d}g$

- Additivity for equiordered support functions. Given two support functions h_1 and h_2 defined on Ω, h_1 is said to be *equiordered* with h_2 if and only if for each pair of information sources S_n and S_m from Ω such that $h_1(S_n) \le h_1(S_m)$, it follows necessarily that $h_2(S_n) \le h_2(S_m)$. If h_1 and h_2 are equiordered support functions, then:

$$\int_C (h_1 + h_2)\,\mathrm{d}g = \int_C h_1 \mathrm{d}g + \int_C h_2 \mathrm{d}g$$

Therefore, although the discrete Choquet integral is not an additive functional, it does have an "additivity property" for equiordered support functions (this property is called *ordered additivity*).

6.4.2 Discrete Choquet Integral for Some Types of Monotone Measures

In this section, some particular monotone measures are considered, and the discrete Choquet integrals associated with these monotone measures are evaluated.

- *Additive monotone measure*

Recall that a monotone measure g is said to be additive if for all disjoint sets A, $B \subseteq \Omega$, we have $g(A \cup B) = g(A) + g(B)$. In this case, g reflects the absence of interaction between disjoint subsets of information sources. In this case,

$$g(A_1) = g(\{S_1\}); g(A_2) = g(\{S_1, S_2\}) = g(\{S_1\}) + g(\{S_2\}); \ldots$$
$$g(A_N) = g(\{S_1\}) + g(\{S_2\}) + \cdots + g(\{S_{N-1}\}) + g(\{S_N\}).$$

Therefore,

$$\int_C h\,\mathrm{d}g = h(S_1)g(\{S_1\}) + h(S_2)g(\{S_2\}) + h(S_3)g(\{S_3\}) + \cdots + h(S_N)g(\{S_N\})$$

This means that the discrete Choquet integral with respect to an additive monotone measure is reduced to the weighted arithmetic mean function where information sources weights are simply the fuzzy densities. It is worthwhile to notice that with the assumption of additivity of the monotone measure g, the interaction between information sources disappears. Moreover, notice that when the monotone measure is a probability measure, Pr, (which is an additive measure), then the discrete Choquet measure coincides with the mathematical expectation of the support function with respect to the probability measure:

$$\int_C h\,\mathrm{d}g = \sum_{n=1}^{N} h(S_n)\,\Pr(\{S_n\})$$

- *Dirac monotone measure*

A Dirac monotone measure g_k centered on S_k is defined as follows:

$$\begin{aligned} g_k:\quad 2^S &\rightarrow [0,1] \\ A &\rightarrow g_k(A) \quad = 1 \text{ if } S_k \in A \\ & \qquad\qquad\quad\; = 0 \text{ if } S_k \notin A \end{aligned}$$

We can easily show that g_k is a monotone measure. To compute the discrete Choquet integral with respect to the Dirac monotone measure g_k, $\int_C h\,\mathrm{d}g_k$, let us assume that S_k is the qth most satisfied information source. This means that $S_k \in A_n$ for all $n \geq q$, and $S_k \notin A_n$ for all $n < q$. Therefore, $g_k(A_n) = 1$ (for $n \geq q$), and $g_k(A_n) = 0$ (for $n < q$). This leads to:

$$\int_C h\,\mathrm{d}g_k = 0 + 0 + \cdots + h(S_k)[1 - 0] + 0 + \cdots + 0 = h(S_k).$$

This means that the discrete Choquet integral with respect to a Dirac monotone measure g_k centered on S_k produces the support value of the information source S_k.

- *Weakest monotone measure on* Ω

$$\text{If} \quad g(A) = 0 \quad \text{for all} A \neq \Omega \text{ and}$$
$$= 1 \quad \text{if} A = \Omega$$

For this monotone measure, we have:

$$\int_C h\,dg = h(S_1)g(A_1) + h(S_2)\,[g(A_2) - g(A_1)] + h(S_3)\,[g(A_3) - g(A_2)]$$
$$+ \cdots + h(S_N)\,[g(A_N) - g(A_{N-1})]$$
$$= h(S_1).0 + h(S_2)[0 - 0] + h(S_3)[\,0 - 0] + \cdots + h(S_N)[1 - 0]$$
$$= h(S_N) = \min_{n=1}^{N} h(S_n)$$

This means that the discrete Choquet integral with respect to the weakest monotone measure on Ω narrows down to the minimum function.

- *Strongest monotone measure on* Ω

$$\text{If} \quad g(A) = 1 \quad \text{for all} A \neq \Phi$$
$$= 0 \quad \text{if} A = \Phi$$

then, $\int_C h\,dg = h(S_1) = \max_{n=1}^{N} h(S_n)$.

This means that for this particular fuzzy measure, the discrete Choquet integral is reduced to the maximum function.

- *Cardinality-based monotone measure*

Recall that a set function $g : 2^{\Omega} \to [0, 1]$ is called a cardinality-based monotone measure if it is a monotone measure for which $g(A) = f(|A|)$ for all subsets $A \subseteq \Omega$ where $|A|$ stands for the cardinality of the subset A. The discrete Choquet integral is given with respect to the three following examples of cardinality-based monotone measures:

$$g_1(A) = 1 \quad \text{if } |A| \neq 0 \;\; (\text{i.e., } A \text{ contains at least one element});$$
$$= 0 \quad \text{if} A = \Phi.$$

$$\Rightarrow \int_C h\,\mathrm{d}g_1 = h(S_1)g_1(A_1) + h(S_2)\,[g_1(A_2) - g_1(A_1)]$$

$$+ \cdots + h(S_N)\,[g_1(A_N) - g_1(A_{N-1})]$$
$$= h(S_1) + h(S_2)\ .\ 0 + h(S_3)\ .\ 0 + \cdots + h(S_N)\ .\ 0$$
$$= \max_{n=1}^{N} h(S_n)$$

This means that the discrete Choquet integral with respect to g_1 is reduced to the maximum aggregation function.

$$g_2(A) = 0 \quad \text{if } |A| < N \text{ (i.e.,} A \neq \Omega);$$
$$= 1 \quad \text{if} A = \Omega.$$

$$\Rightarrow \int_C h\,\mathrm{d}g_2 = h(S_1)g_2(A_1) + h(S_2)\ [g_2(A_2) - g_2(A_1)]$$

$$+ \cdots + h(S_N)\ [g_2(A_N) - g_2(A_{N-1})]$$
$$= h(S_1)\ .\ 0 + h(S_2)\ .\ 0 + \cdots + h(S_N)\ .\ (1 - 0) = h(S_N)$$
$$= \min_{n=1}^{N} h(S_n)$$

This means that the discrete Choquet integral with respect to g_2 is reduced to the minimum aggregation function.

$$g_3(A) = 0 \quad \text{if} A = \Phi;$$
$$= |A|/N \quad \text{if } |A| \leq N.$$

$$\Rightarrow \int_C h\,\mathrm{d}g_3 = h(S_1)[1/N] + h(S_2)[2/N - 1/N] + \cdots + h(S_N)[\,N/N - (N-1)/N]$$

$$= \frac{1}{N}\sum_{n=1}^{N} h(S_n)$$

This means that the discrete Choquet integral with respect to g_3 is reduced to the arithmetic mean aggregation function.

- *Prioritization monotone measure*

Let $\Omega = \{S_1, S_2, \ldots, S_n, \ldots, S_N\}$ denote a finite set of N information sources of interest assumed to be ranked following their decision priorities (i.e., $S_1 > S_2 > \cdots > S_n > \cdots > S_N$), and assume $h(S_{\mathrm{id}(n)}) = x_{\mathrm{id}(n)} \in [0, 1]$ the support degree of that the alternative/decision meets the requirement of the information source $S_{\mathrm{id}(n)}$ where id(n) is the index of the nth most satisfied information source, $n = 1, 2, \ldots, N$.

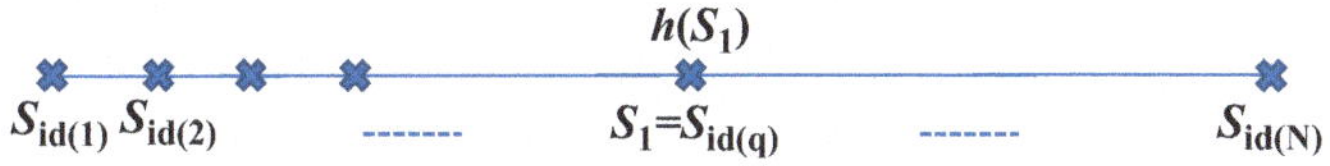

Fig. 6.6 Information sources ordering for the discrete Choquet integral computation

Also, let

$$A_n = \{S_{\mathrm{id}(1)}, S_{\mathrm{id}(2)}, \ldots, S_{\mathrm{id}(n)}\}$$

$n = 1, \ldots, N$ be the subset of the "n" most satisfied information sources. The nested subsets A_n are formed with respect to the information sources support degrees $h(S_{\mathrm{id}(n)})$, $n = 1, \ldots, N$ (Fig. 6.6). In other words, they are not linked with the information sources degrees of priority. Given the definition of the discrete Choquet integral,

$$\int_C h\,\mathrm{d}g_P = \sum_{n=1}^{N} h\left(S_{\mathrm{id}(n)}\right)\left[g_P(A_n) - g_P(A_{n+1})\right]$$

let us consider the three following cases:

- Case where the highest priority is the least supporting information source.

  In this case, $h(S_1) = \min_{n=1}^{N} h(S_n)$. Thus, $g_P(A_n) = 0$ for $n = 1, 2, \ldots, N-1$, and $g_P(A_N) = 1$.

  Therefore,

$$\int_C h\,\mathrm{d}g_P = h(S_1) = \min_{n=1}^{N} h(S_n).$$

This means that $\int_C h\,\mathrm{d}g_P$ produces the smallest support value, and there is no compensation by any other information source.

- Case where the highest priority information source is ranked as being the qth highest supporting information source (Fig. 6.6). In this case, $\{S_1\} \notin A_n$, $n = 1, \ldots q-1$. Thus, $g_P(A_n) = 0$ for $n = 1, \ldots, q-1$, and:

$$\int_C h\,\mathrm{d}g_P = \sum_{n=q}^{N} h\left(S_{\mathrm{id}(n)}\right)\left[g_P(A_n) - g_P(A_{n+1})\right]$$

  Furthermore, for $n = q$ to N, we have $h(S_{\mathrm{id}(n)}) \leq h(S_1)$, and with the fact that:

$$\sum_{n=1}^{N}\left[g_P(A_n)-g_P(A_{n+1})\right]=1, \text{this leads to} \int_C h\,\mathrm{d}g_P \leq h(S_1)$$

- Case where $A_n = B_n$, $n = 1, 2, \ldots, N$. In this case, the information sources supports are ordered the same as the priorities. This leads to $g_P(A_n) - g_P(A_{n-1}) = (1/N)$ for $n = 1, 2, \ldots, N$ (with, $g_P(A-1) = 0$). Therefore,

$$\int_C h\,\mathrm{d}g_P = \frac{1}{N}\sum_{n=1}^{N} h(S_n)$$

(i.e., the arithmetic mean aggregation of the support values).

6.5 Discrete Sugeno Integral

Using the concept of monotone measures and fuzzy integrals, the discrete Choquet integral gave rise to important progress on both mathematical and decision-making points of view. Nevertheless, an important weakness of the discrete Choquet integral concerns its application in qualitative decision-making (i.e., when information sources support and importance are expressed using qualitative, or ordinal, information). In fact, in many applications, information sources support (or satisfaction) and the importance degrees are expressed on a finite ordinal, or a qualitative, scale (e.g., a finite chain {very low, low, moderate, high, etc.}). Turning this ordinal information into cardinal information (i.e., real numbers) and the application of standard arithmetic operators like sum product, difference, etc. is not the optimal solution. Only comparison operations can be done in this case.

In order to face this problem, Sugeno proposed an interesting model of fuzzy integrals as nonlinear functions (generalizing the "max-min" operators) defined with respect to monotone measures.

Let $\Omega = \{S_1, S_2, \ldots, S_n, \ldots, S_N\}$ denote a finite set of N information sources of interest in a decision-making problem, and assume $h(S_n) = x_n \in [0, 1]$ the support degree that the solution meets the requirement of the information source S_n.

Similar to the discrete Choquet integral, the discrete Sugeno integral focuses the attention on the nested subsets $A_1 \subseteq A_2 \subseteq \ldots \subseteq A_N = \Omega$ of the ordered information sources (ordering is realized with respect to the support values) where $A_n = \{S_1, S_2, \ldots, S_n\}$, $n = 1, 2, \ldots, N$ is the subset of the "n" most satisfied information sources. Moreover, the discrete Sugeno integral characterizes each of subset $A_n = \{S_1, S_2, \ldots, S_n\}$, $n = 1, 2, \ldots, N$ by two values (Fig. 6.7):

- The least supporting value of information sources contained in A_n, i.e., $h(S_n)$
- The worth/importance of the subset $A_n = \{S_1, S_2, \ldots, S_n\} : g(A_n)$

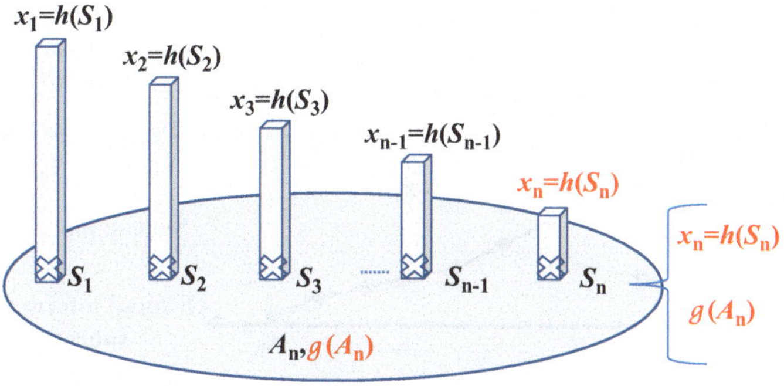

Fig. 6.7 Subset A_n characterization by the discrete Sugeno integral

The decision support of each of the nested subsets $A_n = \{S_1, S_2, \ldots, S_n\}$, $n = 1, 2, \ldots, N$, is defined as the minimal value of:

1. The least supporting information source, S_n
2. The worth associated with the subset A_n : $min\,\{h(S_n), g(A_n)\}$

The discrete Sugeno integral of the support function h with respect to the monotone measure g is, then, defined as the maximum support of different nested subsets A_n, $n = 1, 2, \ldots, N$:

$$\int_S hg = \int_S h\,\mathrm{d}g = \vee_{n=1,\ldots,N}\{\{h(S_n) \wedge g(A_n)\}\}$$

$$= \max_{n=1}^{N} min\,\{h(S_n), g(A_n)\}$$

Therefore, the discrete Sugeno integral of h with respect to the monotone measure g is "simply" a weighted max-min aggregation function showing a trade-off between the degrees of importance of the nested sets of information sources and their worst support in such sets. It is worthwhile to notice that this definition of the discrete Sugeno integral (taking the range [0,1] for the support values and the monotone measure) is valid for other domains. In particular, any linearly ordered scale can be considered instead of the range [0,1].

In fact, the expression of the discrete Sugeno integral involves operations that are consistently defined in an ordinal scale.

Given the fact that information sources degrees of support are arranged in a decreasing order (i.e., $x_1 \geq \cdots \geq x_n \geq \cdots \geq x_N$), and that (due to the use of the fuzzy measure g), the nested subsets A_n, $n = 1, 2, \ldots, N$, degrees of relevance/importance represent an increasing function $(g(A_1) \leq g(A_2) \leq \cdots \leq g(A_N) = 1)$, then (as illustrated in Fig. 6.8) the discrete Sugeno integral resulting value will be the "largest" minimum intersection value between these two curves.

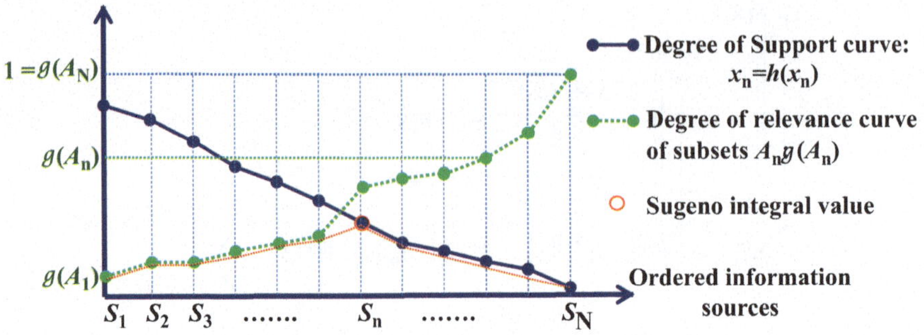

Fig. 6.8 Computation of the possibilistic Sugeno integral of h with respect to g

Example Let us consider the same example considered for computing the discrete Choquet integral where $\Omega = \{S_1, S_2, S_3\}$ is a finite set of three information sources and g_λ the Sugeno λ-measure:

$$g_\lambda(\{S_1\}) = 0.4, \quad g_\lambda(\{S_2\}) = 0.3, \quad g_\lambda(\{S_3\}) = 0.2, g_\lambda(\{S_1, S_2\}) = 0.7446,$$
$$g_\lambda(\{S_1, S_3\}) = 0.6298, \quad g_\lambda(\{S_2, S_3\}) = 0.5223 \text{ and } g_\lambda(\{S_1, S_2, S_3\}) = 1.$$

Assuming the support vector $X = [x_1 = 0.4, x_2 = 1, x_3 = 0.9]$, then information sources ordering and the corresponding nested subsets are illustrated in Fig. 6.5.

Therefore, the discrete Sugeno integral of h with respect to g_λ is given by:

$$\int_S h \, \mathrm{d}g_\lambda = \max_{n=1}^{N} \; min\left\{h(S_n), g_\lambda(A_n)\right\}$$
$$= max\{\, min\{1, 0.3\},\; min\{0.9, 0.5223\},\; min\{0.4, 1\}\}$$
$$= max\{0.3, 0.5223, 0.4\} = 0.5223.$$

6.5.1 Important Properties of the Discrete Sugeno Integral

The discrete Sugeno integral of a support function h (with respect to the monotone measure g) possesses the following interesting properties:

- Idempotent property: If $h(S_n) = \alpha$, $n = 1, \ldots, N$ (α is a constant value), then $\int_S h \, \mathrm{d}g = \alpha$.

  It is worthwhile to notice the following two cases:

- $\alpha = 0 : h(S_n) = 0$ for all S_n (i.e., all information sources have a total non-support level) leads to $\int_S h\,dg = 0$.
- $\alpha = 1 : h(S_n) = 1$ for all S_n (i.e., all information sources have the same highest support level) leads to $\int_S h\,dg = 1\ (= g(\Omega))$.

- $\int_S (h + a)\,dg \leq \int_S h\,dg + a$ for any constant value, a, from the unit interval.
- Monotonicity non-decreasing with respect to the support function:
  If $h_1(S_n) \leq h_2(S_n)$ for all $S_n \in \Omega$ ($h_1 \leq h_2$ in short), then $\int_S h_1 dg \leq \int_S h_2 dg$ (i.e., an increase of the support cannot lead to a decrease of the overall evaluation of the discrete Sugeno integral).
- Monotonicity non-decreasing with respect to the monotone measure:
  If $g_1(A) \leq g_2(A)$ for all $A \in 2^{\Omega}$, then $\int_S h\,dg_1 \leq \int_S h\,dg_2 \forall h : \Omega \rightarrow [0, 1]$.
- $\int_S (h_1 \vee h_2)\,dg \geq \left(\int_S h_1 dg\right) \vee \left(\int_S h_2 dg\right)$ (equality is obtained if and only if h_1 and h_2 are equiordered (i.e., *comonotone*), $\forall n, m : [h_1(S_n) - h_1(S_m)] \cdot [h_2(S_n) - h_2(S_m)] \geq 0$).
- $\int_S (h_1 \wedge h_2)\,dg \leq \left(\int_S h_1 dg\right) \wedge \left(\int_S h_2 dg\right)$ (equality is obtained if and only if h_1 and h_2 are equiordered).
- If $\alpha \in [0, 1]$, then $\forall h : \Omega \rightarrow [0, 1]$:
  - *Max-homogeneity*: $\int_S (\alpha \vee h) = \alpha \vee \int_S h\,dg$
  - *Min-homogeneity*: $\int_S (\alpha \wedge h) = \alpha \wedge \int_S h\,dg$

As it can be noticed, the parallelism between discrete Choquet and Sugeno integrals is complete. In fact, both integrals fit the same formal model, and they only differ in terms of:

- Parameters used to characterize different nested subsets A_n, $n = 1, \ldots, N$
- The used operators (sum and product for the discrete Choquet integral, maximum and minimum in the discrete Sugeno integral)

However, the mathematical features of the used operators give each some functional particular properties that make them useful in different contexts. Also, it is important to notice that the discrete Sugeno integral does not make any sense for support functions values outside the interval [0,1], whereas the discrete Choquet integral makes sense for any nonnegative support function.

6.5.2 Discrete Sugeno Integral for Some Monotone Measures

In this section, some particular monotone measures are considered, and the discrete Sugeno integrals associated with these monotone measures are evaluated.

- *Weakest monotone measure on* Ω

    If $g(A) = 0$ for all $A \neq \Omega$ and $g(A) = 1$ if $A = \Omega$, then $\int_S h\,\mathrm{d}g = h(S_N) = \min_{n=1}^{N} h(S_n)$. This means that the discrete Sugeno integral of h with respect to the weakest monotone measure on Ω narrows down to the minimum function (i.e., produces the least, or minimal, support value).
- *Strongest monotone measure on* Ω

    If $g(A) = 1$ for all $A \neq \Phi$ and $g(A) = 0$ if $A = \Phi$, then $\int_S h\,\mathrm{d}g = h(S_1) = \max_{n=1}^{N} h(S_n)$. This means that for this particular monotone measure, the discrete Sugeno integral of h with respect to g produces the maximal support value.

6.5.3 Twofold Integral

The twofold integral is another type of fuzzy integrals based on the use of both discrete Choquet and Sugeno integrals in the same framework. In fact, let $\Omega = \{S_1, S_2, \ldots, S_n, \ldots, S_N\}$ denote a finite set of N information sources of interest, and assume $h(S_n) = x_n \in [0, 1]$ the support degree of that the solution meets the requirement of the information source S_n. We also assume that different sources are arranged such that their support values are decreasing, i.e., $h(S_1) = x_1 \geq \cdots \geq x_n \geq \cdots \geq x_N = h(S_N)$, where S_1 (resp. S_N) is assumed to show the greatest (resp. lowest) support value. Let g_1 and g_2 denote two monotone measures defined on 2^{Ω}. In this case, the discrete Choquet integral of h with respect to the monotone measure g_2 is given by:

$$\int_C h\,\mathrm{d}g_2 = \sum_{n=1}^{N} h(S_n)\,[g_2(A_n) - g_2(A_{n-1})]$$

which means that $\int_C h\,\mathrm{d}g_2$ is the linear combination of the "additional importance" of each information source weighted by the its support value. The main idea of the twofold integral consists in substituting the information sources support value, i.e., $h(S_n)$, by the discrete Sugeno integral with respect to the monotone measure g_1:

$$\int_{S} h\,dg_1 = \max_{n=1}^{N} \min\{h(S_n), g_1(A_n)\}$$

Therefore, the twofold fuzzy integral is defined by:

$$\int_{TF} h\,dg_1\,dg_2 = \sum_{n=1}^{N} \left\{ \max_{n=1}^{N} \min\{h(S_n), g_1(A_n)\} \right\} [g_2(A_n) - g_2(A_{n-1})]$$

6.6 Possibility Integral

Let us consider a finite set of N information sources $\Omega = \{S_1, \ldots, S_n, \ldots, S_N\}$ and assume $g(.)$ a monotone measure defined on 2^{Ω}. The choice/construction of the monotone measure g which embodies the importance, or relevance, of subsets of information sources is of primary importance for the aggregation of different evidences available through different information sources. Nevertheless, one major difficulty is that there is a curse of dimensionality. In fact, the monotone measure requires defining 2^{N-1} values, where N is the number of information sources. This definition gets laborious when the number of information sources (N) increases. Therefore, reducing the model complexity appears as necessary. As previously shown, one of the most widely used monotone measures is the Sugeno λ-measure for the simple reason that is the decision-maker has just to fix density values, and then, the Sugeno λ-measure recursively computes the measure values on a nested sequence of information sources subsets.

Possibility measures exhibit an even simpler generation capacity than Sugeno λ-measures. Recall that a monotone measure, $g = \Pi$, is called a possibility measure if and only if it fulfills the *maxitive* property, i.e.:

$$\Pi(A \cup B) = \max[\Pi(A), \Pi(B)] \quad \forall A, B \in 2^{\Omega}$$

Therefore, the computation of $\Pi(A)$ is reduced to the computation of the maximal value of basic densities of information sources (i.e., $\pi(S_n) = \Pi(\{S_n\})$) within the subset A. Moreover, the possibility measure is said to be normal if at least one singleton has the highest possibility degree (i.e., $\exists S_n$ such that $\pi(S_n) = 1$). At this level, it is worthwhile to recall that possibility distributions (and possibility measures) are used for two purposes: uncertainty modeling and preference (or priority) modeling. For uncertainty modeling, $\pi(\{S_n\})$ measures the possibility of occurrence of S_n as being the "unique" actual state of the world. Moreover, the normalization condition means that the occurring information source certainly belongs to Ω.

On the other hand, possibility distributions (and possibility measures) are also used for preference, or priority, modeling. In this case, $\pi(\{S_n\})$ measures the importance (relevance, preference, etc.) of the criterion, or information source, S_n. Normalization means that there is at least one information source that is absolutely important (or totally relevant), which implies that the data acquired from this information source is fully credible. This very rigid condition is not easy to satisfy in real-world applications. Obviously, if there were an information source that is fully reliable/credible, there would have been no need to use several information sources and to aggregate the information delivered by these information sources since the decision can be made based on the information obtained from the fully credible information source. As a result, normal possibility measures are not very applicable, and imposing the use of such measures will increase the risk of making decision errors. Therefore, the use of subnormal possibility measures achieves high-fidelity modeling with less risk of making decision errors. In this section, discrete Choquet and Sugeno integrals of a support function with respect to possibility measures are discussed.

6.6.1 Possibilistic Choquet Integral

Let $h : \Omega \to [0, 1]$ be a support function and let Π be a normal possibility measure defined on 2^{Ω}. The possibilistic Choquet integral of $h(.)$ is defined as the discrete Choquet integral with respect to Π:

$$\int_{C} h \mathrm{d}\Pi = \sum_{n=1}^{N} h(S_n)\,[\Pi(A_n) - \Pi(A_{n-1})]$$

An important advantage of the possibilistic Choquet integral is that the measures $\Pi(A_n)$ are easily calculated from the basic densities by the recursive relationship:

$$\begin{aligned} &\Pi(A_1) = \pi(S_1), \text{and} \\ &\Pi(A_n) = \Pi(A_{n-1}) \vee \pi(S_n), n = 2, \ldots, N. \end{aligned}$$

Using this relationship, we get:

$$\begin{aligned} \Pi(A_n) - \Pi(A_{n-1}) &= \Pi(A_{n-1}) \vee \pi(S_n) - \Pi(A_{n-1}) \\ &= 0 \vee [\pi(S_n) - \Pi(A_{n-1})] \\ &= 0 \vee \left[\pi(S_n) - \max_{k=1}^{n-1} \pi(S_k)\right] \end{aligned}$$

This means that the additional importance of the information source S_n (i.e., $\Pi(A_n) - \Pi(A_{n-1})$) is considered as valuable if and only if the possibility degree

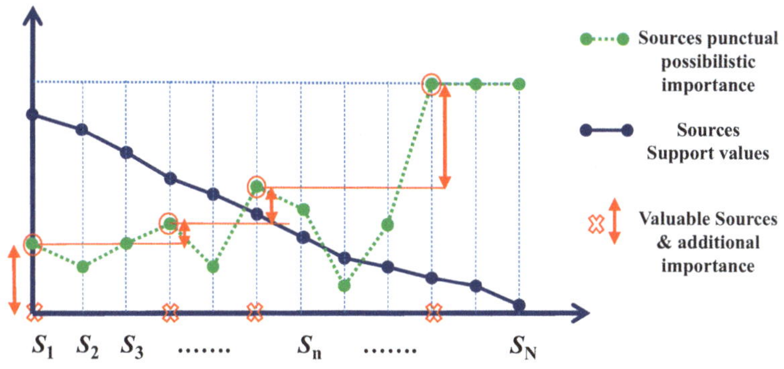

Fig. 6.9 Sources considered for the computation of the possibilistic Choquet integral

$\pi(S_n)$ (i.e., the punctual possibilistic importance of S_n) is greater than all possibility degrees of the first "$n-1$" most supporting information sources (i.e., $\pi(S_n) > \max_{k=1}^{n-1} \pi(S_k)$); see Fig. 6.9.

Notice that if the most supporting information source (i.e., S_1) has the highest importance, i.e.,

$$\pi(S_1) = \max_{n=1}^{N} \pi(S_n),$$

then the possibilistic discrete Choquet integral of h with respect to Π returns the highest supporting value $h(S_1)$. D. Dubois and A. Rico [6] have proposed to rank different information sources with respect to their possibilistic importance (i.e., $S_1 > \cdots > S_n > \cdots > S_N$, where $1 = \pi(S_1) \geq \cdots \geq \pi(S_n) \geq \cdots \geq \pi(S_N)$). In this case, the possibilistic Choquet integral of h with respect to Π can be written as follows:

$$\int_C h \, d\Pi = \sum_{n=1}^{N} [\pi(S_n) - \pi(S_{n+1})] \max_{j=1}^{n} h(S_j)$$

with $\pi(S_{N+1}) = 0$ by convention. Notice that if $\pi(S_1) = \pi(S_2) = \cdots = \pi(S_N) = 1$, then:

$$\int_C h \, d\Pi = \max_{j=1}^{n} h(S_j)$$

This means that the possibilistic Choquet integral of h with respect to Π boils down to the maximum aggregation operator. Moreover, if the conjugate necessity measure (i.e., $\mathcal{N} : 2^{\Omega} \rightarrow [0, 1]$, $\mathcal{N}(A) = 1 - \Pi(A)$, for all $A \in 2^{\Omega}$), as the importance monotone measure, then the corresponding possibilistic Choquet integral of h with respect to $\mathcal{N}$ is given by:

$$\int_C h\, d\mathcal{N} = \sum_{n=1}^{N} [\pi(S_n) - \pi(S_{n+1})] \min_{j=1}^{n} h(S_j)$$

6.6.2 *Possibilistic Sugeno Integral*

Let $h : \Omega \rightarrow [0, 1]$ be a support function and Π a normal possibility measure defined on 2^{Ω}. The possibilistic Sugeno integral, also called *possibility expectation*, of $h(.)$ is defined as the discrete Sugeno integral of h with respect to Π:

$$\int_S h\, d\Pi = \vee_{n=1,\ldots,N}\{h(S_n) \wedge \Pi(A_n)\}$$

$$= \max_{n=1}^{N} \min \{h(S_n), \Pi(A_n)\}$$

where $A_n = \{S_1, S_2, \ldots, S_n\}$ is the subset formed by the "n" most supporting information sources.

Recalling the important advantage of the possibility integrals is that the measures $\Pi(A_n)$ are easily calculated from the basic densities by the recursive relationship:

$$\Pi(A_1) = \pi(S_1)$$
$$\Pi(A_n) = \Pi(A_{n-1}) \vee \pi(S_n), \quad n = 2, \ldots, N$$

The rational of the possibilistic Sugeno integral consists of determining the information source where both the support value $h(S_n)$ and the possibility measure $\Pi(A_n)$ are compatibility large, that is, where the feasibility of the data and the importance of the subset of information sources are jointly optimal. In fact, it is worthwhile to notice that if we define the following possibility distribution (see Fig. 6.10):

$$\pi : \Omega \rightarrow [0, 1]$$
$$S_n \rightarrow \pi'(S_n) = \Pi(A_n)$$

$\pi'(S_n)$ is the worth associated with the subset of the "n" most supporting information sources, and by considering the support values as defining the membership of a fuzzy set $\mu(S_n) = h(S_n)$, $n = 1, 2, \ldots, N$, then:

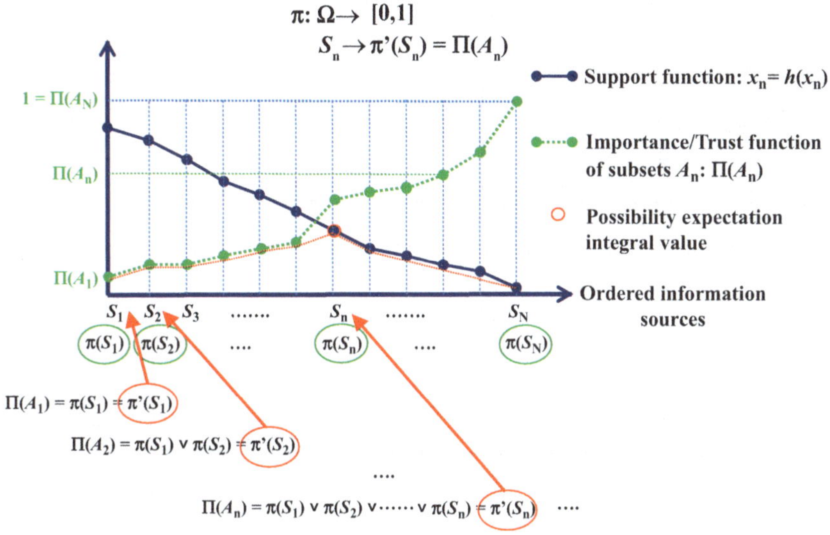

Fig. 6.10 Possibilistic Sugeno integral

$$\int_S h\, \mathrm{d}\Pi = \Pi'(\mu)$$

is the possibility of matching of the possibility distribution π' and the fuzzy set defined by $\mu(S_n) = h(S_n)$, $\forall S_n \in \Omega$ (i.e., measuring the extent to which π' and μ superpose on each other).

It is important to notice that the possibility measure (used in the possibility integral) is not intended to model uncertainty (in the sense that only one information source will occur), since all information sources are assumed to deliver their respective information. In fact, the possibility measure is exploited in terms of its mathematical model and properties, and thus, possibilistic densities are considered as measures of the degrees of importance/preference/relevance of different information sources. Therefore, $\pi(S_n) = 1$ (resp. 0) indicates that the information source S_n is absolutely important (resp. worthless). To understand the way of computing the possibilistic Sugeno integral, let us recall the fact that the information sources are arranged such that their support values are ranked decreasingly (i.e., $h(S_1) \geq \cdots \geq h(S_n) \geq \cdots \geq h(S_N)$) and that the importance values of the nested subsets $A_1 \subseteq A_2 \subseteq \ldots \subseteq A_N = \Omega$ show a monotonic increasing structure (i.e., if $n < m$, then $\Pi(A_n) \leq \Pi(A_m)$). Therefore, and as it is graphically shown in Fig. 6.11, five potential cases can be encountered. Cases (a) and (b) correspond to normal possibility measures and cases (c), (d), and (e) to subnormal possibility measures.

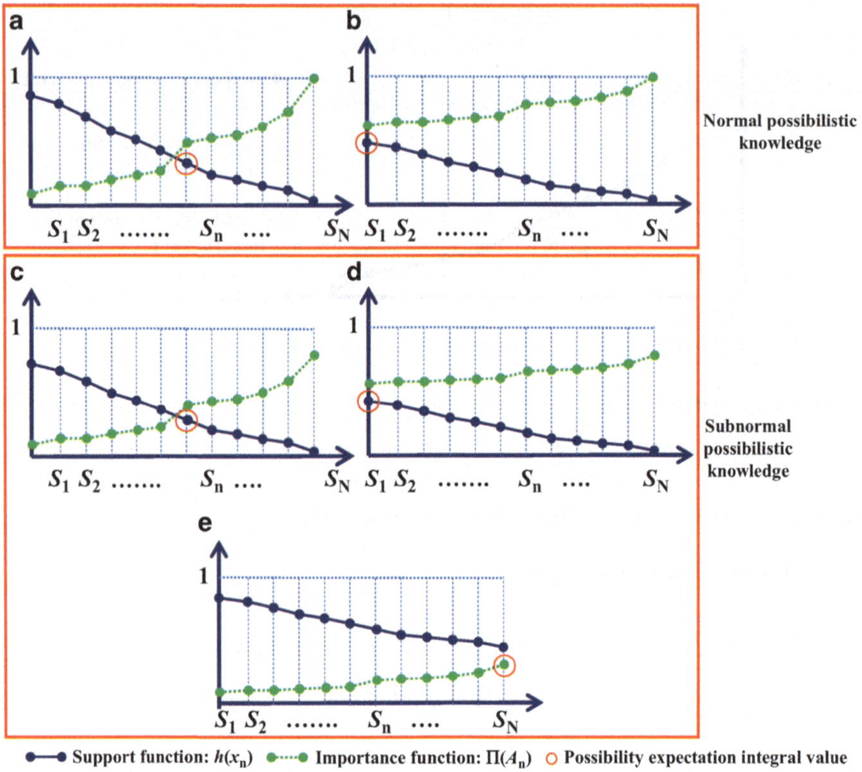

Fig. 6.11 Five potential configurations for the computation of the possibilistic Sugeno integral of h with respect to Π

In cases (a) and (c), the graphical solution of the possibility expectation (i.e., the possibilistic Sugeno integral resulting value) is the lower value of the intersection of the descending support curve and the ascending possibilistic importance values curve of the nested subsets $A_1 \subseteq A_2 \subseteq \ldots \subseteq A_N$. In cases ($b$), ($d$), and ($e$), the possibility expectation is the largest value of the dominated curve (i.e., support values curve in cases (b) and (d) and the possibilistic importance curve in case (e)). Therefore, the information provided by the support values is preserved as much as possible without severe discount by the possibilistic importance measures.

An interesting simplification for the computation of possibilistic Sugeno integral was proposed in [7]. In fact, Vila et al. have shown that if π_1 and π_2 are two possibility distributions defined on the same finite set $\Omega = \{S_1, \ldots, S_n, \ldots, S_N\}$, then:

$$\int_S \pi_1 \, d\Pi_2 = \int_S \pi_2 d\Pi_1 = \vee_{n=1,\ldots,N}\{\pi_1(S_n) \wedge \pi_2(S_n)\}$$

Using this result, the following propositions can be demonstrated straightforward (by considering $\pi_1 = \pi$, and the support function h is simply considered as defining a possibility distribution $\pi_2 = h$).

Proposition 1
Let Π be a possibility measure defined on a finite set of information sources $\Omega = \{S_1, S_2, \ldots, S_n, \ldots, S_N\}$, and let $h : \Omega \rightarrow [0, 1]$ be a support function. The discrete Sugeno integral of h with respect to Π can be written as:

$$\int_S h \mathrm{d}\Pi = \vee_{n=1,\ldots N}\{h(S_n) \wedge \pi(S_n)\}$$

$$= \max_{n=1}^{N} \; \min\{h(S_n), \pi(S_n)\}$$

where $\pi_n(S_n) = \Pi(\{S_n\}), n = 1, \ldots, N$. It is worthwhile to notice that this formulation makes information sources ranking (as a function of the support degrees) no more needed for the computation of the possibilistic Sugeno integral.

Moreover, this important proposition shows that the Sugeno integral corresponds to:

- The possibility degree of matching often used for comparing two ambiguous events:

  1. A possibilistic pattern (called *ambiguous class*), π, an ill-defined possibility distribution representing the information sources importance degrees.
  2. A support profile function (called *a fuzzy profile atom*) representing ill-known support degrees. Both pieces of information are defined over the same set of information sources $\Omega = \{S_1, S_2, \ldots, S_n, \ldots, S_N\}$.

- Prioritized maximum operator (pmax) which is an operator already proposed by D. Dubois and H. Prade [8] and allowing to aggregate the possibility distribution π information with a profile vector given by the support vector $X = [x_1 = h(S_1), \ldots, x_n = h(S_n), \ldots, x_N = h(S_N)] \in [0, 1]^N$.

Proposition 2
Let Π be a possibility measure defined on a finite set of information sources $\Omega = \{S_1, \ldots, S_n, \ldots, S_N\}$, and let be a support function $h : \Omega \rightarrow [0, 1]$.

The discrete Sugeno integral of h with respect to the necessity measure $\mathcal{N}$ associated with Π (i.e., the monotone measure defined by $\mathcal{N} : 2^{\Omega} \rightarrow [0, 1]$ such that $\mathcal{N}(A) = 1 - \Pi(A^C)$) is equivalent to the prioritized minimum operator defined in [8] (where $\pi_n(S_n) = \Pi(\{S_n\})$, $n = 1, \ldots, N$):

$$\int_S h \mathrm{d}\mathcal{N} = \wedge_{n=1,\ldots,N}\{h(S_n) \vee \pi(S_n)\} = \min_{n=1}^{N} \max\{h(S_n), \pi(S_n)\}.$$

6.6.3 Subnormal Possibilistic Sugeno Integral

Let $h : \Omega \to [0, 1]$ be a support function and Π a possibility measure defined on 2^{Ω}. In the framework of possibility integrals, $\pi_n = \Pi(\{S_n\})$ is considered as the credibility of the information source (or criterion) S_n to provide reliable information, in the sense of fidelity, for the considered decision-making problem. Therefore, $\Pi(\Omega) = 1$ implies that at least one information source S_n is reliable enough to provide fully credible information. From a practical point of view, if there were a fully reliable information source S_n, there would have no need to utilize a decisional model based on multi-information sources (since, the decision can be made based on the information delivered by the fully reliable source only!). This means that it is of great importance to revisit the properties of the possibilistic Sugeno integral in the case of subnormal possibility measures. The following properties consider support functions h, h_1, and $h_2(\Omega \to [0, 1])$ and a possibility measure Π defined on 2^{Ω} where:

$$\Pi(\Omega) = \max_{n=1}^{N} \{\pi(S_n)\} \leq 1.$$

The possibility measure Π may be normal (i.e., $\Pi(\Omega) = 1$) or subnormal (i.e., $\Pi(\Omega) < 1$).

1. $0 \leq \int_S h \, d\Pi \leq \text{Height}(h) \wedge \Pi(\Omega)$, where $\wedge$ stands for the minimum operator and $\text{Height}(h) = \max_{n=1}^{N} \{h(S_n)\}$.
2. If $h(S_n) = a \in [0, 1]$, for all S_n (a is a constant value), then $\int_S h \, d\Pi = \min \{a, \Pi(\Omega)\}$.
3. If $h_1(S_n) \leq h_2(S_n)$ for $n = 1, 2, \ldots, N$, then:
   (a) $\int_S h_1 d\Pi \leq \int_S h_2 d\Pi$, if $\Pi(\Omega) > \text{Height}(h_1) = \max_{n=1}^{N} \{h_1(S_n)\}$
   (b) $\int_S h_1 d\Pi = \int_S h_2 d\Pi$, if $\Pi(\Omega) \leq \text{Height}(h_1)$
4. Let $a \in [0, 1]$ denote a constant value and $\text{Height}(h) = \max_{n=1}^{N} \{h(S_n)\}$:

   If $\text{Height}(h) > a$, then:

$$\int_S [a \wedge h] \, d\Pi = a \wedge \int_S h \, d\Pi \quad \text{if } a \leq \Pi(\Omega)$$
$$= \int_S h \, d\Pi \quad \text{if } a > \Pi(\Omega)$$

$$\int_S [a \vee h]\,\mathrm{d}\Pi = [a \wedge \Pi(\Omega)] \vee \int_S h\,\mathrm{d}\Pi \quad \text{if } \Pi(\Omega) < 1$$

$$= a \vee \int_S h\,\mathrm{d}\Pi \quad \in \text{if } \Pi(\Omega) = 1$$

If Height(h) $\leq a$, then:

$$\int_S [a \wedge h]\,\mathrm{d}\Pi = \int_S h\,\mathrm{d}\Pi$$

$$\int_S [a \vee h]\,\mathrm{d}\Pi = \min\{a, \Pi(\Omega)\}$$

5. $\int_S [h_1 \vee h_2]\,\mathrm{d}\Pi \geq \left[\int_S h_1 \mathrm{d}\Pi\right] \vee \left[\int_S h_2 \mathrm{d}\Pi\right]$
6. $\int_S [h_1 \wedge h_2]\mathrm{d}\Pi \leq \left[\int_S h_1 \mathrm{d}\Pi\right] \wedge \left[\int_S h_2 \mathrm{d}\Pi\right]$
7. Let $A \subset \Omega$ (i.e., A is a subset of information sources) and $h_A : \Omega \rightarrow [0, 1]$ where $h_A(S_n) = 1$ if $S_n \in A$ and $h_A(S_n) = 0$ if $S_n \notin A$, then $\int_S h_A \mathrm{d}\Pi = \Pi(A)$.
8. Let $D \subseteq \Omega$ refer to a subset of information sources. If D is considered instead of the whole set of information sources, Ω, the possibilistic Sugeno integral is then denoted as $\int_{S[D]} h\,\mathrm{d}\Pi$.

   Let $A \subseteq B \subseteq \Omega$ (A, B are subsets of sources), then:

   (a)

$$\int_{S[A]} h\,\mathrm{d}\Pi \leq \int_{S[B]} h\,\mathrm{d}\Pi$$

   (b)

$$\int_{S[A \cup B]} h\,\mathrm{d}\Pi = \left[\int_{S[A]} h\,\mathrm{d}\Pi\right] \vee \left[\int_{S[B]} h\,\mathrm{d}\Pi\right]$$

   (c)

$$\int_{S[A \cap B]} h\,\mathrm{d}\Pi = \left[\int_{S[A]} h\,\mathrm{d}\Pi\right] \wedge \left[\int_{S[B]} h\,\mathrm{d}\Pi\right]$$

6.7 Application of the Possibility Integral to Pattern Recognition

The general pattern recognition process can be formulated as follows. Let $\Omega = \{C_m\}$, $m = 1\ldots M$, denote an exhaustive set of M exclusive alternatives (hypotheses, classes, categories, decisions, etc.), and assume that an input pattern is "observed" using a set on N information sources, or sensors, $S = \{S_1, S_2, \ldots, S_N\}$ where each source S_n delivers a feature f_n, $n = 1, 2, \ldots, N$. Each feature f_n (Fig. 6.12) is then injected at the input of a soft labeling algorithm for which the role is to compute, based on the feature f_n, a *soft decision label vector*, $\ell(f_n) = [\ell_n(C_1), \ldots, \ell_n(C_M)] \in [0, 1]^M$, $n = 1, \ldots, N$, where $\ell_n(C_m)$ measures the degree of support, or confidence, according to the nth information source S_n to the hypothesis that the observed pattern comes from the class/decision, or simply decisional alternative, $C_m (m = 1, \ldots, M)$.

Most often, $\ell_n(C_m)$ is an estimate of the *posterior* probability of the decisional alternative C_m given the observed feature $f_n : \ell_n(C_m) = \Pr\{C_m/f_n\}$ or (when the available knowledge is expressed in terms of ambiguous information) the possibility degree related to the occurrence of C_m given the observed feature $f_n : \ell_n(C_m) = \pi_{f_n}(C_m)$. If only one information source is used by the pattern recognition process, then a hard labeling functionality is applied in order to assign an observed object to the mostly supported decisional alternative to C_{m_0} possessing the most important support value, i.e.:

$$\ell_n(C_{m_0}) = \max_{m=1}^{M} \ell_n(C_m).$$

In the case where N information sources are used for observing/recognizing a given pattern, then there is a need for combing the M-dimensional N soft decision output vectors:

$$[\ell_n(C_1), \ldots, \ell_n(C_m), \ldots, \ell_n(C_m)], \quad n = 1, \ldots, N.$$

This combination aims at a higher accuracy than that of any single information source. The general combination scheme is depicted in Fig. 6.13. In fact, for each

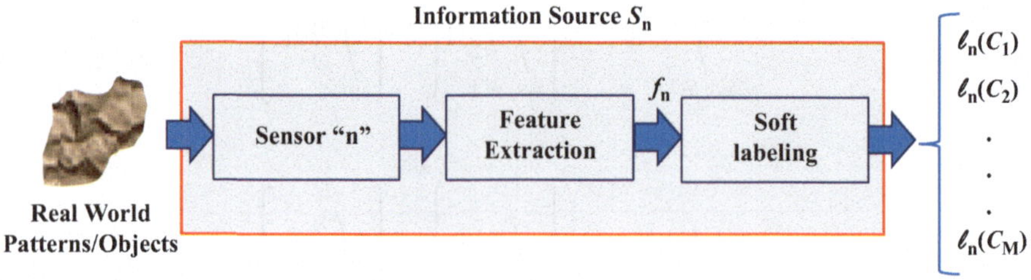

Fig. 6.12 Information source concept

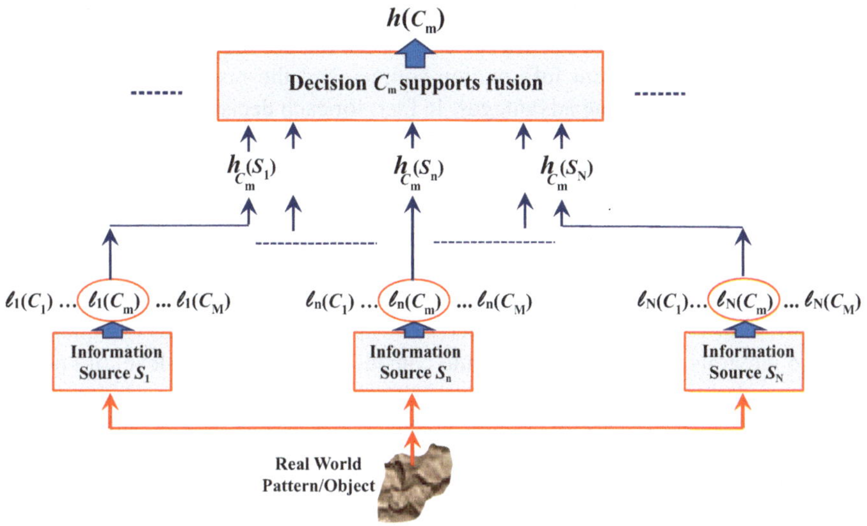

Fig. 6.13 Soft support label vectors combination step

decision C_m, $m = 1, 2, \ldots, M$, once the N soft decision degrees of support/confidence are determined, the soft support label vectors fusion functionality is based on the use of an aggregation function $F : [0, 1]^N \to [0, 1]$ in order to "resume" the N support degrees of each decisional alternative C_m, into a global degree of support (also called the global confidence):

$$h(C_m) = F([\ell_1(C_m), \ldots, \ell_n(C_m), \ldots, \ell_N(C_m)]), \quad m = 1, 2, \ldots, m$$

Similar to the case of using a single information source, to determine to which class an observed pattern belongs, the next functionality in decision-making systems consists of comparing the M global support/confidence values (i.e., $h(C_m)$, $m = 1, \ldots, M$) and to attribute the observed object to the decisional alternative C_{m_0} possessing the most important global support value: i.e., such that:

$$h(C_{m_0}) = \max_{m=1}^{M} h(C_m).$$

As it can be noticed, the soft support fusion leading to obtain a global support value for each decisional alternative is a crucial step in the framework of the decision-making process. It is worthwhile to notice that the "contribution" of the nth information source into the global support value of each decisional alternative C_m (i.e., $h_{C_m}(S_n) = \ell_n(C_m)$) is related to the specific feature f_n acquired by S_n on the observed object and does not reflect the global "informative capacity" of the information source S_n for the discrimination of different decisional alternatives C_m, $m = 1, \ldots, M$.

The use of the discrete Choquet and Sugeno integrals as a soft support label vectors fusion tool (in which the monotone measures can clearly express the interaction among different information sources and the respective importance) has shown some promising advantages. In fact, for each decisional alternative, the degrees of support issued from different information sources, i.e., $\ell_1(C_m), \ldots, \ell_n(C_m), \ldots, \ell_N(C_m)$, are related to the observed features f_n, $n = 1, \ldots, N$, and to the "projection method" of these features by the "individual" sources in order to evaluate the support degrees. It must be noticed that these degrees of support do not reflect the higher level of knowledge which is the capacity, level of importance, or relevance of different information sources for the recognition of patterns issued from different classes. The major interest of applying the discrete Choquet and Sugeno integrals is precisely to take into consideration this higher level of knowledge through the process of estimating the global support degree of each decisional alternative (i.e., class) in terms of the agreement between objective evidence (i.e., degrees of support of different information sources) and the *prior* expectation, or level of importance of different information sources with respect to different individual decisions. First, for each given class C_m, a monotone measure

$$\begin{aligned} g_m &: 2^S \to [0, 1] \\ &A \to g_m(A) \end{aligned}$$

is assumed to measure and to represent the importance/capacity/relevance of any subset $A \in 2^S$ of information sources from S with respect to the given class C_m. Second, for each decisional alternative C_m, the "global" support value is computed as the discrete Choquet and Sugeno integrals of the class evidence vector $H_\mathrm{m} = [\ell_1(C_m), \ldots, \ell_n(C_m), \ldots, \ell_N(C_m)]$ with respect to g_m (Fig. 6.14):

$$h^C(C_m) = \int_C H_m \, \mathrm{d}g_m \text{ and } h^S(C_m) = \int_S H_m \mathrm{d}g_m$$

This means that the support for C_m, i.e., $h^C(C_m)$ or $h^S(C_m)$, can be considered as the "compromise" between the competence (represented by the monotone measure g_m) and the evidence (represented by the class evidence vector H_m and treated as the partial evaluation of the degree of belonging of the considered pattern to the decisional alternative C_m, given by different information sources associated with the different observed features).

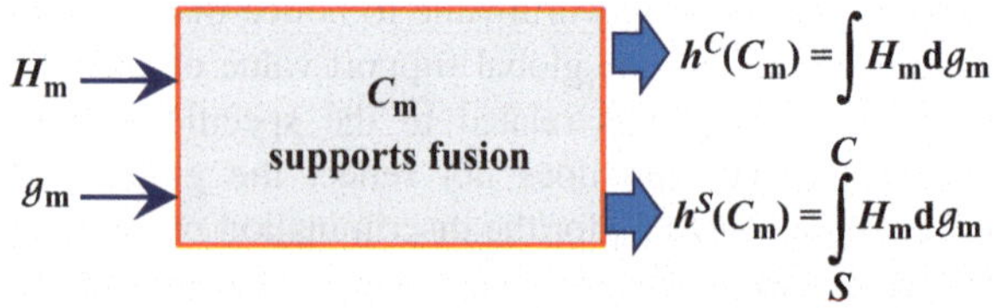

Fig. 6.14 Choquet/Sugeno as soft support label vectors fusion tools at the level of each decisional alternative

Noting that the addition and multiplication operators are used in the discrete Choquet integrals while the maximum and the minimum operators are used in the discrete Sugeno integral, most researchers prefer to use the Choquet integral in pattern recognition problems [9]. One major question when applying the discrete Choquet and Sugeno integrals in multi-information source pattern recognition problems is how to determine the M monotone measures g_m, $m = 1, 2, \ldots, M$. Several approaches based on "learning" the M monotone measures from training samples are already proposed in the literature: using linear/quadratic programing [10], using neural networks [11], using pseudo-gradient [9], and using genetic algorithms [12].

Another important question frequently encountered is how can we interpret the behavior of fuzzy integrals or that of their associated monotone measures? Predefined values $g(A)$ for different subsets $A \in 2^{\Omega}$ of information sources do not immediately give the global importance of subsets $A \in 2^{\Omega}$ nor the degree of interaction among different information sources within A. From a given monotone measure g, information sources importance indicators can be derived in order to enable us interpreting the behavior of the considered monotone measure. Two importance indicators are frequently used: Shapley indices and interaction indices.

1. Shapley Indices

Let $\Omega = \{S_1, S_2, \ldots, S_n, \ldots, S_N\}$ denote a finite set of N information sources of interest, and g is a monotone measure defined on 2^{Ω}. Each information source plays its own respective role in the decision-making process. The overall importance of an information source $S_n \in \Omega$ into a decision problem is not solely determined by the value $g(\{S_n\})$ but also by all $g(A)$ such that $S_n \in A$. Indeed, we may have $g(\{S_n\}) = 0$, suggesting that S_n is unimportant, but we may have some subsets $A \subset \Omega$ such that $g(A \cup \{S_n\})$ is much greater than $g(A)$, suggesting that the association of S_n with the subset of information sources in A is of high importance for the decision-making. In the framework of game theory and based on a set of reasonable axioms, Shapley [13] proposed a definition of a coefficient of importance called *importance index* (or *Shapley index*) of the information source S_n with respect to the monotone measure g:

$$\phi(g, S_n) = \sum_{A \subseteq \Omega \setminus \{S_n\}} \frac{|A|!.(N-|A|-1)!}{N!} [g(A \cup \{S_n\}) - g(A)]$$

This index can be interpreted as the weighted average value of the "additional worth contribution" $g(A \cup \{S_n\}) - g(A)$ of the information source S_n alone in all information sources combinations. It is worth noting the following basic properties of the Shapley indices:

- Shapley indices sum to one: $\sum_{n=1}^{N} \phi(g, S_n) = 1$.
- If the monotone measure g is additive, then $\phi(g, S_n) = g\{S_n\}$ for all $n = 1, \ldots, N$.

The vector $[\phi(g,S_1),\phi(g,S_2),\ldots,\phi(g,S_N)]$ is called the *Shapley value of g*.

2. Interaction Index

Let $\Omega = \{S_1, S_2, \ldots, S_n, \ldots, S_N\}$ denote a finite set of N information sources of interest, and g is a monotone measure defined on 2^{Ω}. Shapley index $\phi(g, S_n)$ measures the average contribution that the information source S_n brings to all possible sources combinations, but it does not explain why the source S_n may have a large importance. In other words, it gives no information on the interaction phenomena existing among different information sources. In fact, the concept of positive interaction (i.e., complementary) of two information sources S_n and S_m means that the satisfaction (i.e., full support) of both information sources S_n and S_m is much more valuable than the satisfaction of them separately. This concept can be translated quantitatively as follows: if A is a subset of information sources from Ω such that $A \cap \{S_n, S_m\} = \Phi$, then:

$$g(A \cup \{S_n, S_m\}) - g(A) \geq [g(A \cup \{S_n\}) - g(A)] + [g(A \cup \{S_m\}) - g(A)]$$

This expression is called the *marginal interaction* between information sources S_n and S_m conditioned to the presence of elements of the combination $A \subseteq \Omega/\{S_n, S_m\}$. This marginal interaction can be rewritten as follows:

$$g(A \cup \{S_n, S_m\}) - g(A \cup \{S_n\}) - g(A \cup \{S_m\}) + g(A) \geq 0$$

Therefore, and similar to the definition of Shapley importance index, the *interaction index* between the two information sources S_n and S_m is obtained by taking the weighted average of the above expression for all possible subsets of information sources $A \subseteq \Omega/\{S_n, S_m\}$:

$$I_{n,m}(g) = \sum_{A \subseteq \Omega \backslash \{S_n, S_m\}} \frac{|A|!.(N-|A|-1)!}{N!} \times [g(A \cup \{S_n, S_m\}) - g(A \cup \{S_n\}) - g(A \cup \{S_m\}) + g(A)]$$

Notice that:

- $I_{n,m}(g) = 0$ means that the two information sources S_n and S_m are independent.
- $I_{n,m}(g) > 0$ means that the two information sources S_n and S_m are complementary.
- $I_{n,m}(g) < 0$ means that the two information sources S_n and S_m are redundant (i.e., the satisfaction of one of the two information sources is sufficient, and, the satisfactory of both sources is useless).

References

1. R.R. Yager, On ordered weighted averaging aggregation operators in multicriteria decisionmaking. IEEE Trans. Syst. Man Cybern. **18**, 183–190 (1988)
2. M. Sugeno, Theory of fuzzy integrals and its applications, Doctoral Thesis, Tokyo Institute of Technology, 1974
3. T. Murofushi, M. Sugeno, An interpretation of fuzzy measures and the Choquet integral as an integral with respect to a fuzzy measure. Fuzzy Sets Syst. **29**, 201–227 (1989)
4. M. Grabisch, C. Labreuche, Fuzzy measures and integrals in MCDA, in *Multiple Criteria Decision Analysis*, (Springer, Cham, 2016), pp. 553–603
5. G. Choquet, Theory of capacities, in *Annales de l'institut Fourier* (1954), pp. 131–295
6. D. Dubois, A. Rico, Axiomatisation of discrete fuzzy integrals with respect to possibility and necessity measures, in *International Conference on Modeling Decisions for Artificial Intelligence* (2016), pp. 94–106
7. M. Vila, M. Delgado, On medical diagnosis using possibility measures. Fuzzy Sets Syst. **10**, 211–222 (1983)
8. D. Dubois, H. Prade, Weighted minimum and maximum operations in fuzzy set theory. Inf. Sci. **39**, 205–210 (1986)
9. Z. Wang, K.-S. Leung, G.J. Klir, Applying fuzzy measures and nonlinear integrals in data mining. Fuzzy Sets Syst. **156**, 371–380 (2005)
10. D.S. Yeung, X.-Z. Wang, E.C. Tsang, Handling interaction in fuzzy production rule reasoning. IEEE Trans. Syst. Man Cybern. B Cybern. **34**, 1979–1987 (2004)
11. W. Jia, W. Zhenyuan, Using neural networks to determine Sugeno measures by statistics. Neural Netw. **10**, 183–195 (1997)
12. R. Yang, Z. Wang, P.-A. Heng, K.-S. Leung, Fuzzified Choquet integral with a fuzzy-valued integrand and its application on temperature prediction. IEEE Trans. Syst. Man Cybern. B Cybern. **38**, 367–380 (2008)
13. L.S. Shapley, A value for n-person games. Contrib. Theory Games **2**, 307–317 (1953)

Chapter 7
Fusion Operators and Decision-Making Criteria in the Framework of Possibility Theory

7.1 Introduction

As already mentioned, uncertainty sources are diverse: incomplete domain knowledge, noisy and conflicting data, incomplete information, linguistic imprecise and ambiguous knowledge, etc. Possibility theory was introduced to allow a reasoning to be carried out in the framework of a vague knowledge. In fact, possibility theory can be described as a collection of techniques centered on the concept of a possibility distribution used for the representation and manipulation of the ambiguous, or vague, knowledge about the encountered uncertainty. Chapter 2 described the fundamental concepts of possibility distributions and theory. This chapter presents different operators allowing the fusion of several possibility distributions. The second issue is related to the decision-making process in the framework of possibility theory. When observing a "real-world situation," decision-making is the process of reducing uncertainty and doubt about several alternatives or decisions, in order to allow a reasonable choice to be made among them. When real-world situations represent patterns observed through sensors, measurements, or other types of data, and when the target is to attribute a label referring to a class or a category to which the observed pattern is assumed to belong, then the decision-making process is called a pattern recognition or a classification process. Fuzzy pattern matching is added as an illustration.

7.2 Possibility Distributions Fusion

Aggregation and fusion of information are basic concerns for all information processing systems. From a general point of view, information fusion has for purpose the simultaneous use of different pieces of information (provided by several

B. Solaiman, É. Bossé, *Possibility Theory for the Design of Information Fusion Systems*, Information Fusion and Data Science,
https://doi.org/10.1007/978-3-030-32853-5_7

sources) in order to reach a conclusion, a decision, or a "better" global information. In the framework of possibility theory, we consider that information is delivered by N possibilistic information sources and encoded by N possibility distributions π_1, $\pi_2 \ldots \pi_N$ defined on the same set of alternatives Ω. These distributions are assumed to use the same "uncertainty scale" to evaluate the degrees of possibility. Three major possibility distributions fusion modes are proposed depending on several factors: the reliability of different information sources (as formulated by some analyst) and our state of knowledge concerning this reliability. There are two "extreme" modes allowing the symmetric fusion: the *conjunctive mode* (when all sources agree and are considered as equally reliable) and the *disjunctive mode* (when sources disagree but at least one of them is reliable but we don't know which one). A third fusion mode, called the *trade-off mode*, allows taking into consideration the individual reliability of different possibilistic information sources when the considered sources have unequal reliabilities (reliability factors are assumed known or can be estimated).

7.2.1 Conjunctive Possibility Distributions Fusion

A conjunctive operator is the counterpart to sets intersection. This corresponds to a severe fusion behavior. The uncertainty in the information resulting from a conjunction is less than the uncertainty of each source alone. Moreover, a conjunctive operator assumes that all sources are reliable and usually result in very precise information. If there is some conflict in the information (i.e., at least one source is not fully reliable), then the result of the conjunction can be poorly reliable, or even empty.

Formally, the conjunctive fusion of N possibility distributions π_1, $\pi_2 \ldots \pi_N$ is given, for all $x \in \Omega$, by:

$$\pi_{T(1:N)}(x) = T_{n=1,2\ldots N}[\pi_n(x)]$$

where T is a triangular fusion norm (for short: *T-norm*).

Recall that a triangular norm T is a two-place real-valued function on the unit square $[0,1] \times [0,1]$ which satisfies the following axioms (for all α, β, γ, α_1, α_2, β_1, and $\beta_2 \in [0,1]$):

1. *Boundary Conditions*: $T(\alpha,1) = T(1,\alpha) = \alpha$, and $T(0,0) = 0$.
2. *Monotonicity*: if $\alpha_1 \leq \alpha_2$ and $\beta_1 \leq \beta_2$, then $T(\alpha_1, \beta_1) \leq T(\alpha_2, \beta_2)$.
3. *Commutativity*: $T(\alpha, \beta) = T(\beta, \alpha)$.
4. *Associativity*: $T(\alpha, T(\beta, \gamma)) = T(T(\alpha, \beta), \gamma)$.

Table 7.1 gives the most widely used T-norm function (case of two possibility distributions, i.e., $N = 2$).

Notice that all T-norm operators fulfill the so-called ***strong zero preservation principle***: *if one information source considers an alternative as fully impossible* (i.e.,

Table 7.1 Most commonly used conjunctive operators (*T*-norms)

T-norm	Name
$T(\alpha,\beta) = \min(\alpha,\beta)$	Zadeh's *T*-norm
$T(\alpha,\beta) = \alpha \cdot \beta$	Probabilistic *T*-norm
$T(\alpha,\beta) = \max(0, \alpha + \beta - 1)$	Lukasiewicz's *T*-norm
$T(\alpha,\beta) = \frac{\alpha\cdot\beta}{\gamma+(1-\gamma)(\alpha+\beta-\alpha\cdot\beta)}$	Hamacher's *T*-norm ($\gamma > 0$)
$T(\alpha,\beta) = \max([\alpha + \beta - 1 + \lambda\alpha\beta)/(1 + \lambda)], 0)$	Weber's *T*-norm ($\lambda > -1$)
$T(\alpha,\beta) = 1 - \min(1, \{[1-\alpha]^p + [1-\beta]^p\}^{1/p})$	Yager's *T*-norm ($p > 0$)
$T(\alpha,\beta) = \frac{\alpha\cdot\beta}{2-\alpha-\beta+\alpha\cdot\beta}$	Einstein's product *T*-norm

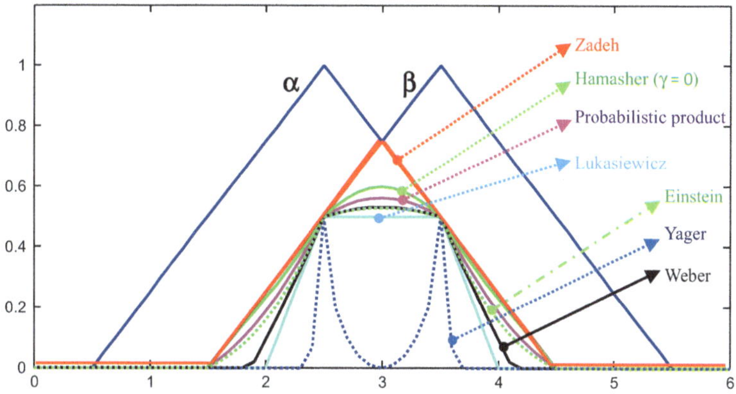

Fig. 7.1 Comparative positioning of different conjunctive operators

$\alpha = 0$ *or* $\beta = 0$), *then the resulting conjunctive fusion produces a zero value* (i.e., *fully impossible output*). The most commonly used *T*-norms are the *minimum* and the *product* operators. The minimum is the largest point-wise *T*-norm and the only one to possess the idempotence property, making it the most conservative (and, therefore, cautious) conjunctive operator in possibility theory. The product is often associated to an assumption of independence between sources. Another important property of the *minimum* operator [1] is that beside the fact that the *minimum* operator assigns the least possibility degree to a given alternative, it constitutes the upper envelope of all the *T*-norm operators: $T_{n=1,2...N}[\pi_n(x)] \leq min_{n=1,2...N}[\pi_n(x)], \forall x \in \Omega$.

In Fig. 7.1, different *T*-norm operators are plotted for the fusion of two triangular possibility distributions (α, resp. β, represents the possibility value of $\pi_1(x)$, resp. $\pi_2(x)$). Moreover, in the framework of possibility theory, the *minimum* operator is considered in order to define conflict and consistency concepts as follows:

- Possibility distributions $\pi_1, \pi_2 ... \pi_N$ are called *in total conflict* when and only when $\pi_{min(1:N)}(x) = 0$ for all $x \in \Omega$.
- Possibility distributions $\pi_1, \pi_2 ... \pi_N$ are called *partially consistent* when and only when $\pi_{min(1:N)}(x) < 1$ and is positive for at least one element $x \in \Omega$.

- Possibility distributions $\pi_1, \pi_2 \ldots \pi_N$ are *totally consistent* when and only when $\pi_{min(1:N)}(x) = 1$ for at least one element $x \in \Omega$.

7.2.2 Disjunctive Possibility Distributions Fusion

Disjunctive fusion corresponds to an indulgent combination behavior assuming that one information source is fully reliable, but this reliable source is "hidden" among the other sources, and we don't know which one it is. In this case, to avoid neglecting any resulting possibility degree, disjunctive fusion operators have a competing behavior in order to keep as much information as possible. The disjunctive fusion of N possibility distributions $\pi_1, \pi_2 \ldots \pi_N$ is given, for all $x \in \Omega$, by $\pi_{C(1:N)}(x) = C_{n=1,2\ldots N}\, \pi_n(x)$, where C is a triangular fusion conorm (for short, T-conorm or C-norm). Recall that a triangular conorm C is a two-place real-valued function on the unit square $[0,1] \times [0,1]$ which satisfies the following axioms (for all $\alpha, \beta, \gamma, \alpha_1, \alpha_2, \beta_1$, and $\beta_2 \in [0,1]$):

- *Boundary Conditions*: $C(\alpha,0) = C(0,\alpha) = \alpha$, and $C(1,1) = 0$.
- *Monotonicity*: if $\alpha_1 \leq \alpha_2$ and $\beta_1 \leq \beta_2$, then $C(\alpha_1, \beta_1) \leq C(\alpha_2, \beta_2)$.
- *Commutativity*: $C(\alpha, \beta) = C(\beta, \alpha)$.
- *Associativity*: $C(\alpha, C(\beta, \gamma)) = C(C(\alpha, \beta), \gamma)$.

Any C-norm, C, can be "generated" from a T-norm T through the *duality transformation*

$$C(\alpha, \beta) = 1 - T(1 - \alpha, 1 - \beta)$$

and conversely.

A conjunctive operator is used when it is believed that all sources are reliable and these sources agree with each other, while a disjunctive operator is applied when it is believed that some sources are reliable, but it is not known which of these sources the reliable ones are. Notice also that all C-norm operators fulfill the so-called *strong possibility preservation principle*: *if one information source considers an alternative as fully possible* (i.e., $\alpha = 1$ or $\beta = 1$), *then the resulting disjunctive fusion produces a unit value* (i.e., *fully possible output*).

The most commonly used C-norms, resumed in Table 7.2, is the maximum C-norm (also called Zadeh's C-norm, $C(\alpha, \beta) = max(\alpha, \beta)$), which is the smallest point-wise C-norm, the dual of the minimum T-norm, and the only C-norm to possess the idempotence property. In fact, the maximum C-norm constitutes the lower envelope of all the C-norm operators:

$$max_{n=1,2\ldots N}[\pi_n(x)] \leq C_{n=1,2\ldots N}[\pi_n(x)], \forall x \in \Omega$$

In Fig. 7.2, different C-norm operators are plotted for the fusion of two triangular possibility distributions (α, resp. β, represents the possibility value of $\pi_1(x)$, resp. $\pi_2(x)$).

Table 7.2 Most commonly used disjunctive operators (C-norms)

C-norm	Name
$C(\alpha, \beta) = \max(\alpha, \beta)$	Zadeh's C-norm
$C(\alpha, \beta) = \alpha + \beta - \alpha \cdot \beta$	Probabilistic C-norm
$C(\alpha, \beta) = \min(\alpha + \beta, 1)$	Lukasiewicz's C-norm
$C(\alpha, \beta) = 1 - \frac{(1-\alpha)\cdot(1-\beta)}{1+(\gamma-1)\,\alpha\cdot\beta}$	Hamacher's C-norm ($\gamma > 0$)
$C(\alpha, \beta) = \frac{\alpha+\beta}{1+\alpha\cdot\beta}$	Einstein's product C-norm

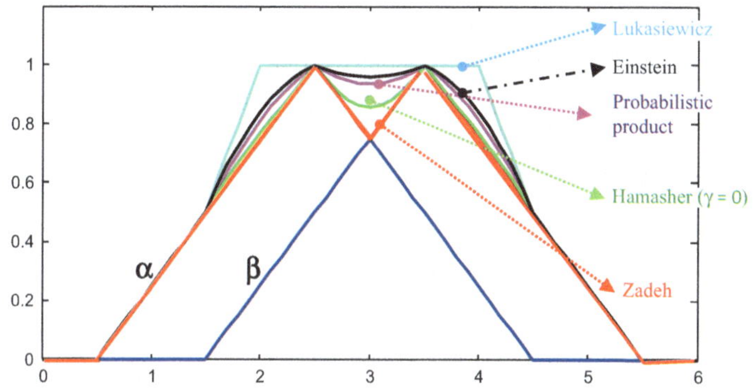

Fig. 7.2 Comparative positioning of different disjunctive operators

7.2.3 Trade-Off Possibility Distributions Fusion

It is worthwhile to notice that the conjunctive fusion mode is meaningful if and only if all possibilistic information sources (encoded as possibility distributions) are considered as equally reliable, whereas the disjunctive fusion mode corresponds to a very weak reliability assumption where there is only one reliable source for sure, but we do not know which one. Therefore, these two modes use general or weak assumptions concerning sources reliabilities producing, thus, a kind of lower and upper envelopes of the resulting possibility degrees of all potential possibilistic fusion operators.

The trade-off possibilistic fusion operators have an intermediate cautious behavior between conjunctive and disjunctive behaviors. In fact, trade-off operators are typically used when sources are partially conflicting. Their objective is to make a trade-off between disjunction and conjunction so as to achieve a good balance between informativeness and reliability.

The following trade-off possibilistic fusion operators allow taking into consideration the individual reliability of different possibilistic information sources having unequal reliabilities (reliability factors are assumed known or can be estimated).

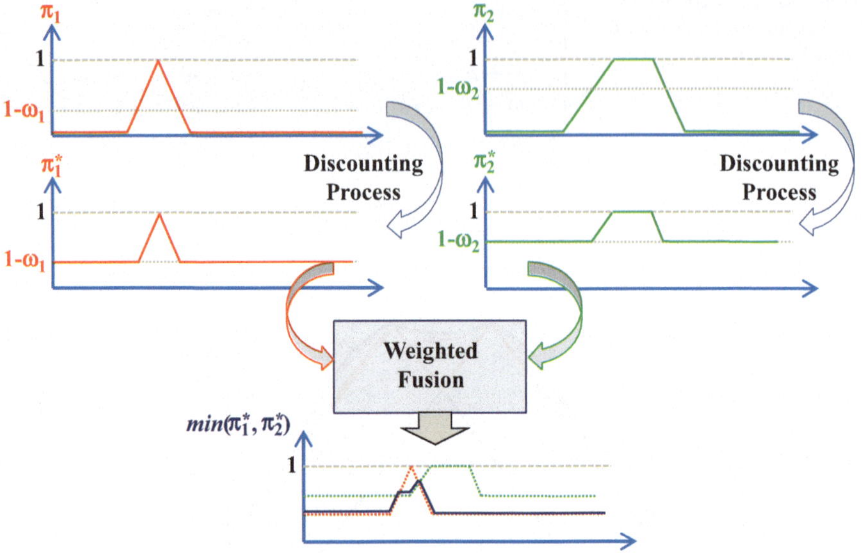

Fig. 7.3 Illustrative example of weighted possibilistic fusion

7.2.3.1 Weighted Possibilistic Fusion Operator

Let π_1, $\pi_2 \ldots \pi_N$ denote N possibility distributions delivered by N possibilistic information sources with unequal reliabilities but where their reliability can be assessed as a number $\omega_n \in [0, 1]$, $n = 1,2\ldots N$, such that $max_{n\,=\,1,\,2\ldots N}\,(\omega_n) = 1$, i.e., at least one possibilistic sources is fully reliable.

The number ω_n is called the ***certainty coefficient*** of the nth possibilistic information source.

Note that $1 - \omega_n$ represents the degree of possibility that the nth possibilistic information source is considered as wrong, or unreliable. D. Dubois [2] proposed to "adjust" each possibility distribution π_n, $n = 1, 2\ldots N$ with certainty coefficient at the degree ω_n (or, equivalently, discounted with a discounting factor $1 - \omega_n$) leading to:

$$\boldsymbol{\pi}_{\boldsymbol{n}}^{*}(x) = max\,(\pi_n(x), 1 - \omega_n) \quad \text{for all}\, x \in \Omega$$

Once discounted, the N possibility distributions $\boldsymbol{\pi}_{\boldsymbol{n}}^{*}$, $n = 1, 2\ldots N$ are considered as being fully reliable and can be combined conjunctively using the *minimum* fusion operator. An illustrative example of this fusion operator is given in Fig. 7.3 in the case of two possibility distributions π_1 and π_2 associated with certainty coefficients at the degrees ω_1 and ω_2.

7.2.3.2 Consistency-Driven Possibilistic Fusion Operator

Weighted possibilistic fusion operator (dealing with unequal reliabilities of the possibilistic sources of information) assumes that reliability degrees (i.e., certainty coefficients) of different sources are a priori known. The question we may raise is that facing unequally reliable sources, how can we merge the information delivered by different information sources when the sources reliability factors are unknown? Dubois et al. [3] have proposed to consider that at least one source, say π_1, for instance, as being fully reliable, i.e., $\omega_1 = 1$. Under this assumption, Dubois et al. proposed a possibilistic information fusion approach, called *consistency-driven fusion operator*.

This operator consists in estimating the certainty coefficients of the remaining unreliable information sources as being the degrees of consistency of the corresponding possibility distributions with the fully reliable possibility distribution, here π_1, i.e., $\omega_n = h(\pi_n \cap \pi_1)$, $n = 2, 3\ldots N$. Using this method of certainty coefficients estimation, the weighting possibilistic fusion operator is then applied. This fusion operator can thus be resumed as follows:

Step1: Estimation of the certainty coefficients for all unreliable sources:

$$\omega_n = h(\pi_n \cap \pi_1) = max_{x\in\Omega}\{\, min\, (\pi_n(x), \pi_1(x))\} \quad n = 2, 3 \ldots N$$

Step2: Possibility distributions discounting using $1 - \omega_n = 1 - h(\pi_n \cap \pi_1)$, $n = 2$, $3\ldots N$, as discounting factors. The resulting discounted possibility distributions are given by:

$$\begin{aligned} \boldsymbol{\pi}_n^*(x) &= max\,(\pi_n(x), 1 - \omega_n) \\ &= max\,(\pi_n(x),\, max_{x\in\Omega}\{\, min\,(\pi_n(x), \pi_1(x))\}) \quad \forall x \in \Omega, n = 2, 3 \ldots N \end{aligned}$$

Step3: The N possibility distributions $\boldsymbol{\pi}_n^*$, $n = 1, 2\ldots N(\boldsymbol{\pi}_1^* = \pi_1)$ are combined conjunctively using the *minimum* fusion operator:

$$\pi(x) = min_{\,n\in\{1,2,\ldots,N\}}\, \boldsymbol{\pi}_n^*(x) \qquad \forall x \in \Omega$$

Notice that if for a given source we have $\omega_n = 0$ (i.e., π_1 and π_n are in a total conflict), then the possibility distribution $\boldsymbol{\pi}_n^*$ will represent the total ignorance situation and will be discarded by the conjunctive fusion operation. On the contrary, if $\omega_n = 1$, then $\boldsymbol{\pi}_n^*$ will be considered as a fully reliable source and will be merged with π_1 in a total conjunctive behavior.

7.2.3.3 Adaptive Fusion Operator

An information fusion operator is called *adaptive* if the fusion result depends on the context. The aim of such an operator is to switch from a conjunctive behavior to a disjunctive behavior according to the increase of conflict (disagreement) among sources. Adaptive operators reduce to disjunction (conjunction) in case of total conflict (agreement) between sources. In-between, they act as trade-off operators. Dubois and Prade [4] have proposed such an adaptive fusion operator allowing the

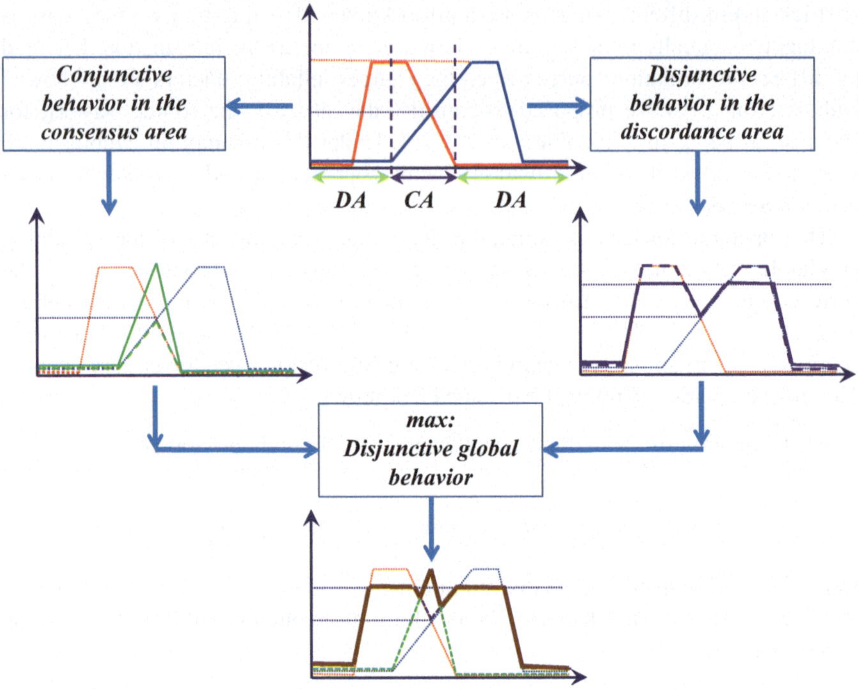

Fig. 7.4 Adaptive possibilistic fusion

gradual transformation from a conjunctive to a disjunctive fusion mode depending on the conflict amount increase between the possibilistic information sources. The conflict between different sources is interpreted in terms of disagreement between different sources. In other words, the set of alternatives is divided into two subsets: $\Omega = \{CA, DA\}$, where CA (standing for *consensus area*) refers to the subset of alternatives for which different sources have a partial agreement (i.e., none of the possibility distributions give to these alternatives a zero possibility degree of occurrence), whereas DA (standing for *discordance area*) refers to the subset of alternatives for which different sources have a total conflict (i.e., one possibility distribution gives to these alternatives strictly positive possibility degree, and all the other sources give a zero possibility degree of occurrence) (Fig. 7.4).

The adaptive fusion operator is based on two "parallel" fusion processes (Fig. 7.4):

- Fusion process in the consensus area: This fusion process uses, first, the minimum conjunctive fusion operator (i.e., $min\{\pi_1(x), \pi_2(x), \forall x \in \Omega\}$ is computed). In a second step, a scaling normalization is applied (i.e., the resulting output on this consensus branch process is given by $[\{\pi_1(x), \pi_2(x)\}]/h, \forall x \in \Omega)$ in order to enforce the possibility degrees resulting from the conjunctive fusion.
- Fusion process in the discordance area: This fusion process uses, first, the maximum disjunctive fusion operator allowing, thus, to capture the possibilistic

knowledge from different "conflictual" sources (i.e., $max\{\pi_1(x), \pi_2(x)\}$, $\forall x \in \Omega$, is computed). In a second step, and in order to keep track the partial conflict, the disjunctive fusion result is discounted by a reliability factor corresponding to the lake of normalization (i.e., amount of conflict): $1 - h$, where h is the consistency index measuring the overlap between different possibility distributions (i.e., the resulting output on this discordance branch process is given by $max[1 - h, max\{\pi_1(x), \pi_2(x)\}]$, $\forall x \in \Omega$).

Finally, the global adaptive fusion behavior is ensured by coupling these two processing branches using the basic maximum disjunctive fusion operator allowing to select the most adequate (i.e., maximum) possibility value issued from the former processing branches. The lake of associativity and the need of an empiric method for the determination of a "global" degree of consistency, h, between several possibilistic sources are certainly an important "cost" that we pay as a counterpart of the adaptive character of this operator. Therefore, the extension of this fusion operator to more than two possibilistic information sources should be "carefully" conducted:

$$\pi(x) = max\left(\left[min_{n\in\{1,2\ldots N\}}\pi_n(x)\right]/h,\ max\left[1 - h,\ max_{n\in\{1,2\ldots N\}}\pi_n(x)\right]\right)\forall x \in \Omega$$

7.3 Decision-Making

As shown in Fig. 7.5, the general pattern recognition process can be formulated as follows. An input pattern $\boldsymbol{p}$ (from a set of patterns ψ) is "observed" using a sensor (or a set of sensors) delivering a feature vector $\boldsymbol{x} \in \Theta$ (Θ is called the *features set*). This feature vector $\boldsymbol{x}$ is then injected into the decision-making system (also called the *classification* or *pattern recognition algorithm*) in order to recognize the most likely decision (hypothesis, alternative, class, category, state of the nature, etc.) from a given exhaustive set $\Omega = \{C_m, m = 1\ldots M\}$ of M exclusive decisions [5].

The development of both pattern recognition algorithm and decision-making criterion is governed by several factors mainly depending on the nature of the feature vector, the nature of the imperfection attached to the observed features, as well as the available knowledge characterizing each decision. Several more global constraints also drive the conception of the global pattern recognition process: the "physical" nature and quality of the measures delivered by the sensors, the categories discrimination capacity of the computed features, and the nature and the quality of the available knowledge used for the development of the decision-making system. However, in much of the literature, the decision-making system is performed by the application of two successive functionalities: the labeling and the hardening functionalities.

The *labeling functionality* [6] uses the available a priori knowledge in order to perform a mapping ℓ between the features set Θ and the decisions set $\Omega(\ell: \Theta \rightarrow \Omega)$. For each feature vector $x \in \Theta$, a *soft decision label vector* $\ell(x) = [\ell_{C_1}(x)\ldots\ell_{C_m}(x)\ldots\ell_{C_M}$

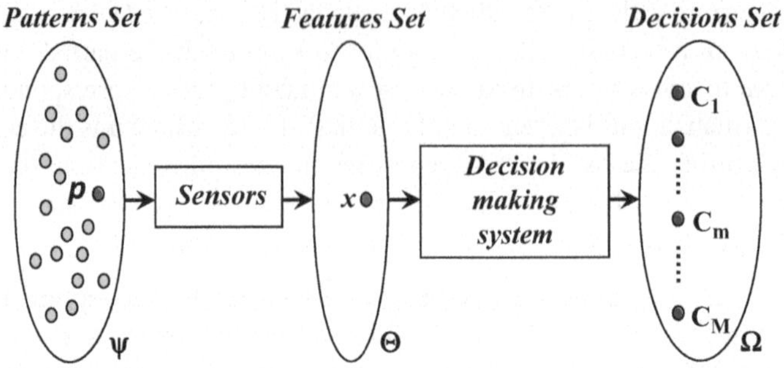

Fig. 7.5 Pattern recognition systems architecture

$(x)]\in[0,1]^M$ is determined in the light of the available knowledge where $\ell_{C_m}(x)$ measures the degree of belief, or support, that we have in the occurrence of the decision C_m. For instance, if the available knowledge allows probabilistic computations, the soft decision label vector is given through $\ell_{C_m}(x) = \Pr\{C_m|x\}$ where Pr $\{C_m|x\}$ represents the a posteriori probability of the decision C_m given the observed feature vector $x\in\Theta$ [7]. When the available knowledge is expressed in terms of ambiguous information, the possibility theory formalism can be used. The soft decision label vector $\ell(x)$ is then expressed with an a posteriori possibility distribution π_x defined on the decisions set Ω. In this case,

$$\ell_{C_m}(x) = \pi_x(C_m)$$

where $\pi_x(C_m)$ represents the possibility degree for the decision C_m to occur, given the observed feature vector $x\in\Theta$.

The second functionality performed by the decision-making system is called the *hardening functionality*. As the ultimate goal of most pattern recognition applications is to select one and only one decision (associated with the pattern "*p*" for which the feature vector $x\in\Theta$ is observed) out of the decisions set Ω, then a hardening mapping has to be applied in order to transform the soft decision label vector $\ell(x)$ into a hard decision label vector for which one and only one decision is selected. The goal is then to make a choice according to an optimality criterion.

In this section, the main focus concerns the case where the knowledge available for the decision-making process is ill-defined in the sense that it is totally or partially expressed in terms of ambiguous information representing limitations in feature values (due to physical, logistical, or other types of constraints) or encoding linguistic expert's knowledge about the relationship between the feature values and different potential decisions. Thus, this available knowledge can be represented as a conditional possibilistic soft decision label vector $\ell(x)$ or π_x, defined on the decisions set Ω such that:

$$\ell_{C_m}(x) = \pi_x(C_m) = \pi(C_m|x)$$

where $\pi(C_m|x)$ represents the possibility degree for the decision C_m to occur, given the observed feature vector $\boldsymbol{x}\in\Theta$ and the underlying pattern $\boldsymbol{p}$. It is worthwhile to notice that possibility theory constitutes the natural framework allowing to tackle this type of information imperfection (i.e., *epistemic uncertainty* imperfection type) when one and only one decision (hard decision) must be selected from the exhaustive decisions set Ω, with incomplete, ill-defined, or ambiguous available knowledge thus encoded as a possibility distribution over Ω.

7.3.1 Possibilistic Decision Criteria

In spite of the fact that possibility theory deals with uncertainty, which means that a unique but unknown elementary decision is to occur and the ultimate target is to determine this decision, it is surprising that only few studies tackle the decision-making issue, in this theory.

In this section, we will investigate existing possibilistic decision-making rules. Two families of rules can be distinguished: rules based on the direct use of the information encapsulated in the possibility distribution and rules based on the use of uncertainty measures associated with this possibility distribution.

Let $\Omega = \{C_1, C_2 ... C_M\}$ be a finite and exhaustive set of M mutually exclusive elementary decisions. Given an observed pattern $\boldsymbol{p}\in\psi$ for which the feature vector $\boldsymbol{x}\in\Theta$ is observed, let $\pi_x(C_m)$ denote the a posteriori possibility distribution $\pi(C_m|x)$ defined on Ω. The possibility $\Pi_x(\{C_m\})$ and necessity $\mathcal{N}_x(\{C_m\})$ measures are obtained as previously mentioned, using the possibility distribution $\pi_x(C_m)$.

7.3.1.1 Decision Rule Based on the Maximum of Possibility

The decision rule based on the maximum of possibility is certainly the most widely used in possibilistic classification—decision-making applications. Indeed, this rule is based on the selection of the elementary decision $C_{m_0} \in \Omega$ having the highest possibility degree of occurrence $\Pi_x(\{C_{m_0}\})$:

$$\text{Decision}[x(p)] \equiv C_{m_0} \text{ if and only if } m_0 = arg\ max_{m=1,\dots,M}\ [\Pi_x(\{C_m\})]$$

A "first" mathematical justification of this "intuitive" possibilistic decision-making rule can be derived from the *Minimax Probabilistic Criterion* (MPC), using a binary cost assignment rule. Indeed, converting a posteriori possibility distributions $\pi_x(.)$ into a posteriori probability distributions $\Pr\{.|x\}$ is assumed to respect the three following constraints [8]:

(a) The consistency principle
(b) The preference ordering preservation
(c) The least commitment principle

The preference ordering preservation, on which we focus the attention here, means that if decision C_{m_1} is preferred to decision C_{m_2}, i.e., $\pi_x(C_{m_1}) > \pi_x(C_{m_2})$, then the a posteriori probability distribution $\Pr\{.|x\}$ obtained from $\pi_x(.)$ should satisfy $\Pr\{C_{m_1}|x\} > \Pr\{C_{m_2}|x\}$. This preference ordering preservation constraint is summed up as:

$$\pi_x(C_{m_1}) > \pi_x(C_{m_2}) \Leftrightarrow \Pr\{C_{m_1}|x\} > \Pr\{C_{m_2}|x\}$$

Therefore, selecting the decision maximizing the a posteriori probability or selecting the decision maximizing the a posteriori possibility decision is identical: using the MPC associated with the binary cost assignment rule or using the maximum possibility decision rule leads to an identical result:

$$\begin{aligned}\text{Decision}[x(p)] \equiv C_{m_0} \text{ if and only if } m_0 &= arg\ max_{m=1,\ldots,M} \Pr\{C_m|x\} \\ &= arg\ max_{m=1,\ldots,M} \pi_x(C_m)\end{aligned}$$

This decision-making approach is called the *Naive Bayes Style Possibilistic Criterion* [8, 9], and most ongoing efforts are oriented into the computation of a posteriori possibility values using numerical data [10, 11]. Notice that this decision rule, strongly inspired from probabilistic decision reasoning, does not provide a decision mechanism when several elementary decisions have the same maximum possibility measure.

7.3.1.2 Decision Rule Based on Maximizing the Necessity Measure

It is worthwhile to notice that the a posteriori measures of possibility Π_x and necessity $\mathcal{N}_x$ coming from the normal a posteriori possibility distribution $\pi_x(.)$ constitute a bracketing for the a posteriori probability distribution $\Pr\{.|x\}$ [12], i.e.,

$$\mathcal{N}_x(\{C_m\}) = 1 - max_{\substack{n=1,\ldots,M \\ n \neq m}} \pi_x(C_n) \leq \Pr\{C_m|x\} \leq \Pi_x(\{C_m\}) = \pi_x(C_m)$$

Therefore, the maximum possibility decision criterion can be considered as an optimistic decision criterion as it maximizes the upper bound of the a posteriori probability distribution. On the contrary, a pessimistic decision criterion based on maximizing the a posteriori necessity measure can be considered as a maximization of the lower bound of the a posteriori probability distribution. This pessimistic decision criterion can thus be expressed as:

$$\text{Decision}[x(p)] \equiv C_{m_0} \text{ if and only if } m_0 = arg\ max_{m=1,\ldots,M} [\mathcal{N}_x(\{C_m\})]$$

The question that we have to raise concerns the "links" between the optimistic and the pessimistic decision criteria. Let us consider the a posteriori possibility distribution $\pi_x(.)$ for which C_{m_1} (resp. C_{m_2}) is the "winning decision" obtained using the maximum possibility (resp. necessity measure) decision criteria as given by:

$$\pi_x(C_{m_1}) = max_{m=1,\ldots,M}\pi_x(C_m) \quad \text{and} \quad \mathcal{N}_x(\{C_{m_2}\}) = max_{m=1,\ldots,M}\mathcal{N}_x(\{C_m\})$$

An important question can be formulated as follows:

"Is the winning decision C_{m_1} (according to the maximum possibility criterion) the same as the winning decision C_{m_2} according to maximum necessity measure criterion?"

First, notice that if several elementary decisions share the same maximum possibility value $v = \pi_x(C_{m_1})$, then the necessity measure becomes a useless decision criterion (since $\mathcal{N}_x(\{C_m\} = 1 - max_{k \neq m}\pi_x(C_k) = 1 - v$ for all the elementary decisions).

Now, suppose that only one decision C_{m_1} assumes the maximum possibility value $v = \pi_x(C_{m_1})$; it is important to raise the question whether the decision C_{m_1} will (or will not) be the decision assuming the maximum necessity measure value. Let us note v', the possibility value for the "second best" decision according to the possibility value criterion. As C_{m_1} is the unique decision having the maximum possibility value v, we have $v' < v$. Therefore, the necessity measure value $\mathcal{N}_x(\{C_m\})$ only gets maximum for the decision C_{m_1} (as $1 - v' > 1 - v$):

$$\mathcal{N}_x(\{C_m\}) = 1 - max_{k \neq m}\pi_x(C_k) = \begin{cases} 1 - v' & \text{if } m = m_1 \\ 1 - v & \text{if } m \neq m_1 \end{cases}$$

As a conclusion, when the maximum necessity measure criterion is useful for application (i.e., only one elementary decision assumes the maximum possibility value), then both decision criteria (maximum possibility and maximum necessity) produce the same winning decision.

In order to illustrate the difference between the maximum possibility and the maximum necessity measure criteria, Table 7.3 presents an illustrative example. In this example, four different a posteriori possibility distributions π_1, π_2, π_3, and π_4, all defined on a five elementary decisions set $\Omega = \{C_1, C_2, C_3, C_4, C_5\}$, are considered. The necessity measures $\mathcal{N}_k(\{C_m\})$ have been computed from the corresponding possibility distribution π_k. The underlined values indicate which decisions result from the maximum possibility decision criterion as well as the maximum necessity measure decision criterion, for the four possibility distributions π_k.

Notice that the necessity measure assumes at most two values whatever the considered possibility distribution. When the a posteriori possibility distribution has one and only one decision having the highest possibility degree, then both decision rules produce the same winning decision. This is the case of the normal possibility distribution π_1 as well as the subnormal possibility distribution π_3.

When several elementary decisions share the same highest possibility degree, then the maximum possibility decision criterion can randomly select one of these potential winning decisions. In this case, the maximum necessity measure decision criterion will affect a single necessity measure degree to all elementary decisions from Ω, and thus, it will be impossible to select any of the potential winning decisions. This behavior can be observed with a normal possibility distribution (π_2) as well as with a subnormal possibility distribution (like π_4). This example

Table 7.3 Comparative example of the maximum possibility and maximum necessity measure decision criteria using four a posteriori possibility distributions

	$\pi_1(C_m)$	$\mathcal{N}_1(\{C_m\})$	$\pi_2(C_m)$	$\mathcal{N}_2(\{C_m\})$	$\pi_3(C_m)$	$\mathcal{N}_3(\{C_m\})$	$\pi_4(C_m)$	$\mathcal{N}_4(\{C_m\})$
C_1	0.1	0	0.1	0	0.1	0.2	0.1	0.2
C_2	0.5	0	0.5	0	0.5	0.2	0.5	0.2
C_3	0.3	0	0.3	0	0.3	0.2	0.3	0.2
C_4	0.8	0	1	0	0.8	0.5	0.8	0.2
C_5	1	0.2	1	0	0	0.2	0.8	0.2

clearly shows the weak (if not similar) decisional capacity of the maximum necessity measure decision criterion when compared to the maximum possibility decision criterion.

7.3.1.3 Decision Rule Based on Maximizing the Confidence Index

Other possibilistic decision rules based on the use of uncertainty measures are also encountered. The most frequently used criterion (proposed by S. Kikuchi et al. [13]) is based on the maximization of the confidence index, *Ind*, previously defined as a combination of the possibility and the necessity measures for each event $A \subseteq \Omega$, given a possibility distribution $\pi(.)$:

$$\begin{aligned} Ind : 2^{\Omega} &\rightarrow [-1, +1] \\ A \rightarrow Ind(A) &= \Pi(A) + \mathcal{N}(A) - 1, \quad \forall A \subseteq \Omega \end{aligned}$$

where 2^{Ω} denotes the power set of Ω, i.e., the set of all subsets from Ω. For an event A, this index ranges from -1 to $+1$:

- $Ind(A) = -1$, if and only if $\mathcal{N}(A) = \Pi(A) = 0$ (the occurrence of A is totally impossible).
- $Ind(A) = +1$, if and only if $\mathcal{N}(A) = \Pi(A) = 1$ (the occurrence of A is totally possible and certain).

Restricting the application of this measure to events A_m having only one decision $A_m = \{C_m\}$ shows that Ind(A_m) measures the difference between the possibility measure of the event A_m (which is identical to the possibility degree of the decision C_m) and the highest possibility degree of all decisions contained in ${A_m}^C$ (the complement of A_m in Ω):

$$\begin{aligned} Ind(A_m) &= \Pi(A_m) + \mathcal{N}(A_m) - 1 \\ &= \pi(C_m) - max_{m \neq n} \pi(C_n) \end{aligned}$$

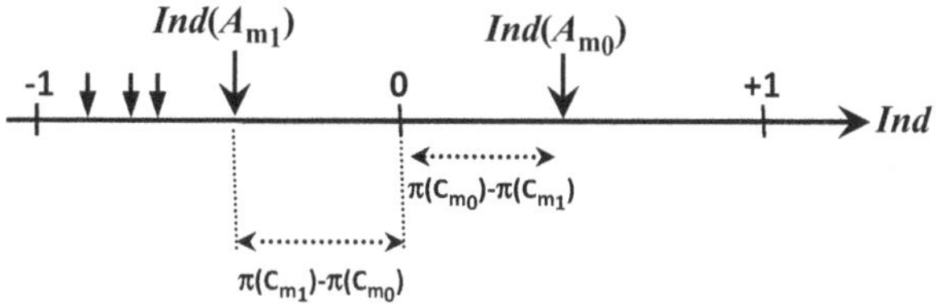

Fig. 7.6 Confidence indices associated with different decisions (A_{m_0}, event having the highest possibility degree; A_{m_1}, event with the second highest possibility degree)

Therefore, if $A_{m_0} = \{C_{m_0}\}$ is the only event having the highest possibility measure value $\pi(C_{m_0})$, then A_{m_0} will be the unique event having a positive confidence index value, whereas all other events will have negative values, as illustrated in Fig. 7.6 (where we assume $\pi(C_{m_0}) > \pi(C_m)$, $\forall m \neq m_0$, and C_{m_1} refers to the decision having the second highest possibility degree).

In the framework of the pattern recognition decision-making process, the decision criterion associated with this index can be formulated as follows:

$$\text{Decision} = A_{m0} \text{ if and only if } Ind(A_{m_0}) = max_{\,m=1,\ldots,M}\,[Ind(A_m)]$$

The main difference between the maximum possibility and the maximum confidence index decision criteria lies in the fact that the maximum possibility decision criterion is only based on the maximum possibility degree, whereas the maximum confidence index decision criterion is based on the difference between the two highest possibility degrees associated with the elementary decisions.

As already mentioned, it is important to notice that the event $A_{m_0} = \{C_{m_0}\}$ having the highest possibilistic value will be the unique event producing a positive confidence index measuring the difference with the second highest possibility degree. All other events $A_m = \{C_m\}$, $\forall m \neq m_0$, will produce negative confidence indices. When several decisions share the same highest possibility degree, their confidence index (the highest one) will be null. This shows the real capacity of this uncertainty measure for the decision-making process. However, this criterion brings the same resulting decisions as the two former ones.

7.4 Fuzzy Pattern Matching (FPM)

Pattern matching is becoming more of a necessity given the needs for such methods for detecting similarities in records, epidemiological patterns, image and information retrieval, etc. Unfortunately, traditional methods for linking or matching records, data fields, etc. rely on certain and precise data matches rather than looking for close matches or patterns. Of course, proximity pattern matches are often necessary when dealing with messy data, data that has inexact, or imprecise, values and/or data with missing key values. A possibilistic pattern (called *ambiguous class*) represents an ill-defined class of objects, while a fuzzy datum (called *a fuzzy pattern atom*) represents an ill-known object whose precise description is not available.

Possibility/necessity degrees of matching are often used for comparing two ambiguous events A (associated with the membership function μ_A) and B (associated with the possibility distribution π_B) defined over the same set of alternatives Ω.

In fact, let us consider the ambiguous event B (as a reference event) defined as an epistemic possibility distribution π_B; then, the possibility and necessity degrees of matching the ambiguous event A with the reference event B are given by:

$$\Pi(A/B) = max_{x \in \Omega} \mu_A \cap \pi_B(x) = max_{x \in \Omega} \left[min \left(\mu_A(x), \pi_B(x) \right) \right]$$

and,

$$\begin{aligned} \mathcal{N}(A/B) &= 1 - \Pi\left(A^C/\pi_B\right) = 1 - \ max_{x \in \Omega} A^C \cap \pi_B(x) \\ &= Inf_{x \in \Omega} \left[max \left(\mu_A \ (x), 1 - \pi_B(x) \right) \right] \end{aligned}$$

$\Pi(A/B)$, resp. $\mathcal{N}(A/B)$, is called the possibility, resp. necessity, matching degree of the ambiguous event A with the ambiguous class B. $\Pi(A/B)$ estimates to what extent it is possible that the "pattern atom" A and the "ambiguous class" B refer to the same alternative x; in other words, $\Pi(A/B)$ represents the degree of compatibility of the fuzzy pattern atom A, with the *prior* possibility knowledge B. On the other hand, the measure $\mathcal{N}(A/B)$ estimates to what extent it is necessary, i.e., certain, that the value to which the ambiguous class B refers is among the one compatible with the pattern atom A; in other words, $\mathcal{N}(A/B)$ is the degree of inclusion of the ambiguous set of possible values of B into the set of values compatible with the pattern atom A.

Note that:

- $\Pi(A/B)$ is symmetrical in both arguments, i.e., $\Pi(A/B) = \Pi(B/A)$ (since both $\Pi(A/B)$ and $\Pi(B/A)$ express to what extent $A \cap B$, defined point-wisely as $\min(\mu_A(x), \pi_B(x))$, is not empty).
- $\mathcal{N}(A/B)$ is not symmetrical in both arguments, i.e.,

$$\mathcal{N}(A/B) = 1 - \Pi\left(A^C/\pi_B\right) \neq 1 - \Pi\left(B^C/\mu_A\right) = \mathcal{N}(B/A)$$

(since $\mathcal{N}(A/B)$, resp. $\mathcal{N}(B/A)$, expresses the degree of inclusion of B into A, resp. A into B).

From a pattern recognition point of view, fuzzy pattern matching process consists in computing the possibility/necessity degrees of matching between a fuzzy pattern atom, A, and the whole set of ambiguous classes $C = \{C_m\}_{m\ =\ 1,2...,M}$ where the available knowledge of each class is of a possibilistic nature and is encoded as a possibility distribution $\pi_m = \pi_{C_m}, m = 1, 2..., M$ (Fig. 7.7). The possibilistic pattern atom π_A and all possibilistic patterns $\pi_m = \pi_{A_m}, m = 1, 2..., M$ are defined over the same set of alternatives Ω. Table 7.4 and Fig. 7.8 resume the computation results of the pair of possibility/necessity degrees of matching: $[\mathcal{N}(A/\pi_{C_m}), \Pi(A/\pi_{C_m})]$, $m = 1, 2, \ldots, M$, where the fuzzy pattern atom and the ambiguous classes are described

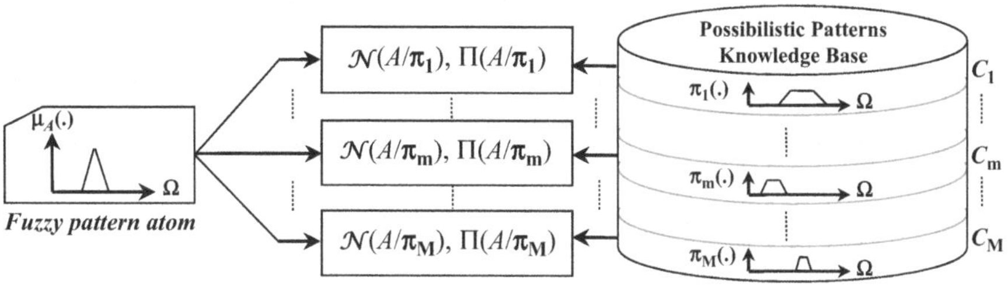

Fig. 7.7 Possibility/necessity degrees of matching determination process

Table 7.4 Pairs of possibility/necessity degrees of matching: $[\mathcal{N}(A/\pi_C), \Pi(A/\pi_C)]$ for different possibilistic information types

Fuzzy pattern atom / Ambiguous class	***Precise*** $\mu_A(x) = 1$ if $x = x_0$; $= 0$ if $x \neq x_0$	***Imprecise*** $\mu_A(x) = 1$ if $x \in [a, b]$; $= 0$ if $x \notin [a, b]$	***Missing data*** $\mu_A(x) = 1$ if $x \in \Omega$; $= 0$ if $x \notin \Omega$	***Ambiguous*** $\mu_A(x)$
Precise $\pi_C(y) = 1$ if $y = y_0$; $= 0$ if $y \neq y_0$	$\mathcal{N}(A/\pi_C) = \Pi(A/\pi_C) = 1$ If $x_0 = y_0$; $= 0$ If $x_0 \neq y_0$	$\mathcal{N}(A/\pi_C) = \Pi(A/\pi_C) = 1$ If $y_0 \in [a,b]$; $= 0$ If $y_0 \notin [a,b]$	$\mathcal{N}(A/\pi_C) = 1$; $\Pi(A/\pi_C) = 1$	$0 \leq \mathcal{N}(A/\pi_C) < \mu_A(x_0)$; $\Pi(A/\pi_C) = \mu_A(x_0)$
Imprecise $\pi_C(y) = 1$ if $y \in [a_1, b_1]$; $= 0$ if $y \notin [a_1, b_1]$	$\mathcal{N}(A/\pi_C) = \Pi(A/\pi_C) = 1$ If $x_0 \in [a_1,b_1]$; $= 0$ If $x_0 \notin [a_1,b_1]$	If $[a,b] \cap [a_1,b_1] = \Phi$: $\mathcal{N}(A/\pi_C) = \Pi(A/\pi_C) = 0$; If $[a,b] \cap [a_1,b_1] \neq \Phi$: $\mathcal{N}(A/\pi_C) = \Pi(A/\pi_C) = 1$	$\mathcal{N}(A/\pi_C) = 0$; $\Pi(A/\pi_C) = 1$	$0 \leq \mathcal{N}(A/\pi_C) \leq \Pi(A/\pi_C)$
Missing data $\pi_C(y) = 1$ if $y \in \Omega$; $= 0$ if $y \notin \Omega$	$\mathcal{N}(A/\pi_C) = \Pi(A/\pi_C) = 1$	$\mathcal{N}(A/\pi_C) = 0$; $\Pi(A/\pi_C) = 1$	$\mathcal{N}(A/\pi_C) = 0$; $\Pi(A/\pi_C) = 1$	$0 \leq \mathcal{N}(A/\pi_C) \leq \Pi(A/\pi_C)$
Epistemic $\pi_C(y)$	$0 \leq \mathcal{N}(A/\pi_C) < \Pi(A/\pi_C)$; $\Pi(A/\pi_C) = \pi_C(y)$	$0 \leq \mathcal{N}(A/\pi_C) \leq \Pi(A/\pi_C)$	$0 \leq \mathcal{N}(A/\pi_C) \leq \Pi(A/\pi_C)$	$0 \leq \mathcal{N}(A/\pi_C) \leq \Pi(A/\pi_C)$

using the following types of possibilistic information elements: precise imprecise, missing data, and ambiguous information type.

For the global matching evaluation process, the second step is to get the best matching, i.e., the most compatible, possibilistic pattern, i.e., ambiguous class, C_{m_0}, with respect to the fuzzy pattern atom A. At this level, we are faced with the problem of ranking different ambiguous classes, each of them being characterized by the pair of possibility/necessity degrees of matching: $[\mathcal{N}(A/\pi_{C_m}), \Pi(A/\pi_{C_m})]$, $m = 1, 2, \ldots, M$.

The Pareto ordering [14] is generally used to rank the pairs of degrees $[\mathcal{N}(A/\pi_{C_m}), \Pi(A/\pi_{C_m})]$, $m = 1, 2, \ldots, M$ as follows: $[\mathcal{N}(A/\pi_{C_m}), \Pi(A/\pi_{C_m})]$ is considered greater than $[\mathcal{N}(A/\pi_{C_k}), \Pi(A/\pi_{C_k})]$ if and only if:

- $\Pi(A/\pi_{C_m}) > \Pi(A/\pi_{C_k})$ and $\mathcal{N}(A/\pi_{C_m}) \geq \mathcal{N}(A/\pi_{C_k})$.
- $\Pi(A/\pi_{C_m}) \geq \Pi(A/\pi_{C_k})$ and $\mathcal{N}(A/\pi_{C_m}) > \mathcal{N}(A/\pi_{C_k})$.

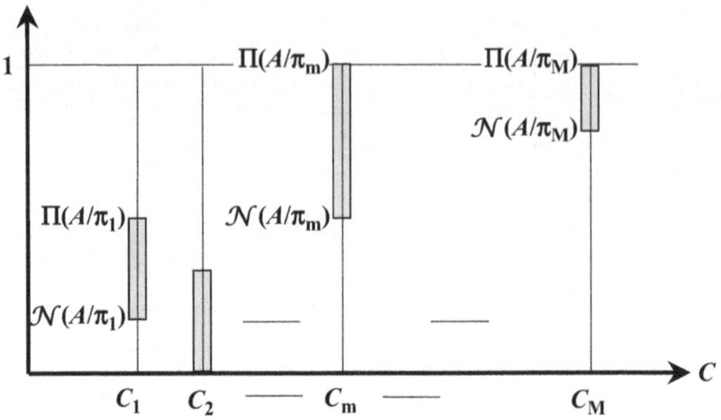

Fig. 7.8 Possibility/necessity degrees of matching visualization

7.4.1 *Confidence Index and Uncertainty Quantification*

Let Ω be a finite set of alternatives and π a normal possibility distribution encoding an ambiguous information defined on the set of alternatives Ω. Possibility theory offers the double possibility/necessity measures to quantify the occurrence uncertainty of a given subset $A \subseteq \Omega$: $\Pi(A)$ and $\mathcal{N}(A)$. As previously discussed, several properties link these two measures. Let us recall the two following properties, for all $A, B \subseteq \Omega$:

$$\mathcal{N}(A) > 0 \Rightarrow \Pi(A) = 1 \text{ and } \Pi(A) < 1 \Rightarrow \mathcal{N}(A) = 0$$

that can be expressed in a single relation: $max(1 - \mathcal{N}(A), \Pi(A)) = 1$. This last property, when observed with $\mathcal{N}(A) \leq \Pi(A)$ (valid for normal possibility distributions), implies two potential configurations of the interval $[\mathcal{N}(A), \Pi(A)]$: $[\mathcal{N}(A) = 0, \Pi(A) < 1]$ and $[\mathcal{N}(A) > 0, \Pi(A) = 1]$ (Fig. 7.9).

It is worthwhile to notice that the variation domain of $(\mathcal{N}(A), \Pi(A))$ is sub-exploited and only reduced to the two "linear" domains $[0, \Pi(A) \in [0, 1]]$ and $[\mathcal{N}(A) \in [0, 1], 1]$.

In order to quantify the occurrence uncertainty of an event $A \subseteq \Omega$, the *confidence index*, *CI*(*A*), was proposed in [15]. This index allows reducing the double necessity-possibility measure $(\mathcal{N}(A), \Pi(A))$, into a single value defined as follows:

$$\begin{aligned} CI : \mathcal{P}(\Omega) &\rightarrow [-1, +1] \\ CI &\rightarrow CI(A) = \mathcal{N}(A) + \Pi(A) - 1 \end{aligned}$$

The function *CI*(.) fulfills the following properties (Fig. 7.10):

$$\text{If } \mathcal{N}(A) = 0, \text{and as } 0 \leq \Pi(A) \leq 1, \text{then} : -1 \leq CI(A) \leq 0, \text{and}$$
$$\text{If } \Pi(A) = 1, \text{and as } 0 \leq \mathcal{N}(A) \leq 1, \text{then} : 0 \leq CI(A) \leq 1.$$

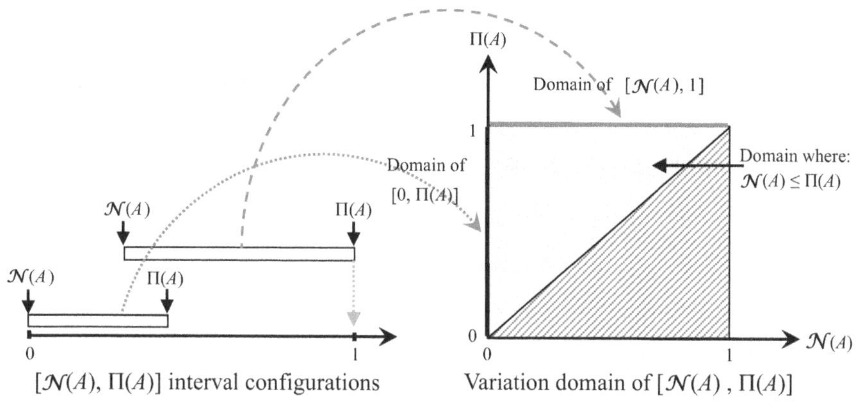

Fig. 7.9 Potential configurations of the interval $[\mathcal{N}(A), \Pi(A)]$

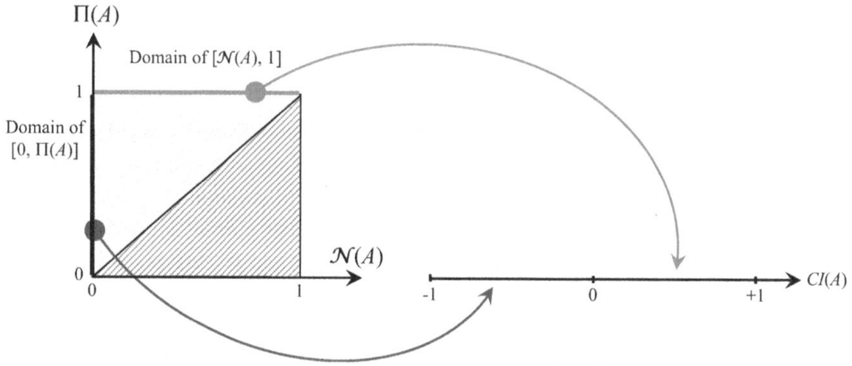

Fig. 7.10 Variation domain mapping from $[\mathcal{N}(A), \Pi(A)]$ to *CI* (*A*)

This means that the $[\mathcal{N}(A), \Pi(A)]$ variation sub-domain $[0, \Pi(A) \in [0,1]]$ is mapped into the variation interval $[-1, 0]$ of $CI(A)$ and the $[\mathcal{N}(A), \Pi(A)]$ variation sub-domain $[\mathcal{N}(A) \in [0,1], 1]$ is mapped into the variation interval $[0, +1]$ of $CI(A)$.

Therefore, the interpretation of the values of $CI(A)$ is as follows:

- If $-1 < CI(A) < 0$, then occurrence of the event A is false to a certain degree.
- If $0 < CI(A) < 0$, then occurrence of the event A is true with strength $CI(A)$.
- $CI(A)$ is a one-to-one mapping, i.e., if $A, B \subseteq \Omega$ for which $CI(A) = CI(B)$, then this implies that $[\mathcal{N}(A), \Pi(A)] = [\mathcal{N}(B), \Pi(B)]$.

Proof Assume $A, B \subseteq \Omega$ for which $CI(A) = \mathcal{N}(A) + \Pi(A) - 1 = CI(B) = \mathcal{N}(B) + \Pi(B) - 1$. We have to show that $A = B$. As π is a normal possibility distribution, then we face two potential situations:

- $\mathcal{N}(A) = 0$. In this case, $\Pi(A) = \mathcal{N}(B) + \Pi(B)$. If $\mathcal{N}(B) > 0$, then certainly we have $\Pi(B) = 1$, and this leads to have $\Pi(A) = \mathcal{N}(B) + \Pi(B) > 1$!.

Therefore, when $\mathcal{N}(A) = 0$ we have the one-to-one mapping $[\mathcal{N}(A), \Pi(A)] = [\mathcal{N}(B), \Pi(B)]$.

- $\Pi(A) = 1$. In this case, $\mathcal{N}(A) = \mathcal{N}(B) + \Pi(B) - 1$. If $\Pi(B) < 1$, then $\mathcal{N}(B) = 0$, Here also, if $\mathcal{N}(B) = 0$, then $\mathcal{N}(A) < 0$!. Thus, $\Pi(B) = 1$ and $\mathcal{N}(A) = \mathcal{N}(B)$.

Therefore, in all cases, we have the one-to-one mapping $[\mathcal{N}(A), \Pi(A)] = [\mathcal{N}(B), \Pi(B)]$.

1. As a consequence of the one-to-one mapping $[\mathcal{N}(A), \Pi(A)] \leftrightarrow CI(A)$, the inverse transformation $CI(A) \rightarrow [\mathcal{N}(A), \Pi(A)]$ is determined as follows:
   - If $CI(A) \in [-1,0]$, then $\mathcal{N}(A) = 0$ and $\Pi(A) = CI(A) + 1$.
   - If $CI(A) \in [0,+1]$, then $\Pi(A) = 1$ and $\mathcal{N}(A) = CI(A)$.
2. Extreme situations: The correspondence $[\mathcal{N}(A), \Pi(A)] \leftrightarrow CI(A)$ is given as follows in extreme situations:

$[\mathcal{N}(A), \Pi(A)] \leftrightarrow$	$CI(A)$	***Meaning***
(0, 0)	−1	Event *A* is certainly false
(0, 1)	0	Total ignorance
(1, 1)	+1	Event *A* is certainly true

An interesting use of the confidence index is in the framework of a pattern recognition application. In fact, consider $\Omega = \{x_1, x_2 ..., x_M\}$ a finite set of alternatives, or actions, open to a decision-maker where one of these alternatives must be selected; and let $\pi: \Omega \rightarrow [0, 1]$ denote a possibility distribution defined on Ω and encoding our state of knowledge concerning the possible occurrence of different elementary alternatives. The situation that we call "*basic decision-making situation*" consists in choosing the true alternative depending on the only available knowledge, i.e., π. At this level, an important concept has to be discussed: *decision granularity*. In fact, recall that we are operating under the uncertainty type of imperfection which means that one and only one alternative occurs. Therefore, the expected output of any decision-making system is the most credible, plausible, reasonable, probable, true alternative. Nevertheless, if we analyze the situation from the decision-maker point of view, he may be interested by a less restrictive than a single alternative decision; he may be interested to know that the true alternative belongs to a crisp or a vague subset of alternatives.

For instance, in a situation faced by a procurement manager who knows that his/her budget is \$10,000 for an item, a vague quote from a vendor that the item may cost somewhere between \$2000 and \$4000 is adequate for the manager to determine that the item can be procured. Although, the quote is vague and uncertain, the current proposition that "the item can be purchased" can be answered with full certainty. In other words, even though we are operating under uncertainty and that one and only one alternative occurs, the decision granularity is fixed by the decision-maker.

Therefore, the decision-maker is generally more interested by a *decision-making support system* (DMSS) than by a *decision-making system* (DMS).

In the case where a single alternative is to be chosen, the decision-maker interest is to choose one and only one alternative out of the set of alternatives $\Omega = \{x_1, x_2 ...,$

$x_M\}$ where the only available knowledge to him is encoded by the possibility distribution π: $\Omega \to [0, 1]$. The possibility and necessity measures can be used:

$$\Pi(\{x_m\}) = \pi(x_m), m = 1, 2 \ldots M$$
$$\mathcal{N}(\{x_m\}) = 1 - \Pi\left(\{x_m\}^C\right) = 1 - \max_{x_n \, [n \neq m]} \pi(x_m)$$

These two measures can be combined to form the *confidence index*, *CI*, as a combination of $\Pi(\{x_m\})$ and $\mathcal{N}(\{x_m\})$. $\Pi(\{x_m\})$ and $\mathcal{N}(\{x_m\})$ represent the force, or momentum, to support $\{x_m\}$, optimistically and conservatively, respectively. *CI* $(\{x_m\})$ is defined as:

$$\begin{aligned} CI(\{x_m\}) = \Pi(\{x_m\}) - \Pi\left(\{x_m\}^C\right) &= \Pi(\{x_m\}) - [1 - \mathcal{N}(\{x_m\})] \\ &= \Pi(\{x_m\}) + \mathcal{N}(\{x_m\}) - 1 \end{aligned}$$

The confidence index value ranges in the interval $[-1, +1]$, with $CI(\{x_m\}) = -1$ being pure invalidation of the alternative $\{x_m\}$ and $CI(\{x_m\}) = +1$ being pure validation of the alternative $\{x_m\}$ (i.e., x_m is fully possible and fully necessary, meaning that $\{x_m\}^C$ is impossible). These "limit" cases represent completeness of information and, hence, the absence of uncertainty. Between these two limit cases, we have a whole range of uncertain situations. When $CI(\{x_m\}) = 0$, $\{x_m\}$ is completely possible, and at the same time, $\{x_m\}^C$ is also completely possible. In other words, $CI(\{x_m\}) = 0$ represents the total ignorance (i.e., anything is possible yet nothing is necessary). Using the confidence index, the uncertainty is then defined as the marginal demand for additional information; this is because the greater the confidence index is, the less the uncertainty, and vice versa. Therefore, the measure of uncertainty in a system, A, is defined, as a function of $CI(A)$, as:

$$U_A = -\log\left(|CI(A)|\right)$$

V. Perincherry et al. [16] have further extended the confidence index to incorporate the attitude of the decision-maker. In fact, if the decision-maker is totally optimistic, then his confidence level will be based on the possibility measure $\Pi(\{x_m\})$; and if he is totally pessimistic, then his confidence level will be based on the necessity measure $\mathcal{N}(\{x_m\})$. Let $\alpha \in [-1,+1]$ denote the decision-maker *attitude factor* ($\alpha > 0$ for optimistic attitude and $\alpha < 0$ for pessimistic attitude). The proposed *α-confidence index* is given by:

$$CI^{\alpha}(\{x_m\}) = \begin{cases} CI(\{x_m\}) + \min\{[CI(\{x_m\}) + 1]\alpha, \ 1\} & \text{for } \alpha > 0 \\ CI(\{x_m\}) - \max\{[CI(\{x_m\}) - 1]\alpha, \ 0\} & \text{for } \alpha < 0 \end{cases}$$

The variations of $CI^{\alpha}(\{x_m\})$ with $\Pi(\{x_m\})$, $\mathcal{N}(\{x_m\})$, and α are shown in Fig. 7.11 where five different decision-makers with attitude factors $\alpha_4 < \alpha_3 < \alpha_0 = 0 < \alpha_1 < \alpha_2$ are considered. They are in order of very optimistic (α_2) to very pessimistic (α_4).

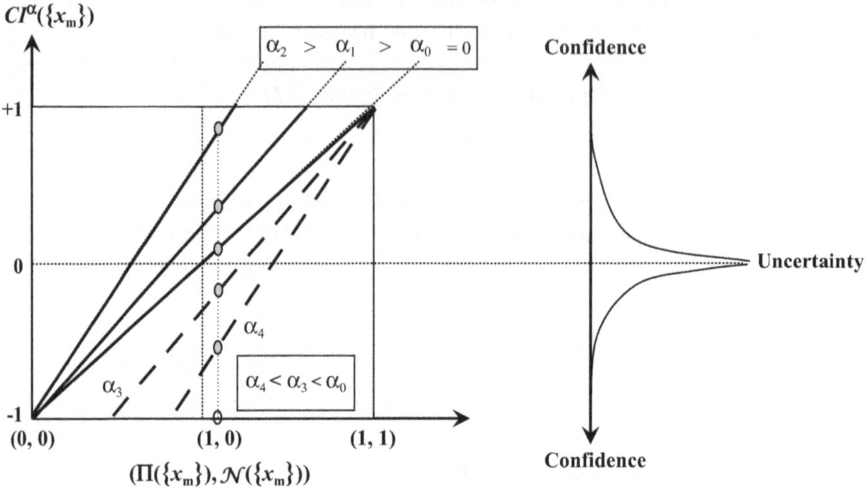

Fig. 7.11 Variation of CI^{α} with $\Pi(\{x_m\})$, $\mathcal{N}(\{x_m\})$ and α measure of uncertainty as a function of confidence index

Notice that for the same values of $(\Pi(\{x_m\}), \mathcal{N}(\{x_m\}))$, the optimists show a higher level of confidence than the pessimists. Also, the variation of the uncertainty is shown on the same figure as a function of the confidence index. When the confidence index equals +1 or -1, the decision-maker is completely confident (positively or negatively); therefore, additional knowledge has no utility, and thus $U_{A=}\{x_n\} = 0$. At a zero confidence index, where the decision-maker has no confidence, additional information has infinite utility and thus uncertainty $U_{A=}\{x_m\} = \propto$.

References

1. A. Kaufmann, *Bibliography on Fuzzy Sets and Their Applications* (BUSEFAL, LSI Lab., Université Paul Sabatier, Paris, 1980)
2. D. Dubois, H. Prade, Combination of fuzzy information in the framework of possibility theory. Data Fusion Rob. Mach. Intell. **12**, 481–505 (1992)
3. D. Dubois, H. Prade, Possibility theory in information fusion, in *Proceedings of the third international conference on information fusion*, vol 1 (2000), pp. PS6–P19
4. D. Dubois, H. Prade, Possibility theory and data fusion in poorly informed environments. Control. Eng. Pract. **2**, 811–823 (1994)
5. R.O. Duda, P.E. Hart, D.G. Stork, *Pattern Classification* (John Wiley & Sons, New York, 2012)
6. C. Frélicot, On unifying probabilistic/fuzzy and possibilistic rejection-based classifiers, in *Joint IAPR International Workshops on Statistical Techniques in Pattern Recognition (SPR) and Structural and Syntactic Pattern Recognition (SSPR)* (1998), pp. 736–745
7. J.T. Tou, R.C. Gonzalez, *Pattern Recognition Principles* (Addison-Wesley, London, 1981)
8. S. Benferhat, K. Tabia, An efficient algorithm for naive possibilistic classifiers with uncertain inputs, in *International Conference on Scalable Uncertainty Management* (2008), pp. 63–77

9. B. Haouari, N.B. Amor, Z. Elouedi, K. Mellouli, Naïve possibilistic network classifiers. Fuzzy Sets Syst. **160**, 3224–3238 (2009)
10. M. Bounhas, K. Mellouli, H. Prade, M. Serrurier, From Bayesian classifiers to possibilistic classifiers for numerical data, in *International Conference on Scalable Uncertainty Management* (2010), pp. 112–125
11. M. Bounhas, H. Prade, M. Serrurier, K. Mellouli, Possibilistic classifiers for uncertain numerical data, in *European Conference on Symbolic and Quantitative Approaches to Reasoning and Uncertainty* (2011), pp. 434–446
12. D. Dubois, H. Prade, When upper probabilities are possibility measures. Fuzzy Sets Syst. **49**, 65–74 (1992)
13. S. Kikuchi, V. Perincherry, Handling uncertainty in large scale systems with certainty and integrity, MIT Engineering Systems Symposium, Cambridge, MA, (2004)
14. T.D. Nielsen, N.L. Zhang, in *Symbolic and Quantitative Approaches to Reasoning with Uncertainty: 7th European Conference, ECSQARU 2003, Aalborg, Denmark, July 2–5, 2003. Proceedings* vol 2711 (Springer, 2004)
15. J.F. Geer, G.J. Klir, A mathematical analysis of information-preserving transformations between probabilistic and possibilistic formulations of uncertainty. Int. J. Gen. Syst. **20**, 143–176 (1992)
16. V. Perincherry, S. Kikuchi, Y. Hamamatsu, Uncertainties in the analysis of large-scale systems, in *1993 (2nd) International Symposium on Uncertainty Modeling and Analysis* (1993), pp. 216–222

Chapter 8
Possibilistic Concepts Applied to Soft Pattern Classification

8.1 Introduction

This chapter presents three applications of possibilistic concepts in the domain of soft pattern classification. The first one is on pixel-based image classification where an approach referred as Iterative Refinement of Possibility Distributions by Learning (IRPDL) is presented. The IRPDL approach is based on the use of possibilistic reasoning concepts exploiting expert knowledge sources as well as "ground" possibilistic seeds learning. Full details concerning the approach are available in Alsahwa et al. [1]. Here, we summarize the method and present results obtained using synthetic images. In [1], the performance of this approach is demonstrated with synthetic images as well as with real images. The performance is compared to three relevant reference methods [2, 3]: region growing, semi-supervised fuzzy pattern matching (detailed in Chapter 7), and Markov random fields.

The second application is on spatial unmixing based on possibilistic similarity [4]. The approach presented here exploits possibilistic concepts to provide flexibility for the integration of both contextual information and a priori knowledge. Possibility distributions are first obtained using a priori knowledge given in the form of learning areas delimitated by an expert. These areas serve for the estimation of the probability density functions of different thematic classes also called endmembers. The resulting probability density functions are then transformed into possibility distributions using Dubois-Prade's probability-possibility transformation (detailed in Chapter 5). The pixel unmixing is then performed based on the possibilistic similarity between a local possibility distribution estimated around the considered pixel and the obtained possibility distributions representing the predefined endmembers in the analyzed image. Several possibilistic similarity measures have been evaluated to improve the discrimination between endmembers. Results [4] show that the proposed approach represents an efficient estimator of the proportion of each endmember present in the

B. Solaiman, É. Bossé, *Possibility Theory for the Design of Information Fusion Systems*, Information Fusion and Data Science,
https://doi.org/10.1007/978-3-030-32853-5_8

pixel (i.e., abundances) and achieves higher classification accuracy. Performance analysis has been conducted using synthetic as well as real images in [4], but here, we only present results obtained using synthetic images. We refer the reader to [4] for full details.

Finally, the third application is about image segmentation based on possibilistic concepts [5]. The aim of a segmentation process is to distinguish homogeneous regions within an image that belong to different objects. The image segmentation approach is based upon the use of possibility theory concepts and imitating human reasoning scheme when analyzing an observed image in terms of its constituting homogeneous regions. Based on the observed features of gray level intensity, color, texture, patterns, shapes, etc., a human understands image by first segmenting the whole image into regions/objects. This initial segmentation task is realized by "visually" locating subregions, called "reference seed." Reference seeds are homogenous small regions having semantic visual significance or satisfying some simple similarity criteria: feature-based homogeneity, membership to a known structure, or a thematic class. Then, a second "iterative" step follows. It consists of spreading the semantic homogeneous seeds into their spatial context using a set of similarity criteria. Additional sources of knowledge are then used to enlarge the initial homogenous regions. This process leads to a global segmented image over which other interpretation tasks may be conducted.

8.2 Pixel-Based Image Classification

A crucial task in image classification is the representation of the thematic classes contained in an analyzed image. An important difficulty related to this task stems from the inability to have a representative knowledge of these different thematic classes. This is mainly due to the lack of solid models capturing the representative constraints of the available knowledge. The quality of this "representativity" can significantly influence the performance of the classifier to be used for the image representation.

Starting with limited initial *prior* knowledge, an efficient classifier is assumed to have the capacity to extract additional knowledge, with a high degree of confidence, while preserving the previously acquired knowledge. This process is called incremental learning [6]. In learning-based systems, initial *prior* knowledge is generally expressed by the expert as a set of samples or learning areas, called ground seeds, representing the "certain" knowledge expressed by an expert about the different classes in the image. Additional knowledge can be extracted by locating new learning areas for different classes in the analyzed image. This additional knowledge is "appealed" to offer a better representation (or discrimination capability) for classification purpose.

The description given by the expert about the thematic classes is often imprecise and ambiguous. For example, according to medical experts on digitized mammograms, contrast is the most relevant characteristic for extracting the over-density regions of interest in an analyzed image. Indeed, over-density areas can have various contrast levels according to the nature of the breast (dense or clear). This leads to an ambiguous description of these areas and brings imprecision in the delimitation of their contours. On the other hand, the description given by the expert can be vague or uncertain. For instance, on digitized mammograms, normal tissue may have a contrast level close to that of the cancerous tissue, especially in noisy or low contrast images.

Various models can be used to represent different forms of knowledge imperfections. It is well established that the probability theory possesses the necessary toolset for optimal decision-making in classification problems dealing with uncertainty. However, in several applications, it is generally impossible to get the required knowledge to attain this optimality such as in conditions of poor data or incomplete knowledge environment. Belief functions, or Dempster-Shafer, theory has been used to express imprecise and uncertain knowledge. However, for incomplete, vague as well as the knowledge expressed by human experts, possibility theory constitutes a more appropriate and flexible tool. In fact, possibility theory is useful to represent uncertainty when the available knowledge is affected by different forms of imperfections: imprecision, incompleteness, vagueness, etc. It is worth noticing that even when the previously acquired knowledge is perfect, the additional knowledge extracted through any incremental learning process will likely be affected by those different forms of imperfection.

A set of possibility distributions can be used to represent thematic classes present in an image. Defining these distributions in terms of standard shapes with their associated parameters is generally not an easy task, especially with limited *prior* knowledge. Class representation using possibility theory is a nontrivial task due to the lack of consensus on the definition and interpretation of possibility distributions as well as the lack of measures to evaluate their goodness and their correctness or to attach a physical meaning to those distributions. Several methods based on possibility theory suffer from the choice of too many parameters that has a great influence on the classification performance. Therefore, especially in noisy environments, the representation of the thematic classes remains an important challenge in image classification.

The approach proposed in [1] is referred as Iterative Refinement of Possibility Distributions by Learning (IRPDL). *Prior* knowledge is assumed to be given as an initial and small set of learning ground seeds areas extracted from the considered image. These ground seeds are supposed to correspond to an expert point of view about an initial definition of the M considered classes. The first step in the proposed IRPDL approach consists of enriching the initial set of the learning areas and class representation with an incremental learning process exploiting spatial knowledge. The second step in IRPDL approach aims at reducing the number of parameters used in the mathematical representation of different classes exploiting probability-possibility transformations (shown in Chapter 5).

8.2.1 Pixel-Based Methods within the Context of Limited Prior *Knowledge*

In pixel-based image classification, several methods have been proposed in the literature to cope with limited initial *prior* knowledge. Among them, the method of image segmentation by region growing is to select manually the "seeds" designating anchor points for the initialization of the segmentation process: partitioning the image into multiple regions. According to a criterion of similarity measure, a region grows iteratively by merging adjacent pixels similar to the initial starting seeds. It is known that the semantic image segmentation approaches suffer from two main problems: the automatic selection of seeds and the sorting and ordering of pixels (for labeling). These approaches are called "semantic" because they involve high-level knowledge of image components in the seeds selection procedure as in the case of the proposed IRPDL approach.

In the possibilistic framework, class representation is done via a set of possibility distributions encoding knowledge about the real world: the thematic classes of interest. Several methods have been proposed in the literature in order to estimate the possibility distributions from a limited *prior* knowledge to represent the thematic classes. These methods can be divided into two categories.

The first category reproduces fuzzy set theory concepts by using the standard and predefined membership functions. They apply Zadeh's postulate for which possibility values numerically duplicate the membership ones. In fact, Zadeh's postulate transforms membership degrees to a fuzzy set describing an ambiguous concept into possibility degrees that represent the uncertainty concept. This first category is well adapted to the case where the available expert's knowledge is expressed using an ambiguous description over the set of thematic classes that can be modeled by the standard membership forms.

The second category of methods is based on the use of statistical data methods such as probability-possibility transformations, histogram-based methods, neural networks, etc. A representative example of this second category is the histogram-based method of Roux and Desachy [7]. They simply normalize the class histograms in order to obtain the possibility distribution for pixel-based image classification without taking into consideration neither the contextual information nor the decision confidence.

The IRPDL approach takes into account both contextual information and decision confidence. IRPDL also aims at reducing the number of parameters used in the representation process. To this specific end, the probability-possibility transformation method [8] is adopted. The Kernel Density Estimation (KDE) approach [9, 10] is first used to estimate the M "poor" probability density functions which then are transformed into M possibility distributions using Dubois-Prade transformations. The novelty of IRPDL does not reside in the use of the $\Pr \rightarrow \pi$ transformation but rather in the integration of an incremental learning process through the application of possibilistic reasoning concepts on the M possibilistic maps. Follows the estimation refinement of the M possibility distributions that characterize the ambiguity between classes in the analyzed image, a confidence criterion is used to measure the degree of

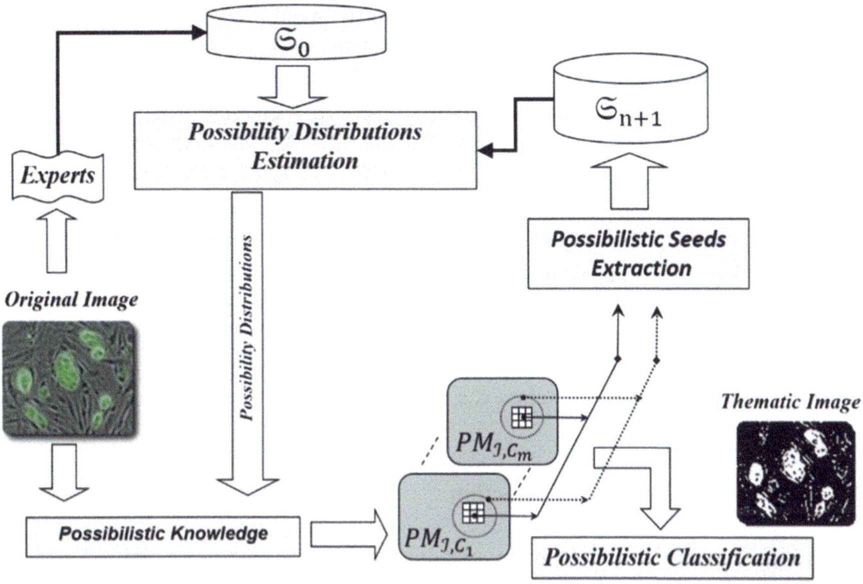

Fig. 8.1 The IRDPL approach. (Adapted from Ref. [1])

certainty that each sample belongs to a class. The extraction of new learning samples is, then, conducted using possibilistic spatial information applied to different possibilistic maps. This extraction process is thus highly inspired from the region growing image segmentation approaches that select similar spectral values from representative locations. The extraction process is then iteratively repeated until there are no more new learning samples that can be extracted and added to the incremental process.

8.2.2 The IRPDL Method

The IRPDL approach consists of two main steps. The first step consists of enriching the initial set of the expert's knowledge learning areas and class representation samples set $\mathfrak{S}_0$ (as illustrated in Fig. 8.1) with an incremental learning process exploiting the spatial knowledge. This enrichment is conducted by means of possibility distributions in order to deal with both vagueness and uncertainty in the expert's description. *Prior* knowledge is assumed to be given as an initial and small set of learning ground seeds areas extracted from the considered image. These ground seeds are supposed to correspond to an expert's point of view about an initial definition of the *M* considered classes. At iteration "*n*," the application of the possibilistic seeds extraction rule produces the additional set of seeds $\mathfrak{S}_{n+1}$. This seeds set is then used to enrich the samples set $\mathfrak{S} = \bigcup_{k=0}^{n+1} \mathfrak{S}_k$ used for the possibility distributions estimation. The seeds enrichment process is then iteratively repeated until no more seeds are added to the set $\mathfrak{S}_{n+1}$.

The second step in IRPDL approach aims at reducing the number of parameters used in the mathematical representation of different classes. This is done via probability-possibility transformations. In the IRPDL method, as shown in Fig. 8.1, the transformation process starts with the Kernel Density Estimation (KDE) approach to estimate the M probability density functions which are then transformed into M possibility distributions using Dubois-Prade transformations (see Chapter 5). The KDE approach is a nonparametric estimator used particularly in the case of a small size sample set. The application of the M class possibility distributions onto the considered image $\mathcal{I}$ will lead to M possibilistic maps $PM_{\mathcal{I},C_m}$, $m = 1, \ldots, M$. A possibilistic map, $PM_{\mathcal{I},C_m}$, encodes the possibility degree of different image pixels belonging to a thematic class C_m, $m = 1, \ldots, M$.

8.2.2.1 Possibilistic Seeds Extraction

This section describes the possibilistic seeds extraction process presented in Fig. 8.1. Each pixel P_0 from the analyzed image $\mathcal{I}$ can be associated with a spatial possibility distribution defined on the set of M thematic classes $\Omega = \{C_1, C_2, \ldots, C_M\}$ according to:

$$\overline{\pi}_{P_0} = [\overline{\pi}_{P_0}(C_1), \overline{\pi}_{P_0}(C_2), \overline{\pi}_{P_0}(C_3), \ldots, \overline{\pi}_{P_0}(C_M)]$$

where $\overline{\pi}_{P_0}(C_m), m = 1, 2, \ldots, M$ represents the possibility degree, for the pixel P_0, belonging to the class C_M. This possibility degree is extracted by the application, on the mth possibilistic map, of an operator Φ (e.g., mean operator) as follows:

$$\overline{\pi}_{P_0}(C_m) = \underbrace{\Phi}_{P \in V(P_0)} (PM_{\mathcal{I},C_m}(P))$$

where $V(P_0)$ refers to the considered spatial neighborhood of the pixel P_0 and $PM_{\mathcal{I},C_m}$ is the mth possibilistic map, $m \in \{1, \ldots, M\}$. Using the spatial possibility distribution $\overline{\pi}_{P_0}$, a spatial confidence index $\overline{Ind}$ can be computed for each class C_M according to:

$$\overline{Ind}(A_M) = \overline{\pi}_{P_0}(C_m) - \max_{n \neq m} \overline{\pi}_{P_0}(C_n)$$

$$\text{where } A_m = \{C_m\}_{m=1,\ldots,M}.$$

In the case of $\Phi = \mathcal{I}_{P_0}$ (identity operator and $V(P_0) = \{P_0\}$), the spatial possibility distribution is restricted to the following possibility distribution:

$$\overline{\pi}_{P_0} = [\overline{\pi}_{P_0}(C_1), \overline{\pi}_{P_0}(C_2), \ldots, \overline{\pi}_{P_0}(C_M)] = [\pi_{P_0}(C_1), \pi_{P_0}(C_2), \ldots, \pi_{P_0}(C_M)]$$

where $\pi_{P_0}(C_m) = PM_{\mathcal{I},C_m}(P_0)$ is the possibility degree, for the pixel P_0 without taking into consideration the spatial context belonging to class C_m. The pixel P_0 is called a *C_m-possibilistic seed* if the state of the available possibilistic knowledge allows to confirm, with high certainty, that this pixel is a candidate for belonging to the class C_m. In [11], using Φ as a mean operator Φ_{mean}, the behavior of $\overline{Ind}$ as well as *Ind* in seeds extraction has been studied as a function of a threshold $S \in [0, 1]$. In the case of low discrimination complexity, seeds are extracted without errors for a threshold $S = 0.5$. Accordingly, P_0 is considered as *C_m-possibilistic seed* if its highest confidence index value, $\overline{Ind}(A_M)$, belonging to the class C_m, exceeds the threshold $S = 0.5$. In order to deal with high discrimination complexity in noisy environment, or low contrast images, the IRPDL approach suggests the use of two operators, Φ_{mean} and $\mathcal{I}_{P_0}$, on each possibilistic map, $PM_{\mathcal{I},C_m}$. Hence, the pixel P_0 is considered as a *C_m-possibilistic seed* if and only if $\overline{Ind}(A_M)$ and *Ind* exceed the threshold $S = 0.5$.

8.2.2.2 Possibilistic Knowledge Projection and Decisions

The application of possibility theory into pixel-based image classification is conducted in two successive steps: possibilistic knowledge projection and decision mechanisms (Fig. 8.1). The step, *possibilistic knowledge projection*, considers the expert's predefined set of M thematic classes contained in the analyzed image, $\Omega = \{C_1, C_2, \ldots, C_M\}$ as well as the set of M possibility distributions:

$$\pi_{C_m} : D \rightarrow [0, 1] \text{ and } x(P) \rightarrow \pi_{C_m}(x(P))$$

where D refers to the definition domain of the observed feature $x(P)$ of the pixel P (e.g., $D = [0{,}255]$). For each class C_m, $\pi_{C_m}(x(P))$ associates each pixel $P \in \mathcal{I}$, observed through a feature $x(P) \in D$, with a possibility degree of belonging to the class C_m, $m = 1, \ldots, M$. Given an input image $\mathcal{I}(H \times W)$, the application of these M possibility distributions $\pi_{C_m}(\cdot)$, $m = 1, \ldots, M$ into the image $\mathcal{I}$ allows the possibilistic expert's knowledge projection leading to M possibilistic images, called *possibilistic maps* $PM_{\mathcal{I},C_m}(H \times W)$ and written as:

$$PM_{\mathcal{I},C_m}(i, j) = \pi_{C_m}(x(P(i, j)))$$

The second pixel-based image classification step consists of the application of *decision mechanisms* (e.g., $R1$ or $R2$) on the M possibilistic maps in order to compute the thematic image where each pixel is associated with a class label. It is worth mentioning that the expert tends generally to describe the different thematic classes by the delimitation of small learning sample sets belonging to these classes with high certainty. Once this knowledge (i.e., learning sample sets) is injected into the pixel-based possibilistic image classification process, the expert can only appreciate the resulting thematic image.

8.2.2.3 Possibilistic Knowledge Diffusion (PKD) Process

The extracted seeds from the refining process enrich the current knowledge representation. Indeed, these seeds form a set of "benchmarks" which guide the discrimination process between classes in the analyzed scene. In any method based upon the knowledge diffusion principle like "region growing," neighboring pixels that have similar properties to those seeds (gray level, texture measurement, etc.) are added. At the end of the diffusion operation, the segmented image will consist of a set of homogeneous regions. Note that the knowledge diffusion principle establishes a kind of analogy with human reasoning in the perceptual analysis of an image. Indeed, the human interpretation of image content (in terms of its segmentation into homogeneous regions) is to identify reference areas of each class and then enrich these areas in an incremental way. This is done by analyzing their spatial neighborhood in order to refine the possibilities of belonging to different classes. This enrichment process becomes more important for the discrimination between classes in the context of high discrimination complexity.

In order to evaluate the refining process, a possibilistic knowledge diffusion (PKD) step is conducted after the seeds extraction process has been completed. This process is inspired from the region growing method and is applied to the space of possibilistic knowledge representation (i.e., possibilistic maps) instead of the space of gray levels. In this step, the possibilistic knowledge available for each pixel on the image is analyzed in order to reduce the uncertainty encountered in the classification decision.

The diffusion method used in IRDPL is based on the Nagao filter [12] since it produces relatively low error rates when considering the maximum possibilistic classification for almost all levels of contrast and noise. In addition, this filter allows the preservation of objects contours. This is implemented through a set of nine (9) new "sub-windows or cliques" $f = \{f_1, f_2, f_3, \ldots, f_9\}$ [1] in the neighborhood of each pixel considered in each possibilistic map $PM_{\mathcal{J},C_m}(P_0)$ at the end of the refining process. Secondly, the winning clique f_g among the nine cliques is the clique that has the maximum possibilistic mean value. Hence, the knowledge diffusion is performed by a Nagao filter at iteration $n + 1$ as $PM_{\mathcal{J},C_m}(P_0, n+1)$. This process is repeated until stabilization of the possibilistic maps. The algorithm used is listed in [1].

8.2.3 The Performance Evaluation of the Proposed IRPDL Approach

The performance evaluation of the IRPDL approach has been conducted for two main cases: low discrimination and high discrimination complexity between classes in the analyzed image. The study is conducted using both a set of synthetic images

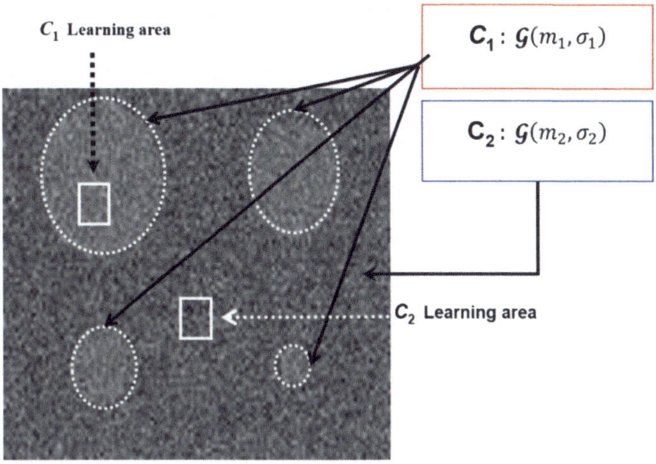

Fig. 8.2 Synthetic image composed of two classes generated using Gaussian distributions

and real images. The performance of the IRPDL approach is being compared to the following three reference methods of pixel-based classification: region growing [13], semi-supervised fuzzy pattern matching (FPM) [9], and Markov random field (MRF) [14]. The performance is evaluated by computing the average of maximum possibilistic pixel-based recognition rate for the three reference methods and comparing these rates with the IRPDL recognition rate.

Complete results are given in [1] including those with real medical images. In this section, we present only the results of the possibilistic refining process obtained with synthetic images.

For the experimental evaluation purpose, a 96 × 128 pixel synthetic image, given in Fig. 8.2, is generated. This synthetic image is composed of two thematic classes $\{C_1, C_2\}$. Class C_1 is assumed to represent circular-form objects of interest of four different sizes; class C_2 constitutes the image background. Pixels from C_1 and C_2 are generated as two Gaussian distributions $\mathcal{G}(m_1, \sigma_1)$ and $\mathcal{G}(m_2, \sigma_2)$. 10 × 10 pixel learning areas positioned by the expert (as being representative areas of the considered thematic classes) are also illustrated on the generated images. Using the learning areas, the initial estimation of the class probability density functions is established using the Kernel Density Estimation (KDE) approach. The application of the $\Pr \rightarrow \pi$ Dubois-Prade's transformation allows obtaining the initial possibility distributions.

The quality of the refined possibility distributions is evaluated using the synthetic image generated using two Gaussian distributions and given in Fig. 8.2. This image is considered for two cases: first, with means $m_1 = 140$ and $m_2 = 100$ and standard deviations $\sigma_1 = 15$, $\sigma_2 = 20$ (i.e., low discrimination complexity) and, second, the same mentioned parameters except for mean $m_1 = 120$ (i.e., high discrimination complexity). The intentional low contrast between these two thematic classes is

Table 8.1 Recognition rate in four cases: low complexity

Cases	Recognition rate (%)	Standard deviation
Before refining without PKD	88.1917	0.3988
Before refining with PKD	94.4889	1.2958
After refining without PKD	89.2077	2.4134
After refining with PKD	**97.0998**	0.0927

Table 8.2 Recognition rate in four cases: high complexity

Cases	Recognition rate (%)	Standard deviation
Before refining without PKD	76.7972	2.1228
Before refining with PKD	80.3589	2.5826
After refining without PKD	72.1368	3.8513
After refining with PKD	**89.2578**	1.3963

assumed to reproduce difficult pixel-based classification situations. As the probability density function of the original generated image is a priori known, the possibility distribution obtained by the application of $Pr \to \pi$ Dubois-Prade's transformation is considered the *reference possibility distribution* to be estimated through the refining process. This assumption is of great importance since it gives place to a "physical interpretation." The class possibility distribution as being the Dubois-Prade's symmetric probability distribution has the "physical significance of the frequency of occurrence."

The possibilistic pixel-based classification, using the maximum rule, is applied to this synthetic image in four cases: before refining without PKD, before refining with PKD, after refining without PKD, and, finally, after refining with PKD. The synthetic image is generated 100 times. The mean and standard deviation of the recognition rates are used as evaluation criteria of the refining process (Tables 8.1 and 8.2).

The increase in the recognition rate of about 3% (case of low discrimination complexity, Table 8.1) after the refining process and the possibilistic knowledge diffusion (PKD) compared to the one before refining shows the importance of refining process in order to characterize the classes in the analyzed image. An improvement of about 10% (case of high discrimination complexity, Table 8.2) in the recognition rate is obtained after the refining process compared to that obtained before the refining process. These results raise the important question of the physical significance of the possibility distribution used for pixel-based image classification and constructed through the iterative statistical/spatial possibilistic reasoning approach. In fact, in the first situation (low discrimination complexity), the refined possibility distributions encapsulate a characterization significance of different classes, whereas in the second situation (multi-class relatively complex to be discriminated), the significance of the refined possibility distributions can be interpreted as being class discrimination-based possibility values.

8.3 Spatial Unmixing Based on Possibilistic Similarity

This section proposes an approach for pixel unmixing based on possibilistic similarity. The approach exploits possibilistic concepts to provide flexibility in the integration of both contextual information and a priori knowledge. Possibility distributions are first obtained using a priori knowledge given in the form of learning areas delimitated by an expert. These areas serve for the estimation of the probability density functions of different thematic classes also called endmembers. The resulting probability density functions are then transformed into possibility distributions using Dubois-Prade's probability-possibility transformation. The pixel unmixing is then performed based on the possibilistic similarity between a local possibility distribution estimated around the considered pixel and the obtained possibility distributions representing the predefined endmembers in the analyzed image. Several possibilistic similarity measures have been tested to improve the discrimination between endmembers. Results show that the proposed approach represents an efficient estimator of the proportion of each endmember present in the pixel (abundances) and achieves higher classification accuracy. Performance analysis has been conducted using synthetic and real images.

8.3.1 Possibilistic Knowledge Representation

Let us consider an exclusive and exhaustive universe of discourse $\Omega = \{C_1, C_2, \ldots, C_M\}$ formed by M mutually exclusive elementary decisions (e.g., decisions, thematic classes, endmembers, hypothesis, etc.), and let $\{\pi_{C_m}\}$, $m = 1, \ldots, M$ be a set of M possibility distributions defined on Ω. Exclusiveness means that one and only one decision may occur at one time, whereas exhaustiveness states that the occurring decision certainly belongs to Ω. Possibility theory is based on the notion of possibility distribution denoted by π, which maps elementary decisions from Ω to the interval [0, 1], thus encoding "our" state of knowledge or belief, on the possible occurrence of each element $C_m \in \Omega$. The value $\pi(C_m)$ represents to what extent it is possible for C_m to be the unique occurring decision. This value $\pi(C_m)$ encodes our state of knowledge, or belief, about the real world and represents the possibility degree for C_m to be the unique occurring element. In this context, two extreme cases of knowledge are given:

- Complete knowledge: $\exists\,!\; C_m \in \Omega$, $\pi(C_m) = 1$ and $\pi(C_n) = 0, \forall\, C_n \in \Omega, C_n \neq C_m$.
- Complete ignorance: $\forall C_m \in \Omega$, $\pi(C_m) = 1$ (all elements from Ω are considered as totally possible).

$\pi(\cdot)$ is called a normal possibility distribution if at least one element C_{m_0} from Ω such that $\pi(C_{m_0}) = 1$.

8.3.2 Possibility Distributions Estimation Based on Pr → π Transformation

A crucial step in possibility theory applications is the determination of possibility distributions. Recall that a possibility distribution encodes our state of knowledge about the real world. Nevertheless, the appropriate estimation of the possibility distribution shape and shape's parameters is a difficult task. Two approaches are generally used for the estimation of a possibility distribution. The first approach consists of using standard forms predefined in the framework of fuzzy set theory for membership functions (i.e., triangular, Gaussian, trapezoidal, *S*-shape, etc.) and tuning the shape-parameters using a manual or an automatic tuning method. The second estimation approach of possibility distributions is based on the use of statistical data and is conducted in the following two consecutive steps:

- Using statistical data, an uncertainty function describing the uncertainty inherent to the statistical data is estimated first (e.g., histogram, probability density function, basic belief function, etc.).
- The estimated uncertainty function is then transformed into a possibility distribution. In the framework of possibility theory, the probability-possibility transformations (Pr $\rightarrow \pi$ transformations) are frequently used for the implementation of this step.

In this study, the available expert's knowledge is expressed through the definition of small learning areas representing different endmembers, i.e., statistical data. The second estimation approach will then be used. Several Pr $\rightarrow \pi$ transformations are proposed in the literature. Dubois et al. [15] suggested that any Pr $\rightarrow \pi$ transformation of a probability density function, Pr, into a possibility distribution, π, should be guided by the two following principles:

- The probability-possibility *consistency principle*. This principle is expressed by Zadeh [16] as "*what is probable is possible.*" Dubois and Prade formulated this principle by indicating that the induced possibility measure Π should encode upper probabilities:

$$\Pi(A) \geq \Pr(A), \quad \forall A \subseteq \Omega$$

- The *preference preservation principle* ensuring that any Pr $\rightarrow \pi$ transformation should satisfy the relation:

$$\Pr(A) < \Pr(B) \Leftrightarrow \Pi(A) < \Pi(B), \quad \forall A, B \subseteq \Omega$$

Verifying these two principles, a Pr $\rightarrow \pi$ transformation turning a probability distribution Pr, defined by probability values $\Pr(\{C_m\})$, $C_m \in \Omega$, $m = 1, 2, \ldots, M$, into a possibility distribution π, defined by $\Pi(\{C_m\})$, $C_m \in \Omega$, $m = 1, 2, \ldots, M$ has been suggested by Dubois et al. [15]. This transformation, called *symmetric* Pr $\rightarrow \pi$ *transformation*, is defined by:

$$\pi(C_m) = \Pi(\{C_m\}) = \sum_{j=1}^{M} min\left[\Pr(\{C_m\}), \Pr(\{C_j\})\right]$$

This transformation is being used in our study to transform the probability distributions into possibility distributions. The reason standing behind this choice is due to good performance of the symmetric transformation which provides pattern recognition and classification [4, 9].

8.3.3 Possibilistic Similarity

The concept of similarity is a very important topic for many applications. Any system that needs to analyze or organize automatically a set of data or knowledge must use a similarity operator to estimate relations and resemblances that exist in data [17]. The issue of comparing imperfect pieces of information depends on the way they are represented. In the case of possibility theory, comparing uncertain pieces of information is to compare their possibility distributions. Hence, a similarity measure is a quantification of the amount of similarity between two possibility distributions.

Considering the expert's predefined set of M endmembers contained in the analyzed image, $\Omega = \{C_1, C_2, \ldots, C_M\}$, a set of M possibility distributions can be defined as follows:

$$\begin{aligned} \pi_{C_m}&:D_m \rightarrow [0, 1] \\ x(P) &\rightarrow \pi_{C_m}(x(P)) \end{aligned}$$

where D refers to the definition domain of the observed feature $x(P)$ of the pixel P. For each class C_m, $\pi_{C_m}(x(P))$ associates each pixel $P \in \mathcal{I}$, of an image $\mathcal{I}$ observed through a feature $x(P) \in D$, with a possibility degree of belonging to the class C_m, $m = 1, \ldots, M$. Different possibilistic similarity and distance functions "Sim" can be defined between the two possibility distributions π_{C_m} and π_{C_n} of two endmembers C_m and C_n of the set Ω. The behavior of these functions can be studied in order to obtain a better discrimination between classes C_m and C_n. To this end, calculating a similarity matrix S informs on such inter-classes behavior and helps in the choice of a suitable measure in a given context:

$$S = \begin{bmatrix} Sim(\pi_{C_m}, \pi_{C_m}) & Sim(\pi_{C_m}, \pi_{C_n}) \\ Sim(\pi_{C_n}, \pi_{C_m}) & Sim(\pi_n, \pi_{C_n}) \end{bmatrix}$$

8.3.3.1 Possibilistic Similarity Functions

This subsection reviews some existing possibilistic similarity and distance functions that are the most frequently used in literature (see Chapter 4). Recall that one considers an exclusive and exhaustive universe of discourse $\Omega = \{C_1, C_2, \ldots, C_M\}$ formed by M mutually exclusive elementary decisions, and let $\{\pi_{C_m}\},\ m = 1, \ldots, M$ be a set of M possibility distributions defined on Ω.

- *Information closeness*: this similarity measure was proposed by Higashi and Klir [18] based on the information variation measure G:

$$G(\pi_{C_m}, \pi_{C_n}) = g(\pi_{C_m}, \pi_{C_m} \vee \pi_{C_n}) + g(\pi_{C_n}, \pi_{C_m} \vee \pi_{C_n})$$

where $g(\pi_{C_m}, \pi_{C_n}) = U(\pi_{C_n}) - U(\pi_{C_m})$. The operator $\vee$ is taken as the maximum operator, and U is the non-specificity measure. Given an ordered possibility distribution π such that $1 = \pi_1 \geq \pi_2 \geq \cdots \geq \pi_M$, the U of π is formulated as:

$$U(\pi) = \left[\sum_{i=1}^{M} (\pi_i - \pi_{i+1}) \log_2 i\right] + (1 - \pi_1)\log_2 M$$

where $\pi_{M+1} = 0$ by convention. Hence, the similarity measure based on the *information closeness* is given by:

$$Sim_G(\pi_{C_m}, \pi_{C_n}) = 1 - \frac{G(\pi_{C_m}, \pi_{C_n})}{G_{\text{Max}}}$$

- *Minkowski distance*: since possibility distributions are often represented as vectors, the most popular metrics for possibility distributions are induced by the *Minkowski* norm (L_p) which is used in vector spaces:

$$L_p(\pi_{C_m}, \pi_{C_n}) = \sqrt[p]{\sum_{i=1}^{|p|} |\pi_{C_m}(x_i) - \pi_{C_n}(x_i)|^p}$$

Two particular cases are often investigated: L_1-norm (*Manhattan distance*) and L_2-norm (*Euclidean distance*). They are given by the following expressions:

$$L_1(\pi_{C_m}, \pi_{C_n}) = \sum_{i=1}^{|p|} |\pi_{C_m}(x_i) - \pi_{C_n}(x_i)|$$

$$L_2(\pi_{C_m}, \pi_{C_n}) = \sqrt{\sum_{i=1}^{|p|} |\pi_{C_m}(x_i) - \pi_{C_n}(x_i)|^2}$$

These cases of *Minkowski* distance can be transformed into similarity measure by the following:

$$Sim_p(\pi_{C_m}, \pi_{C_n}) = 1 - \frac{L_p(\pi_{C_m}, \pi_{C_n})}{\sqrt[p]{|D|}}$$

- *Information affinity*: this similarity measure was proposed by Jenhani et al. [19].

$$Sim_{\text{IA}}(\pi_{C_m}, \pi_{C_n}) = 1 - \frac{\kappa L_p(\pi_{C_m}, \pi_{C_n}) + \lambda Inc(\pi_{C_m}, \pi_{C_n})}{\kappa + \lambda}$$

where $\kappa > 0$ and $\lambda > 0$, $Inc(\pi_{C_m}, \pi_{C_n})$ represents the inconsistency degree between π_{C_m} and π_{C_n} defined as follows:

$$Inc(\pi_{C_m}, \pi_{C_n}) = 1 - max(min(\pi_{C_m}, \pi_{C_n}))$$

- *Similarity index* [19]:

$$Sim_{\text{SI}}(\pi_{C_m}, \pi_{C_n}) = min\{\alpha(\pi_{C_m}, \pi_{C_n}), \alpha(1 - \pi_{C_m}, 1 - \pi_{C_n})\}$$

where

$$\alpha(\pi_{C_i}, \pi_{C_j}) = \frac{\sum_{n=1}^{N} \pi_{C_i}(X_n) \times \pi_{C_j}(X_n)}{\sum_{n=1}^{N} \{max(\pi_{C_i}(X_n), \pi_{C_j}(X_n))\}^2}$$

It is worth noticing that this list should not be considered as complete depending on the application or the type of images that have to be analyzed. In the following subsection, a process allowing the selection of the most "suitable" similarity measure is proposed and evaluated using synthetic images.

8.3.3.2 Evaluation of the Similarity between Two Classes

A 100×100 synthetic image composed of two thematic classes has been generated in order to evaluate the similarity between two classes. The intensity of the pixels from C_1 and C_2 is generated as two Gaussian distributions $G(m_1 = 110, \sigma_1 = 10)$ and $G(m_2 = 120, \sigma_2 = 10)$. The evaluation principle of the similarity between the two classes selects the possibilistic similarity function for which the similarity matrix is the closest to the identity matrix I_2 in terms of the Euclidean distance. In the considered case, where only two classes are involved, this distance D is summarized by the following measure $(i, j \in \{0, 1\})$:

$$D = \sqrt{\sum_{i,j} [S(i,j) - I_2(i,j)]^2}$$

The lower is the distance value D, the better is the discrimination power (between classes) of the considered similarity function. The distance value D is computed for each similarity function by, first, varying the mean value of the generated pixels of class C_2 and, then, the standard deviation while maintaining a fixed value for the mean and standard deviation of class C_1.

8.3.4 A Possibilistic Approach of Pixel Unmixing by Endmembers

The approach, based on possibilistic similarity for pixel unmixing by endmembers, exploits possibilistic concepts to provide flexibility, on one side, in representing limited information (contextual information and a priori knowledge) and, on the other side, in the integration of both contextual information and a priori knowledge. Alsahwa et al. [1] present also a method to enrich the limited initial *prior* knowledge. This method is based on transforming different class probability density functions into possibility distributions using the Dubois-Prade Pr $\to \pi$ transformation and on using spatial context to evaluate the decision of attributing a new pattern to the different predefined classes and, hence, updating their possibility distributions. Possibility distributions are first obtained using a priori knowledge given in the form of learning areas delimitated by an expert. These areas serve for the estimation of the probability density functions of different endmembers. The resulting probability density functions are then transformed into possibility distributions using Dubois-Prade's probability-possibility transformation [8, 15]. The pixel unmixing is then performed based on the possibilistic similarity between a local possibility distribution estimated around the considered pixel and the obtained possibility distributions representing the predefined endmembers in the analyzed image (Fig. 8.3).

The estimation of these M possibility distributions forms the first step in the proposed approach. The second step consists in determining the pixels' similarity to the M predefined classes of the analyzed image $\mathfrak{I}$ by first estimating the local possibility distribution around the pixel of interest P_0 and secondly, by measuring the similarity Sim_{SI} between this local possibility distribution and each of the M estimated possibility distributions. All the measured similarity values are transformed into percentages as the following:

$$a_k = \frac{Sim_{\mathrm{SI}}(\pi_{C_k}, \pi_{P_0})}{\sum_{m=1}^{M} Sim_{\mathrm{SI}}(\pi_{C_m}, \pi_{P_0})}$$

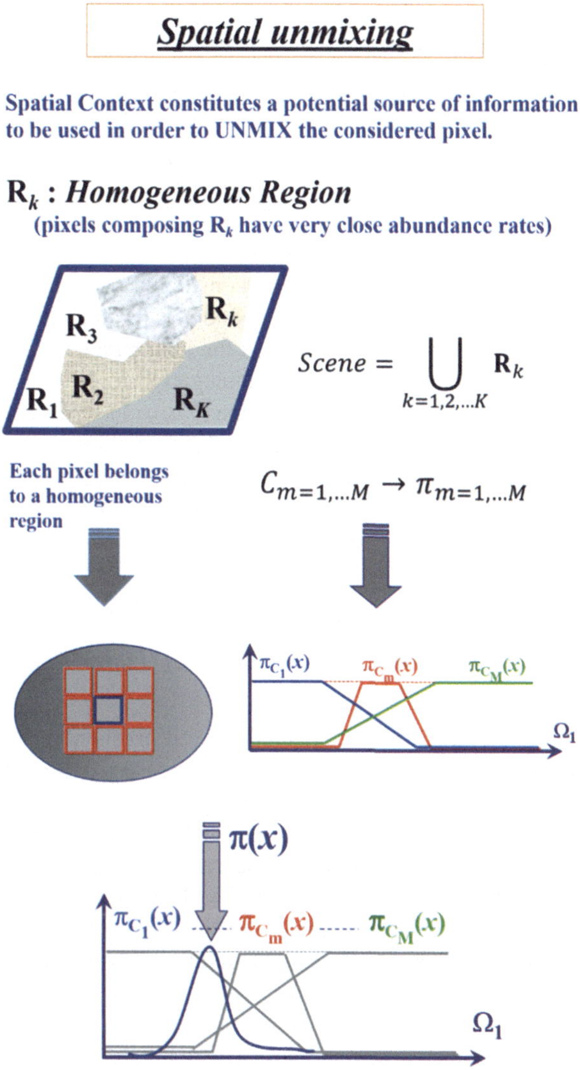

Fig. 8.3 A possibilistic approach of pixel unmixing. (Adapted from Ref. [4])

where a_k is supposed to be the "abundance rate" of the kth predefined thematic class in the considered pixel P_0 and $\sum Sim_{SI}$ serves as a normalizing factor. It is worth noticing that high overlapping case (high discrimination complexity) between the predefined thematic classes can be treated with this approach. In the case of low overlapping (i.e., low discrimination complexity), the "abundance rate" of a predefined thematic class in the considered pixel P_0 is roughly inversely proportional to the distance between the pixel vector and the mean of that class [20].

The simplest and most widely used approach, i.e., the linear mixture model [21], is used in the proposed unmixing approach. This model is based on the assumption that a linear combination exists between the pixel brightness and the M predefined thematic class. The spectral reflectance of a pixel is the sum of the spectral reflectance from the predefined endmembers weighted by their relative "abundance rates":

$$B = \sum_{i=1}^{M} a_i \times B_i$$

where B is the brightness value (i.e., the realization of the random variable measured by the sensor) of the considered pixel P_0, B_i is the brightness value of the ith predefined thematic class, and a_i is its abundance rate in the considered pixel P_0. There are two constraints on the abundance rates that should be satisfied: the abundance rates must all be nonnegative to be meaningful in a physical sense $a_i \geq 0$ [22] and must sum up to one ($\sum a_i = 1$). A classification step is conducted at the end of the proposed approach. This step consists in the process of assigning a class to the considered pixel P_0 by determining the nearest class via the similarity function $\mathrm{Sim}_{\mathrm{SI}}$ used to measure the similarity between this pixel's local possibility distribution and the possibility distributions of each of the M classes.

8.3.5 *Performance on Synthetic Data*

In many applications, collecting mixed pixels and determining their exact abundance rates of the predefined thematic classes is a very difficult task. Therefore, a 550 × 550 pixel synthetic image, given in Fig. 8.4, is generated. This image is composed of 11 sectors. The first and second sectors are assumed to contain two "pure" thematic classes generated by two Gaussian distributions $G(m_1 = 100, \sigma_1 = 15)$ and $G(m_2 = 150, \sigma_2 = 15)$. Pixels of sectors from 3 to 11 are generated as a linear mixture of the first and second pure classes. The abundance rates of class C_1 and class C_2 in these mixed pixels are varying incrementally by 10%. For instance, the abundance rate of class C_1 in the third sector is 10% (resp. abundance rate of class C_2 is 90%), in the fourth sector 20% (resp. abundance rate of class C_2 is 80%), etc. 20 × 20 pixel learning zones positioned by the expert (as being representative areas of the considered thematic classes) are also illustrated on the generated image. The effectiveness evaluation of the proposed approach is studied through two perspectives (next two sections): an estimation of classes' abundance rates in the mixed pixels and the evaluation of the improvement in overall classification accuracy.

8.3.5.1 Estimation of Classes' Abundances in the Mixed Pixels

Using the learning zones, the initial estimation of the class probability density functions is established based on the KDE (Kernel Density Estimation) approach.

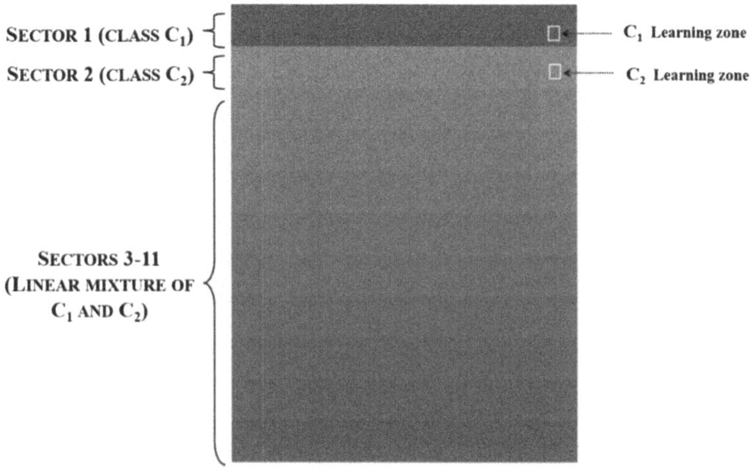

Fig. 8.4 Synthetic image composed of two classes and their learning zones

The application of the Pr $\rightarrow \pi$ Dubois-Prade's transformation allows obtaining the possibility distributions for each class in the analyzed image. A 3×3 pixel window centered on each pixel is considered as the local spatial possibilistic context, and, then, local probability density functions are established based on the KDE approach. The application of the Pr $\rightarrow \pi$ Dubois-Prade's transformation allows obtaining the local possibility distributions.

Abundance rates of the predefined endmembers in each sector, from 3 to 11, can be estimated from the possibilistic similarity values. In each of these sectors, the proposed approach, using the possibilistic similarity measure, $\mathrm{Sim}_{\mathrm{SI}}$, is applied on all its pixels, and their possibilistic similarity values of each endmember are calculated. The results obtained, in terms of abundance maps, are given in Fig. 8.5. A visual inspection of the abundance maps shows a linear variation in the abundance values corresponding to both classes. A quantitative analysis step of abundances maps is conducted to confirm the visual inspection. The mean and the standard deviation of the possibilistic similarity values for each class are given in Table 8.3.

Results show that the abundance rates of the predefined endmembers in the mixed pixels can be estimated with a reasonable accuracy from possibilistic similarity values. This estimation is in conformity with the values used in synthetic image generation. For instance, it can be estimated that the fifth sector contains about 28% of class C_1 and 72% of class C_2, while the used values in synthetic image generation are 30% of class C_1 and 70% of class C_2.

The small values of standard deviation constitute another indication confirming that this estimation is quite consistent to the values used in synthetic image generation.

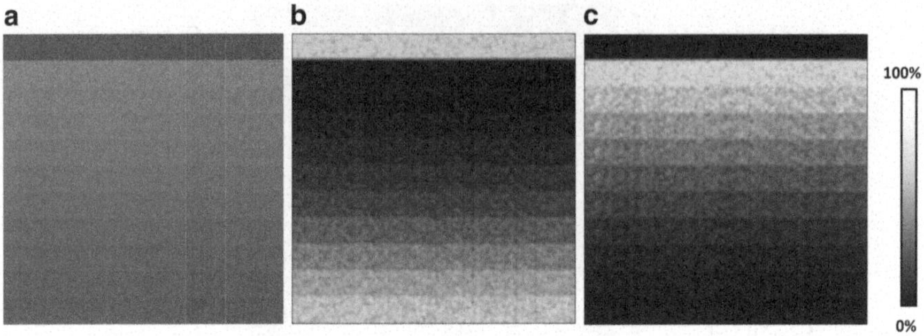

Fig. 8.5 (**a**) Synthetic image composed of two classes, (**b**) abundance map of class C1, and (**c**) abundance map of class C2

Table 8.3 Abundance rates of the predefined endmembers in each sector

	C_1 (10%) C_2 (90%)	C_1 (20%) C_2 (80%)	C_1 (30%) C_2 (70%)	C_1 (40%) C_2 (60%)	C_1 (50%) C_2 (50%)	C_1 (60%) C_2 (40%)	C_1 (70%) C_2 (30%)	C_1 (80%) C_2 (20%)	C_1 (90%) C_2 (10%)
Mean (C_1)	0.14	0.20	0.28	0.39	0.50	0.61	0.72	0.79	0.87
Std (C_1)	0.07	0.08	0.09	0.10	0.10	0.09	0.08	0.08	0.06
Mean (C_2)	0.86	0.80	0.72	0.61	0.50	0.39	0.28	0.21	0.13
Std (C_2)	0.06	0.07	0.09	0.10	0.10	0.10	0.09	0.07	0.06

8.3.5.2 Evaluation of the Improvement in Overall Classification Accuracy

The above synthetic image (Fig. 8.5) is classified using the possibilistic approach and the conventional Bayesian approach, respectively. The classification recognition rate is then calculated in order to compare the classification results of the two approaches.

Results (Table 8.4) show an overall improvement in terms of classification accuracy using the possibilistic approach. This improvement has reached 17% in some cases (e.g., C_1(40%) and C_2(60%)). In addition to this improvement in terms of the classification accuracy, the estimation of the classes' abundances in the mixed pixels enables the assessing of the classification accuracy which, in turns, may contribute to the interpretation of the analyzed scene. For instance, the classification of the third sector is 100% class C_1 with a small deviation of the assignment to its pixels (about 14% of the class C_2), while the classification result of the fourth sector is also about 100% class C_1 but with a bigger deviation of the assignment to its pixels (about 20% of class C_2). It is important to note that this assessment of accuracy cannot be obtained using the conventional pixel-based

Table 8.4 Classification recognition rate of the predefined thematic classes in each sector calculating first by the proposed approach and second by the Bayesian approach

	Recognition rate (%)								
	C_1 (10%) C_2 (90%)	C_1 (20%) C_2 (80%)	C_1 (30%) C_2 (70%)	C_1 (40%) C_2 (60%)	C_1 (50%) C_2 (50%)	C_1 (60%) C_2 (40%)	C_1 (70%) C_2 (30%)	C_1 (80%) C_2 (20%)	C_1 (90%) C_2 (10%)
Our approach (C_1)	0	1	1	11	49	93	99	100	100
Our approach (C_2)	100	99	99	89	51	7	1	0	0
Bayesian approach (C_1)	2	4	12	27	51	28	88	95	99
Bayesian approach (C_2)	98	96	88	73	49	72	12	5	1

images classification systems. We invite the reader to see the experimental results using real medical images that are presented in [4].

8.4 Blind Image Segmentation Using a Possibilistic Approach

This section presents an approach for image segmentation based on possibilistic concepts. The aim of a segmentation process is to distinguish homogeneous regions within an image that belong to different objects. For instance, the segmentation process can be based on finding the maximum homogeneity in gray levels within the identified regions, i.e., the basic information delivered at the output of an imaging sensor. Regions resulting from a segmentation process have to display a uniform behavior with regard to a considered set of features. First, when can we consider that a behavior is uniform with respect to a feature or a set of features? Second, given that objects have uniform boundaries, then we are faced with the boundary location problem. In fact, for marking the region boundaries and for locating uniform ones, the uniformity degree of neighboring pixels shall be considered. Moreover, several challenges related to image segmentation require attention.

One important challenge is related to the selection of the suitable approach to isolate different objects from their background. Another important challenge involves measuring the performance of a segmentation approach and assessing its impact on the global image and scene interpretation.

Image segmentation issues have been approached from a wide variety of perspectives, and different techniques have been developed to perform that task. These techniques are generally grouped into four major categories: (a) clustering-based and histogram thresholding approaches; (b) edge-based approaches; (c) region growing-based approaches; and finally (d) hybrid approaches combining both region growing- and edge-based techniques. The four image segmentation approaches (a, b, c, and d) implement different "understandings" of the human-based image segmentation process. Nevertheless, in all cases, the major encountered difficulty is related to the fact that all image segmentation approaches mainly depend on the "nature" of the spatial knowledge to be used and its spatial diffusion as well as the set of similarity criteria which are strongly based on the physical parameters measured by the imaging sensor in action: consequently, other sources of knowledge are not being integrated.

The image segmentation approach is based upon the use of possibility theory concepts imitating the human reasoning scheme when analyzing an observed image in terms of its constituting homogeneous regions. Based on the observed features of gray level intensity, color, texture, patterns, shapes, etc., a human understands image by first segmenting the whole image into regions/objects. This initial segmentation task is realized by "visually" locating subregions, called "reference seed." Reference seeds are homogenous small regions having semantic visual significance or satisfying some simple similarity criteria: feature-based homogeneity, membership to a known structure, or a thematic class. Then, a second "iterative" step follows. It consists of spreading the semantic homogeneous seeds into their spatial context using a set of similarity criteria. Additional sources of knowledge are then used to enlarge the initial homogenous regions. This process leads to a global segmented image over which other interpretation tasks may be conducted.

Applications presented in Sects. 8.2 and 8.3 above use IRPDL but from different aspects. In [5], the focus is on the spatial knowledge diffusion process that is the keystone of the IRPDL approach. The main reason behind the use of possibility theory is that this theory allows adequate semantic knowledge modeling without huge constraints. Possibilistic concepts offer simple means for modeling human reasoning related to spatial similarity and to contextual knowledge diffusion. The proposed possibilistic region growing image segmentation approach consists of the possibilistic modeling of different semantic classes present in the observed scene. The observed image is then projected into the classes' possibilistic representation space forming, thus, possibilistic class maps. Finally, the region growing process is conducted at the possibility maps level.

8.4.1 Region-Based Approaches in Image Segmentation

An image represents a partial view of an observed scene obtained by interpreting physical measures from sensors. Region growing-based image segmentation methods aim at the identification of the "constituting elements" contained within the image by grouping image pixels into homogeneous regions formed by linked pixels. Each region is assumed to meet a homogeneity criterion according to a common property: its constituting pixels belong to a same semantic entity or a same thematic class. The result from this process is "synthesized" as an image segmentation map formed by the identified homogeneous regions and provides a higher semantic description level of the analyzed image: the label of each detected region, the shape of each region (size, orientation, etc.), the spatial distribution (connections, neighborhoods, etc.) in different regions, etc.

Two main approaches are used for the region growing process. The first one, called *sensor-level information diffusion* approach, is based on grouping pixels to form homogeneous regions at the measured physical parameter level (i.e., the observed gray level or some derived low-level feature). The pixel grouping step consists, first, on the application of an image filtering operation for smoothing the observed image and wiping off noisy pixels. This filtering operation is usually conducted in an iterative manner so that the resulting obtained regions are more and more homogeneous, facilitating thus, in a second step, the pixel grouping operation using some similarity measures. A labeling step is finally conducted in order to attribute a semantic label leading to transform the segmented map into a semantic, or a thematic, map. The second segmentation approach, called *semantic-level information diffusion* approach, is based on the application, first, of a pixel-based classification process leading to a per-pixel classification map. The objective of this per-pixel classification process is to transform the physical measured parameter into semantic labels.

It is worthwhile to notice that both segmentation approaches are based upon the use of some similarity measures for pixel grouping or for pixel classification. Both approaches operate over image gray levels: physical parameters measured by sensors. In all cases, an iterative filtering operation is generally applied either to reduce the noise effects due to the statistical nature of physical parameters or to facilitate the pixel grouping process by increasing the discrimination power of different pixels and improving, thus, their classification quality. The iterative filtering operation aims at exploiting the spatial contextual information at the physical measured parameter level, for the sensor-level information diffusion approach, or at the semantic label level, for the semantic-level information diffusion approach. It represents a sort of spatial knowledge diffusion both at the sensor level and at the semantic level.

The iterative filtering operation is frequently used in most existing image segmentation methods. Some are listed in [5] such as the iterative estimation of segmentation parameters; active contour-based segmentation methods; Markovian segmentation methods; iterative cutting (top-down cutting) image segmentation methods; region growing methods (bottom-up); and hybrid methods of image segmentation. In all these methods, different iterations operate over either the sensor level or the semantic level. Some of the iterative filtering methods, used in this study, are briefly described in [5]: *mean filtering diffusion method, Nagao filtering diffusion method, altered Nagao filtering diffusion method, anisotropic diffusion method*, and *Gabor diffusion method.*

8.4.2 Possibilistic Region Growing Approach

The only use of the pixel's gray level for image informative content description appears to be insufficient for the segmentation process. Some iterative methods based on the gray level use tend to solve this problem by exploiting the pixels spatial context. However, these methods often propagate information of imprecise nature, which increases the ambiguity when classifying pixels. For example, the use of Markov random fields to overcome this problem leads to exploit the pixel's occurrence probability in a given class. The real deal in the use of these methods is to establish the probability law in real time during an iterative segmentation process. Other methods employ fuzzy sets theory often associated to decision rules derived from the possibility theory. Such methods produce segmented regions more homogeneous than those obtained by the methods previously mentioned due to a better exploitation of spatial knowledge contained in the image. However, most of these methods are similar to the statistical segmentation since the pixel's belonging degree to different classes is never re-evaluated, especially in terms of its spatial meaning.

A current trend is to develop "hybrid" methods to exploit the two levels of image processing: the low level (i.e., numerical measures from physical sensors) and the high level (i.e., the semantic content from processed information). An important challenge associated to that exploitation is a lack of a general formalism for integrating numerical and semantic levels. It is then important to develop a method inspired from human reasoning to allow such integration. The possibility theory's formalism offers flexibility in the representation of information and its different forms of imperfection within the same framework. Moreover, this theory allows the development of computer-based reasoning mechanisms very similar to those of a human expert.

The possibilistic approach for image segmentation exploits to a larger extent the spatial information contained within an image. The possibilistic approach, strongly inspired from region growing methods, consists, first, in representing available diverse knowledge sources in the possibility theory formalism. The second step consists of applying a possibilistic knowledge diffusion process based on contextual information in a similar manner that of a region growing operation. The proposed segmentation approach is illustrated in Fig. 8.6 showing the IRPDL approach. This architecture consists of four constituting subsystems: (1) knowledge projection, (2) decision-making, (3) possibilistic diffusion, and finally (4) evaluation and updating.

To detail these subsystems, the following assumptions are made: the input images of the segmentation system are $R \times Q$ sized; each pixel at the position (i,j), , $0 \leq i \leq R - 1; 0 \leq j \leq Q - 1$ is observed at L gray levels, $(i,j) \in \{0, 1, \ldots, L - 1\}$; analyzed images include M thematic classes of universe $\Omega = \{C_1, C_2, \ldots, C_M\}$; we assume operating under the closed world assumption (i.e., the M thematic classes are the only classes found in the considered images).

8.4.2.1 The Knowledge Projection Subsystem

The first information processing stage of the proposed segmentation approach (Fig. 8.6) consists of the application of a "knowledge projection" of the image sensor-level information into the possibilistic knowledge representation level. The possibilistic knowledge level creates an "intermediate" level between the sensor

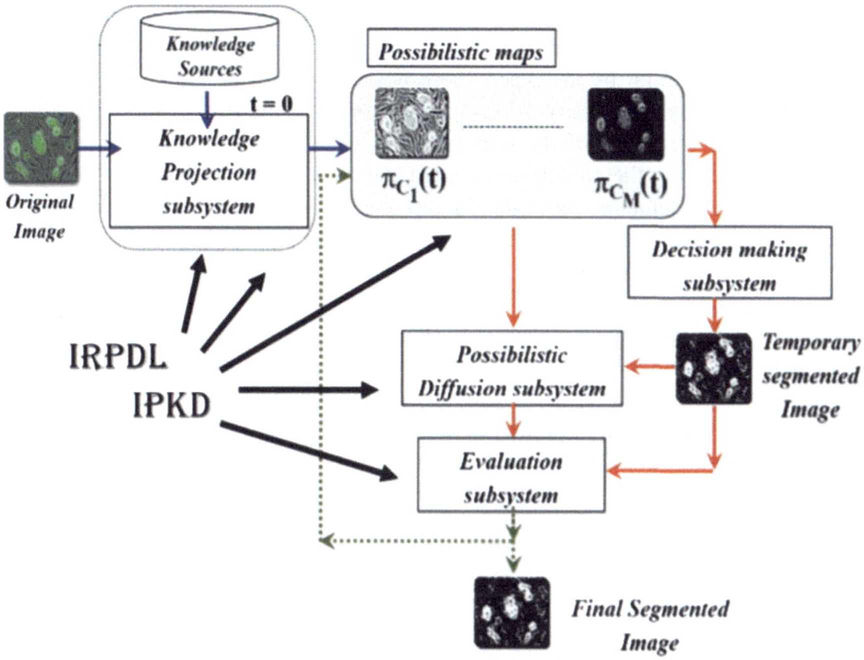

Fig. 8.6 The possibilistic image segmentation approach. (Adapted from Ref. [1, 5])—also named as Iterative Possibilistic Knowledge Diffusion (IPKD) [5] and Iterative Refinement of Possibility Distributions by Learning (IRPDL) [1]

and the semantic knowledge level. This projection is enriched by the expert knowledge injected through the possibility distributions of the thematic classes. This process is achieved by estimating, for each pixel at the position (i,j), an "initial" possibility distribution defined on the universe $\Omega = \{C_1, C_2, \ldots, C_M\}$. It represents our degree of belief that the considered pixel belongs to different thematic classes, formalized as:

$$\pi_{(t=0),(i,j)} \,\vdots\, \Omega \to [0,1], \;\; C_m \to \pi_{(t=0),(i,j)}(C_m)$$

Let $y \in \Theta$ denotes the considered primitive at the pixel's level and where Θ denotes the primitive's universe. Using a priori sources of knowledge, let $\{\pi_m\}$, $m = 1, 2, \ldots, M$ be a set of M predefined classes' possibility distributions defined on Θ where $\pi_m(y)$ represents the possibility degree for a pixel (with the primitive value y) belonging to the thematic class C_m. The initial estimation of the possibility distribution $\pi_{(t\,=\,0),\,(i,j)}$ is thus achieved as follows:

$$\pi_{(t=0),(i,j)}(C_m) = \pi_m\left(y_{(i,j)}\right) \quad m = 1, 2, \ldots, M$$

where $y_{(i,j)}$ represents the primitive value of the pixel at the position (i,j).

Two approaches are generally considered for the definition of the M predefined thematic class possibility distributions $\{\pi_m\}_{m = 1, 2, \ldots, M}$. The first approach consists of empirically defining by an expert these possibility distributions using existing standard set of possibility distributions (i.e., Gaussian, triangular, S-shape, etc.). The second approach consists of the expert's identification of relevant class training zones (characterizing the predefined classes) and then using some existing possibility distribution estimation methods. In this paper, the role of the expert in the image segmentation process is limited to the definition of these possibility distributions representing the M thematic classes' according to one of the two mentioned methods above.

Several approaches have appeared in literature to infer possibility distributions from other representation models, such as Dubois-Prade's transformations [15]; Klir's transformation [23]; and fuzzy membership function, owing to Zadeh's postulate [16]. Once these possibility distributions are determined, the preprocessing subsystem generates M possibilistic images, called initial possibilistic maps, $\pi_{(t = 0)}(C_m)$, $m = 1, 2, \ldots, M$ defined as follows:

$$\pi_{(t=0)}(C_m) = \left[\pi_{(t=0),(i,j)}(C_m)\right]_{0\le i\le R-1;0\le j\le Q-1 \text{ and } m=1,2,\ldots,M}$$

The sensor-related information contained within the input image is spread into M new knowledge sources that are the M possibilistic maps. Each possibilistic map realizes a kind of "knowledge focus" on one and only one thematic class.

8.4.2.2 Decision-Making Subsystem

The decision-making subsystem plays the role of the local per-pixel decision-making expert process. For a given pixel, and based on analyzing the local possibilistic values in different possibilistic maps, this subsystem aims to deciding if a pixel belongs, or not yet, with a high degree of certainty, to one of the different thematic classes. The expected output of this subsystem is, thus, a temporary classification/rejection map called temporary segmentation map indicating the class label of each pixel.

Several strategies can be adopted to evaluate the appropriateness of making a classification decision for each analyzed pixel. In this chapter, we have adopted a realistic decision-making rule expressed as follows:

If	$\pi_{t,m_0}\left(x_{(i,j)}\right) \geq 0,95$, for some m_0 .And. $\pi_{t,\ m}(x_{(i,j)}) \leq 0,\ 05$, for all $m \neq m_0$
Then	Pixel (i,j) is labeled as being a *reference seed* of the class m_0
Else	Pixel (i,j) is labeled as rejected

This decision strategy consists of considering that labeling a pixel as being a *reference seed* of a class is an irreversible decision. Therefore, when a pixel (i,j) is considered as a reference germ, i.e., a seed, for some thematic class at a given iteration t, then its per-pixel possibility value $\pi_{t,\,m}(x_{(i,j)})$, $m = 1, \ldots, M$ is appealed to remain unchanged through following iterations $t'(t' > t)$.

8.4.2.3 Possibilistic Diffusion Subsystem

The possibilistic diffusion subsystem constitutes the heart of the IRPDL proposed segmentation approach. This subsystem offers a sort of analogy with human perceptual image interpretation. Indeed, the human interpretation of an image content into homogeneous labeled regions consists, first, on "identifying reference areas" of each thematic class and, in a second step, enriching these reference areas in an incremental way by analyzing their spatial neighborhood in order to spatially "extend" (or, "enlarge") these areas.

The sensor-level information diffusion approach focuses on the per-pixel enrichment process at the radiometric level (i.e., the measured gray level). In fact, the sensor-level approach realizes, first, an image smoothing operation in order to transform the analyzed image into "homogeneous regions." Region labeling is then realized in a second step. The semantic-level information diffusion approach for image segmentation reproduces the human interpretation process of the image content by considering reference areas as being the "seeds" for the scene interpretation. The per-pixel enrichment process (i.e., the region growing process) is realized by extending the knowledge of these seeds into their neighboring pixels using a semantic-based similarity measure (for instance, having the same label) or a sensor-based similarity measure (by analyzing the similarity between adjacent pixels at the sensor output physical data measure level).

The proposed knowledge diffusion subsystem consists of considering the temporary segmented image as encapsulating two types of information. The first information, called the seed zone information, concerns the labeled pixels that can be considered for the definition of the "initial" reference areas. The second information concerns the unlabeled or rejected pixels for which the available knowledge, at time "t," does not allow a classification decision. The main focus of the approach consists of the use of contextual possibilistic knowledge in order to update the per-pixel possibilistic knowledge of all pixels from the ambiguous zones. Therefore, this knowledge updating intends to reduce the ambiguity for the classification decision-making process. It intends to perform the region growing process at the possibilistic knowledge representation level rather than at the sensor-based grayscale or at the semantic-based decision knowledge levels.

The proposed approach exploits the temporary segmented image as a "constraint" to restrict the possibilistic knowledge updating process. The process is done in an iterative manner by updating the temporary segmented image and possibilistic maps. The role of sensor data is, thus, limited to the "initial" determination of the per-pixel possibilistic knowledge, whereas, the knowledge updating is realized at the possibilistic knowledge level using the spatial context of different unlabeled pixels, and not at the sensor or the semantic knowledge levels. Different image filtering techniques are used to achieve this knowledge diffusion on different possibilistic maps.

8.4.2.4 Evaluation Subsystem

The evaluation subsystem has the role of assessing if the process requires further iterations. The expected output of the evaluation subsystem is a segmented/classified image in which each pixel is associated with a class label or a rejection label. The set of rejection pixels must be analyzed and explained (open world assumption, pixels including classes' mixtures, invalid initial knowledge, etc.). As shown in Fig. 8.6, two information sources are assumed to feed the evaluation subsystem: the temporary segmented image and the set of different possibilistic maps. The simplest criterion consists of a priori fixing the number of possibilistic knowledge diffusion iterations. Another interesting criterion consists of evaluating the "convergence" of the possibilistic maps through successive iterations. This convergence can be appreciated by computing the mean squared error (MSE) between successive possibilistic maps obtained for each thematic class.

8.4.3 Empirical Results from Synthetic Images

This section presents a quantitative and qualitative evaluation of the Iterative Possibilistic Knowledge Diffusion (IPKD) approach using a synthetic image. The synthetic test image (illustrated in Fig. 8.7) is composed of a basic miniature image containing four circular disks (class C_1) and the image background (class C_2). The use of several sizes constitutes a first indicator allowing to appreciate the capacity of

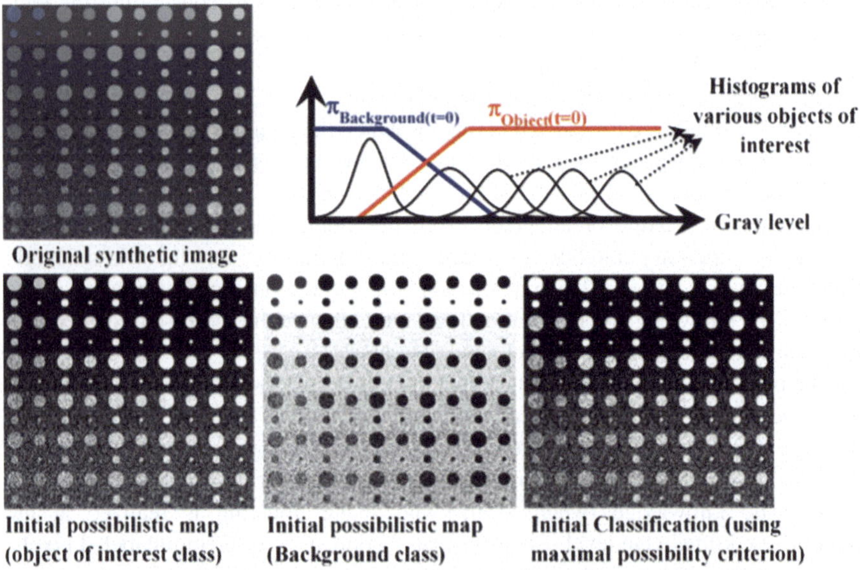

Fig. 8.7 Initial estimation of possibilistic maps. (Source: Ref. [5])

the proposed approach to "locate/detect" objects of interest at various sizes. Pixels from C_1 and C_2 are generated as two Gaussian distributions $\mathcal{G}(m_{object}, \sigma_{object})$ and $\mathcal{G}(m_{Background}, \sigma_{Background})$. Moreover, on the horizontal axis, the basic miniature image is duplicated by increasing the contrast of objects relative to the image background. The contrast is defined as being the ratio between the difference and the sum of the gray level averages, i.e., $(m_{object} - m_{Background}/m_{object} + m_{Background})$.

Five contrast levels are represented in the synthetic image. These five contrast levels take values of 0.3, 0.43, 0.5, 0.56, and 0.6. Six levels of additive white Gaussian noise are progressively introduced into the image corresponding to six increasing values of the noise standard deviation. The constructed synthetic image is very representative of contexts encountered in different applications for object detection and classification. The use of this image allows the analysis under three important aspects: object size, contrast level, and noise level.

The first phase of the experimentation is to use this synthetic image. A first step is to estimate the initial possibilistic maps for both classes C_1 and C_2. A second step is to perform the various iterative strategies for the possibilistic knowledge diffusion and, finally, make a comparison with classical gray level knowledge diffusion.

8.4.3.1 Possibilistic Map Initial Estimation

The knowledge formulated by an expert concerning different semantic classes, or objects, is related to their "visual appreciation" in the scene framework. This visual appreciation is generally expressed using fuzzy sets defined over the universe of the observed physical parameter, i.e., gray levels, like bright, very bright, dark, etc. Therefore, the estimation of the initial possibilistic maps consists of estimating the membership functions of the fuzzy sets expressed by the expert that is considered as possibility distributions of the classes. The trapezoidal model is used [5] for the estimation of different membership functions.

The application of the simulated annealing algorithm is then conducted in order to automatically determine the parameters defining these models. The fuzzy entropy is considered as being the objective function to be maximized by the simulated annealing algorithm. The application of this approach on the two classes (C_1, C_2), synthetic image, is illustrated in Fig. 8.7.

Different histograms of objects of interest corresponding to the five considered contrast levels, the background histogram as well as the two classes (C_1, C_2) initial possibility distributions, are plotted on this figure. Since we are using a synthetic image, the thematic class of each pixel is a priori known, so the classification error rate for the various levels of contrast and noise can be easily evaluated.

A per-pixel misclassification rate using the initial possibility distributions has been achieved for 30 basic miniature images corresponding to the 5 contrast and the 6 noise levels. For a high noise level and a low contrast level, the per-pixel misclassification rate obtained with the classification according to the per-pixel maximum possibility classification criterion reaches values of the order of 30%.

Finally, the global mean per-pixel misclassification rate, obtained with this initial classification, is of 13%. The poor quality resulting per-pixel classification rates are quite natural due to the weak knowledge used for the classification process (i.e., the basic initial possibility distributions).

8.4.3.2 Evaluation of Different Knowledge Diffusion Techniques

The second evaluation step consists of considering the initial possibilistic maps (obtained in the first step) and, then, achieving the possibilistic knowledge diffusion applying the various iterative approaches to give the corresponding possibilistic maps and per-pixel error rates. The various diffusion approaches that are considered are Gabor filtering diffusion; anisotropic filtering diffusion; mean filtering diffusion; Nagao filtering diffusion; and Nagao altered filtering diffusion. Detailed results can be found in [5].

All the considered approaches converge in less than four diffusion iterations with respect to error rates. All, they allow the detection, even partially, of small-sized objects for all contrast and noise levels. The possibilistic knowledge diffusion using Nagao filter produces the lowest error rates and this, whatever the considered iteration is and practically for all contrast and most reasonable noise levels. In accordance with these results and the fact that Nagao filter allows object edge preserving, this filter can be considered as the most promising possibilistic knowledge diffusion approach.

8.4.3.3 Efficiency Assessment of Possibilistic Knowledge Diffusion

An important question raised here is what is the added value of making knowledge diffusion at the possibilistic maps level rather than at the sensor-data physical level like in classical diffusion methods? To tackle this important question, an experimental scheme is tested on the proposed synthetic image. On one setup, the realization of the sensor data knowledge projection into possibilistic maps followed by possibilistic knowledge diffusion using Nagao filter is achieved.

The final thematic map is obtained by the maximum possibility per-pixel classification process. A second setup is used to create homogeneous regions, with the iterative application of Nagao filter on the sensor data (i.e., the gray level observed image). This iterative process is stopped when the resulting image reaches convergence. Possibilistic knowledge projection is then applied to be followed by the maximum possibility per-pixel classification process leading to the final thematic map. Results obtained in [5] show a potential to get better performance when achieving the knowledge diffusion process in the possibilistic space rather than the sensor data representation space.

Experimental results using synthetic images as well as mammographic images from MIAS (Mammographic Image Analysis Society) database are performed in order to assess the efficiency of the proposed segmentation method according to the visual criterion as well as some quantitative criteria. IPKD's performance (in terms of recognition rate, 94.37%, and global predictive rate, 92.18%) is compared with three relevant reference methods: level-set, fuzzy *C*-mean, and region growing methods.

The IPKD approach outperforms the other three methods, respectively, at the recognition rates of 89.77%, 84.43%, and 88.11% and at the global predictive rates of 87.86%, 89.72%, and 84.04%. Noise-sensitivity experiments have been conducted on synthetic as well as on real images. The proposed IPKD approach outperforms the three reference methods and, in addition, exhibits a desired stability behavior.

References

1. B. Alsahwa, B. Solaiman, S. Almouahed, E. Bosse, D. Gueriot, Iterative refinement of possibility distributions by learning for pixel-based classification. IEEE Trans. Image Process. **25**, 3533–3545 (2016)
2. S.S. Nath, G. Mishra, J. Kar, S. Chakraborty, N. Dey, A survey of image classification methods and techniques, in *2014 International Conference on Control, Instrumentation, Communication and Computational Technologies (ICCICCT)*, (2014), pp. 554–557
3. D. Lu, Q. Weng, A survey of image classification methods and techniques for improving classification performance. Int. J. Remote Sens. **28**, 823–870 (2007)
4. B. Alsahwa, B. Solaiman, É. Bossé, S. Almouahed, D. Gueriot, A method of spatial unmixing based on possibilistic similarity in soft pattern classification. Fuzzy Inf. Eng. **8**, 295–314 (2016)
5. I.K. Kallel, S. Almouahed, B. Solaiman, É. Bossé, An iterative possibilistic knowledge diffusion approach for blind medical image segmentation. Pattern Recogn. **78**, 182–197 (2018)
6. A. Vailaya, A. Jain, Incremental learning for bayesian classification of images, in *Proceedings 1999 International Conference on Image Processing (Cat. 99CH36348)*, (1999), pp. 585–589
7. L. Roux, J. Desachy, Satellite image classification based on multi-source information-fusion with possibility theory, in *Proceedings of IGARSS'94–1994 IEEE International Geoscience and Remote Sensing Symposium*, (1994), pp. 885–887
8. D. Dubois, L. Foulloy, G. Mauris, H. Prade, Probability-possibility transformations, triangular fuzzy sets, and probabilistic inequalities. Reliab. Comput. **10**, 273–297 (2004)
9. M.S. Mouchaweh, Semi-supervised classification method for dynamic applications. Fuzzy Sets Syst. **161**, 544–563 (2010)
10. F. Wang, Fuzzy supervised classification of remote sensing images. IEEE Trans. Geosci. Remote Sens. **28**, 194–201 (1990)
11. T. Nakashima, G. Schaefer, Y. Yokota, H. Ishibuchi, A weighted fuzzy classifier and its application to image processing tasks. Fuzzy Sets Syst. **158**, 284–294 (2007)
12. M. Nagao, T. Matsuyama, Edge preserving smoothing. Comput. Graphics Image Process **9**, 394–407 (1979)
13. A. Mencattini, G. Rabottino, M. Salmeri, R. Lojacono, E. Colini, Breast mass segmentation in mammographic images by an effective region growing algorithm, in *International Conference on Advanced Concepts for Intelligent Vision Systems*, (2008), pp. 948–957
14. O. Demirkaya, M.H. Asyali, P.K. Sahoo, *Image Processing with MATLAB: Applications in Medicine and Biology* (CRC Press, Boca Raton, 2008)
15. D. Dubois, H. Prade, S. Sandri, On possibility/probability transformations, in *Fuzzy logic*, (Springer, Berlin, 1993), pp. 103–112
16. L. Zadeh, Fuzzy sets as the basis for a theory of possibility. Fuzzy Sets Syst. **1**, 3–28 (1978)

17. G. Bisson, La similarité: une notion symbolique/numérique. in, *Apprentissage symbolique-numérique*, ed. by Moulet, Brito, Cepadues edn. (2000)
18. M. Higashi, G.J. Klir, On the notion of distance representing information closeness: Possibility and probability distributions. Int. J. Gen. Syst. **9**, 103–115 (1983)
19. I. Jenhani, N.B. Amor, Z. Elouedi, S. Benferhat, K. Mellouli, Information affinity: A new similarity measure for possibilistic uncertain information, in *Symbolic and Quantitative Approaches to Reasoning with Uncertainty*, (Springer, Berlin, 2007), pp. 840–852
20. F. Wang, Fuzzy supervised classification of remote sensing images. Geosci. Remote Sens. IEEE Trans. On **28**, 194–201 (1990)
21. J.B. Adams, M.O. Smith, P.E. Johnson, Spectral mixture modeling: A new analysis of rock and soil types at the Viking Lander 1 site. J. Geophys. Res.: Solid Earth (1978–2012) **91**, 8098–8112 (1986)
22. N. Keshava, A survey of spectral unmixing algorithms. Lincoln Lab. J. **14**, 55–78 (2003)
23. G.J. Klir†, B. Parviz, Probability-possibility transformations: A comparison. Int. J. Gen. Syst. **21**, 291–310 (1992)

Chapter 9
The Use of Possibility Theory in the Design of Information Fusion Systems

9.1 Introduction

Information overload and complexity are core problems to most organizations of today. The advances in networking capabilities have created the conditions of complexity by enabling richer, real-time interactions between and among individuals, objects, systems, and organizations. Fusion of Information and Analytics Technologies (FIAT) are key enablers for the design of current and future decision support systems to support prognosis, diagnosis, and prescriptive tasks in such complex environments. Hundreds of methods and technologies exist, and several books have been dedicated to either analytics or information fusion so far. This chapter is a sort of conclusion for this book on the use of possibility theory in the design of information fusion systems. It presents the overall picture of FIAT-based design in which possibility theory can be of practical use.

9.2 The General Context for the Design of a FIAT-Based System

All organizations have a tendency of generating more and more information that obviously challenges decision-makers and analysts with a deluge of potentially valuable data. The data deluge is a result from a multitude of advanced sensors and sources capable of generating a diversity and a volume of data never envisioned before. Decision-makers and analysts cannot cope with this flow of data, without any impact on decisions quality and on actions efficiency; so, there is a need of FIAT-based systems to support decisions and actions. On the other hand, decision-makers can foresee value in this data deluge which can translate to technological opportunities, for instance, the Internet of Things [1] and the Big Data [2]. More and more,

B. Solaiman, É. Bossé, *Possibility Theory for the Design of Information Fusion Systems*, Information Fusion and Data Science,
https://doi.org/10.1007/978-3-030-32853-5_9

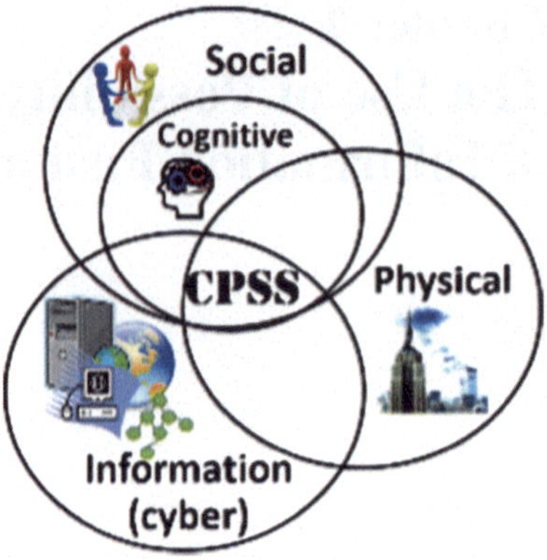

Fig. 9.1 Cyber-Physical and Social Systems (CPSS). (Adapted from [5])

IoT and Big Data are perceived as two sides of the same coin where Big Data would be a subset of IoT. Big Data is evidently contextual to Cyber-Physical and Social Systems (CPSS) [3, 4]. CPSS emerge from the interrelation of social, cognitive, information/cyber, and physical worlds as pictured in Fig. 9.1. Social and cognitive dimensions interface with the physical world through the cyber world. An ever-increasing rich literature (books and journals) now exists both on Big Data and IoT.

CPSS demand multidisciplinary contributions from human and social sciences, physics, engineering, and computer sciences to meet system overall challenges such as 24/7 availability, 100% connectivity, predictability and repeatability, real-time, etc. The advances in Information and Communications Technologies (ICT) in particular smart ICT, to which FIAT belongs, although providing a lot of benefits to improve dependability, efficiency, and trustworthiness in systems, have also increased tremendously the networking capabilities so creating the conditions of complexity by enabling richer, real-time interactions between and among entities that compose CPSS. As a result, events that may once have had isolated consequences can now generate a cascade of consequences that affect badly system dependability and trustworthiness. The question is, thus: How to assemble a set of FIAT that would support decision-makers in such complex systems?

9.2.1 What Is FIAT?

At this point, the question is what is FIAT? The answer is an assemblage of techniques and methods in order to analyze (analytics) and to synthesize (fusion) information from multiple sources to support a decision cycle (e.g., OODA [11]). In Fig. 9.2 an assemblage of FIAT is proposed around three main categories to be considered in a holistic fashion for the design of FIAT-based support systems [5]: (1) multi-agent systems (MAS) theories to formalize the distributed system aspect and the notion of autonomy; (2) generalized information theory (GIT) for

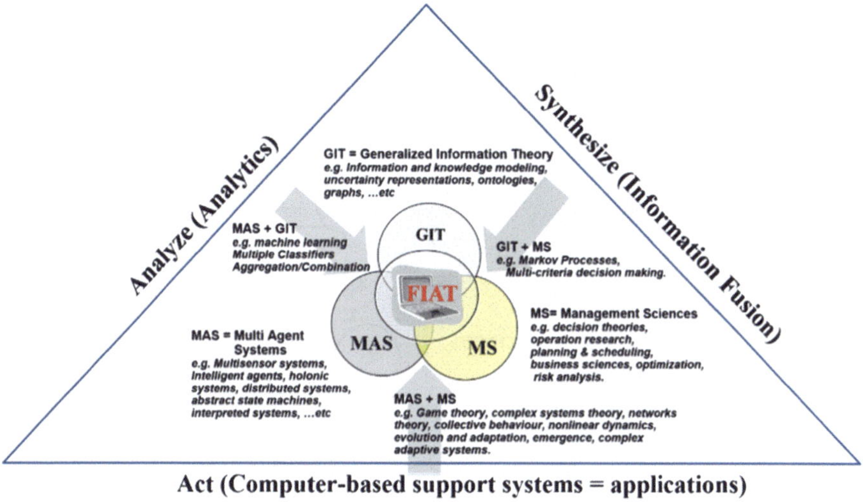

Fig. 9.2 Main categories of FIAT. (Adapted from [5])

knowledge, information, and uncertainty representation; and (3) decision theories (represented by management sciences (MS) in Fig. 9.2) in order to explicitly account for actuation (e.g., decisions-actions and their impact). We assume here that MS also includes dynamic decision-making methods and techniques. The challenge is to assemble an appropriate set of techniques and methods that will support measuring, organizing, reasoning, understanding, deciding, and acting about/upon situations in complex environments such as CPSS in conditions of data overload and complexity.

9.2.2 *Situation Awareness*

Situation awareness (SAW) [6, 7] is a concept around dynamic human decision-making (DM) in both military and civilian complex environments. Situation analysis (SA) is defined as the process that sustains a state of situation awareness for the decision-maker(s). The SA process is the provision of decision quality information to the decision-maker, thereby enabling timely situation awareness. A recent state of the art on this topic is provided in chapter 2 of [7]. Endsley [8] provides a theoretical model of SAW based on its role in dynamic human decision-making. SAW is defined along three main processes: "

1. The perception of the elements in the environment, within a volume of time and space;
2. The comprehension of their meaning; and
3. The projection of their status in the future."

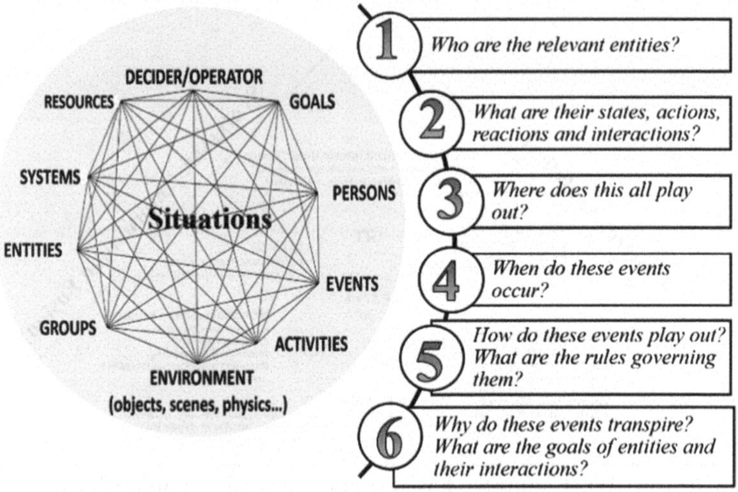

Fig. 9.3 What is a situation? (Adapted from [11])

Llinas [9] raises issues involving interdependencies among the situation analysis processes (e.g., analytics, fusion, sense-making) and decision-making processes for complex decision-making environments. He advocates that a critical examination of inter-process (i.e., situation analysis and decision-making) interdependencies is needed to design decision support systems (DSS) for optimum performance. Therefore, Llina recommends an integrated, multidisciplinary approach (cognitive sciences, human factors, decision sciences, and computer sciences); otherwise DSS designs will remain disconnected and suboptimal. The interconnected processes identified by Llinas [9] are (1) automated FIAT-based situation analysis process; (2) sense-making and information foraging; and (3) decision-making, a semi-automated process that operates in an analytic, an intuitive, or a "hybrid/mixed" decision-making mode to support actuation onto the real-world situation.

Roy [10] defines situation analysis (SA) as "a process, the examination of a situation, its elements, and their relations, to provide and maintain a product, i.e., a state of situation awareness, for the decision maker" and a situation as "A specific combination of circumstances, i.e., conditions, facts, or states of affairs, at a certain moment." The SA process is concerned with understanding the world. The situation can be defined in terms of events, entities, scenes, systems, people, etc. and their mutual interactions as pictured in Fig. 9.3. Entities and events are quite important elements of concern for SA.

A FIAT-based support system can provide answers, in whole or in part, to basic epistemic questions such as "What are the objects of interest? Where are they? How are they moving? Where have they been? Where might they be going?" Objects could refer to either physical objects, such as vehicles, or symbolic objects such as terrorist events, system faults, or business plans.

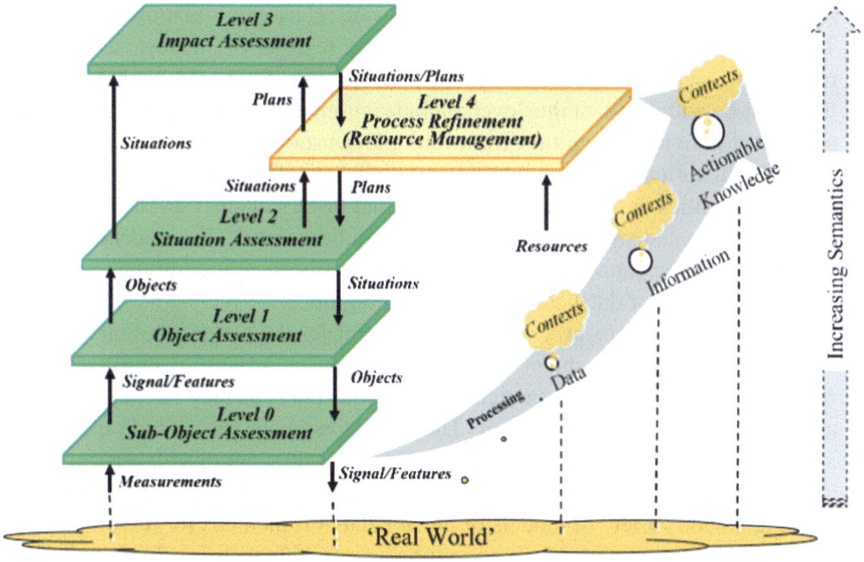

Fig. 9.4 The JDL data information fusion model. (Source: Slightly modified from [5])

9.2.3 Analytics and Information Fusion

9.2.3.1 Information Fusion

The data fusion model maintained by the Joint Directors of Laboratories' Data and Information Fusion Group (JDL DIFG) [12] is the most widely used approach for categorizing data fusion-related functions. The JDL distinction, among fusion "levels" in Fig. 9.4, provides a valuable way of differentiating between data fusion processes that relate to the refinement of "objects," "situations," "threats," and "processes." The definitions are as follows:

Level 0—Sub-Object Data Assessment: Estimation and prediction of signal/object observable states on the basis of pixel/signal level data association and characterization. Level 0 assignment involves hypothesizing the presence of a signal (i.e., detection) and estimating its state.

Level 1—Object Assessment: Level 1 data fusion combines data from single or multiple sensors and sources to provide the best estimate of objects and events in the battlespace in terms of their position, kinematics (e.g., tracks), identity, or identification features. Level 1 involves tracking and identification that includes reliable location, tracking, combat ID, and targeting information. Such information may be of high quality particularly when it leverages multiple sensors to provide robustness to spoofing or deception, reliability in case of sensor malfunctions, and extended space time coverage due to the diversity of observations.

Level 2—Situation Assessment: Level 2 data fusion focuses on situation assessment. This requires object/entity recognition in the regions of interest, as well as recognizing activities of these objects, and inferring their relationships. Issues that must be addressed, at this level, include automated target/object recognition; automated activity recognition from multiple sensors and reports collected and stored for historical analyses; inferring relationships of objects in the scene based on their identities or coordinated behaviors and historical analyses; the capability for the automated system to estimate certainties about object identities and activities; and the capability to request human assistance or additional information from sensors or databases to resolve ambiguities.

Level 3—Impact Assessment: Level 3 fusion estimates the "impact" of an assessed situation, i.e., inferring intent of objects/entities, or groups of objects, in the regions of interest and/or the outcome of various plans as they interact with one another and with the environment. The impact estimate can include likelihood and cost/utility measures associated with potential outcomes of a player's planned actions. Issues that need to be addressed include methods for constructing and learning a wide variety of models of threat behavior; methods for reasoning with uncertain and incomplete information for assessing threats from object activities; and methods for efficient data mining of databases.

Level 4—Process Refinement: Adaptive data acquisition and processing to support mission objectives. Level 4 processing involves planning and control, not estimation. This level, also, involves assigning resources to tasks.

In most defense applications, data fusion processing tends to be hierarchical in nature due to the inherent hierarchies built into defense organizations and operations. As a result, the fusion process also progresses through a hierarchical series of inferences at varying levels of abstraction (data-information-knowledge as shown in Fig. 9.4). The exploitation of contexts in the inference processes targets an increase in semantics that is only obtained by "what one can do with the information?", i.e., "actionable" knowledge. There is an important technical literature available that presents more definitions, explains concepts in detail, and develops mathematical techniques and models related to Fig. 9.4.

9.2.3.2 Analytics

Eckerson [13] defines analytics as "everything involved in turning data into insights into action." This is a quite broad definition that could include "data and information fusion" but does not help understanding analytics from an applied point of view. Unlike information fusion community, the analytics community did not benefit from a well-structured organization like JDL to fix the terminology in order to ease communication among communities. The definition of analytics associated to Big Data becomes more and more confusing with various vendors, consultants, and trade publications defining and offering new technologies.

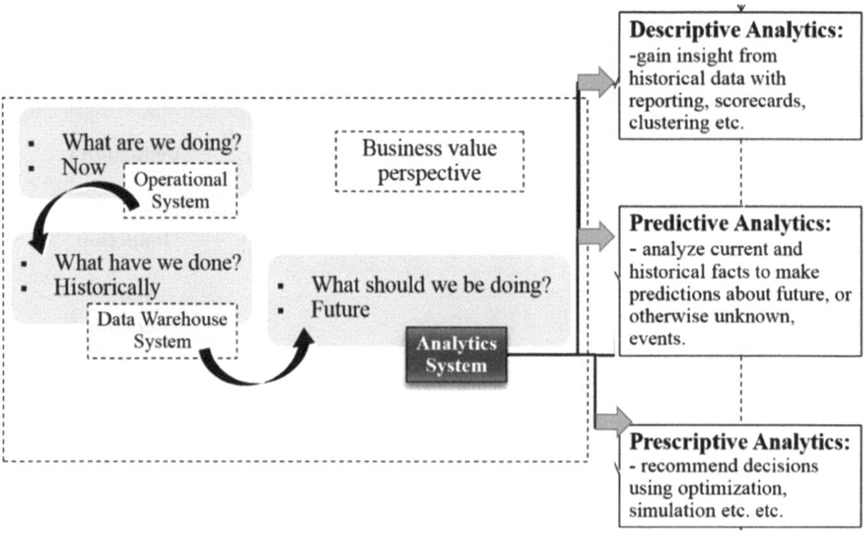

Fig. 9.5 Sheikh's proposal definition of analytics. (Source: Adapted from [5])

Analytics is not new and originates from business intelligence. It has been rejuvenated with Big Data. Figure 9.5 illustrates Sheikh's proposal [14] to define analytics based upon a business and a technical implementation perspective.

The business value perspective looks at data in motion as it is generated through normal conduct of business. For this data, there are three variations of value: the present, the past, and the future, in the exact order as represented on Fig. 9.5. When data is created, referenced, modified, and deleted during the course of normal business activities, it lives in an operational system. The operational system at any given time can tell us what we are doing now. The data from day-to-day operations (e.g., selling merchandise, reviewing applications, etc.) is accumulated for record keeping and starts to build history in a data warehouse. Reporting can then be done to help understanding how a business did (e.g., total sales) in the last month, quarter, or year. These analysis reports provide managers the tools to understand the performance of their departments. This leads into the question that "analytics" should help to answer: "What should we be doing within departments and business units to improve business performance?". Any tools, technologies, or systems that help with this question can qualify to be in the analytics space.

The technical implementation perspective described in Sheikh is also the tangent adopted by Das [15] in his recent book on computational business analytics. Das' book describes the characteristics of analytics in terms of the techniques used to implement analytics solutions. In the analytics literature, three general types of analytics make consensus even though terminology may differ. This is reflected on Fig. 9.5: descriptive, predictive, and prescriptive. A lot

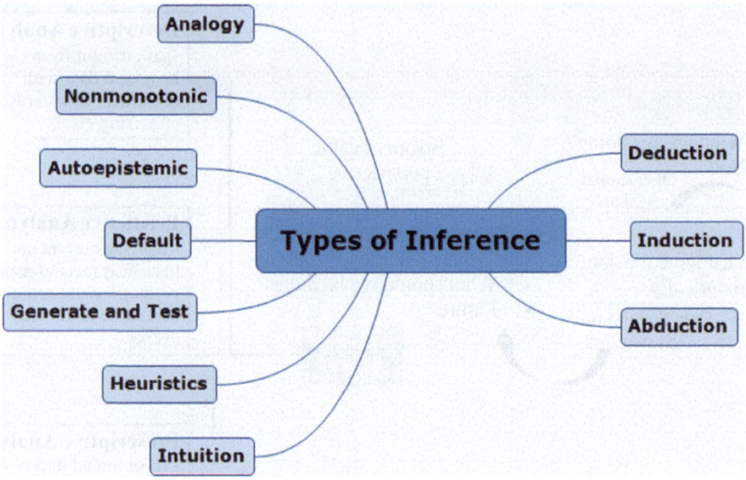

Fig. 9.6 Different types of inference

of good reference books are available to describe these techniques particularly under data mining and machine learning fields.

Das [15] discusses a model-based approach to analytics that contributes to define and understand analytics. He describes computational models as a combination of symbolic, sub-symbolic, and numerical representations of a problem. This is obtained through a series of inferences (e.g., inference types are illustrated in Fig. 9.6) for description, prediction, and prescription: the three kinds of analytics. Structured input in the form of transactional data is fed into an inference cycle for the model to produce analytical results. If the input is unstructured text, structured sequences need to be extracted. The knowledge base on which implicit facts can be inferred is composed of computational models in the form of rules and structure relational data in the form of facts.

Das [15] explains how to build these computational models by mimicking human reasoning. One of the steps is to represent by expressive graphical constructs and linguistics variables, business analysts' internal mental models of things they observe and with which they interact. Computational models can also be viewed as patterns that are embedded within huge volumes of transactional data continuously generated by many business processing systems. Such models can therefore be extracted or learned via automated learning methods.

Finally, as a conclusion on FIAT, an assemblage of techniques and methods guided by a sort of transcendence principle (or an integrating framework) show a better potential to address the multifaceted problems present in the real world. This assemblage is "Analytics and Information Fusion": a processing chain that "transforms data into actionable knowledge." Meaning transcends from actuation (contexts of actions). Semantics grows as the transformations (FIAT) progress data toward actionable knowledge (Fig. 9.4).

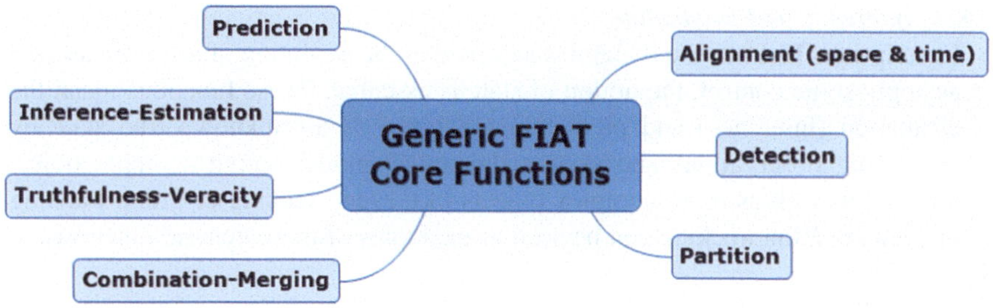

Fig. 9.7 FIAT generic functions

9.2.4 *Generic FIAT Core Functions*

The following set of generic functions or high-level requirements are being proposed as a basic composition for a FIAT-based processing chain [5, 16] as illustrated in Fig. 9.7.

1. *Alignment* (time-space):
   Description: The world is observed in time and space. To make sense out of observations one need to understand the context. Common referencing under time-space is necessary. Registration in time and space is well known in image processing.
2. *Detection*:
   Description: Applying the term "detection" to the levels of the data-information-knowledge processing chain. This generic function refers also to the field of data mining or knowledge discovery, i.e., the process of automatically searching large volumes of data for complex patterns.
3. *Partition*:
   Description: Facing the complexity of the world, partitioning eases the understanding of that world. Clustering and classification are examples of partitions. Image segmentation is also partitioning.
4. *Combination-Merging*:
   Description: After the information elements have been well characterized, this generic function combines those elements to gain better awareness. Aggregation, integration are other terms that can perform that function of combination or merging.
5. *Veracity-Truthfulness*: (true, true, degree of truth) elements, objects, relations
   Description: The action of putting forward some statement or proposition as true. The statement, or proposition, can also be put forward not as assertion, but as a supposition or hypothesis, as possibly true, and so on. Assertion using logical approaches is an example of that function. Reliability and relevance are also notions that can be calculated under that function.

6 & 7. *Inference and Prediction*

Description: To track the dynamics and predict the evolution of a system as well as apply some control, the notion of state is essential. These functions mean the estimation (inference) and prediction (analysis) of an unknown true state by combining observations and system dynamics (model output). Simple object tracking as well as more complex objects like event tracking or group tracking or even situation tracking can be seen as examples of these generic functions.

8. *Association*

Description: Identification and characterization of any link or relation between objects and information elements considering all dimensions. Correlation between information elements, association between detections and object tracks, application such as link analysis, or relations in social networks are all examples of implementation of that generic function.

A plethora of methods and techniques do exist to implement these FIAT functions as listed in Fig. 9.8. Possibility theory is one of them. Building a FIAT-based support system with only the tools provided by the possibility theory is not very wise since the characteristics of the information sources might be quite diverse. It is often required to hybrid the methods of information representation. Examples of such hybridization are fuzzy evidence theory [11] and fuzzy rough sets [13] to name a few.

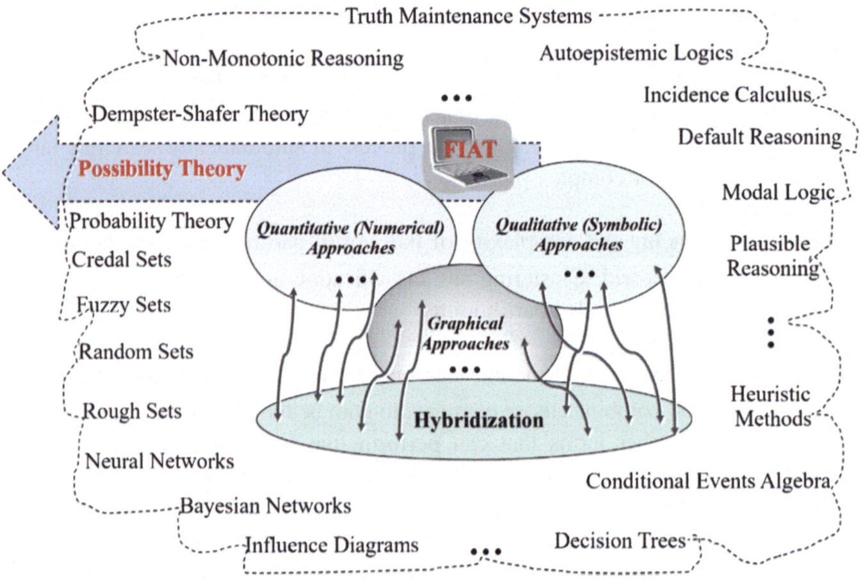

Fig. 9.8 Methods and techniques to implement FIAT generic functions

9.2.5 An Integrating Framework to Support the Design of FIAT-Based System

The assemblage of this plethora of techniques and methods represented in Fig. 9.8 into a processing chain that "transform data into actionable knowledge," i.e., to a specific goal (a support system), must be guided by an integrating framework. The development of a framework in which knowledge, information, and uncertainty can be represented, organized, structured, and processed to achieve a specific objective is a core requirement to cope with the complications inherent with the design of such a FIAT-based support system.

This integrating FIAT framework should:

1. Provide means to represent knowledge through well-defined notions of situation and awareness
2. Support the modeling of uncertainty, belief, and belief update
3. Provide the key "computational model" for FIAT and linkage to actions (users and machines)
4. Provide practical support for system design through modularization, refinement, validation, and verification
5. Offer a good compromise between operational and functional modeling in capturing systems behavior
6. Enable rapid prototyping and experimental validation of fairly abstract models
7. Support modeling of multi-agent systems

Taken individually, the techniques in Fig. 9.8 can only resolve aspects of the problems associated with complex systems. A very limited number of information fusion frameworks have been proposed in the literature that offer partial fulfillment of the above requirements. A number of significant and powerful "analytics" tools and techniques have also been offered to scientific and engineering communities for organizing, integrating, and visualizing large volumes of data [17]. In the domain of information fusion alone, noticeable efforts have recently been dedicated to define an integrating fusion framework [5], namely a promising one based on archetypal dynamics theory.

Sulis [18] introduced archetypal dynamics as:

> Archetypal dynamics is a formal framework for dealing with the study of meaning laden information flows within complex systems.

> This is a formal framework for dealing with the study of the relationships between systems, frames and their representations and the flow of information among these different entities. The framework consists of a triad of: semantic frame (representation), realizations (system) and interpretation (agent/user). Real systems relate to semantic frames through one of the dimensions of that triad. The viewpoint of archetypal dynamics is that meaning is tight with actions. A semantic frame is an organizing principle that ascribes meaning in a coherent and consistent manner to phenomena that have been parsed into distinct entities, mode of being, modes of behaving, modes of acting and interacting.

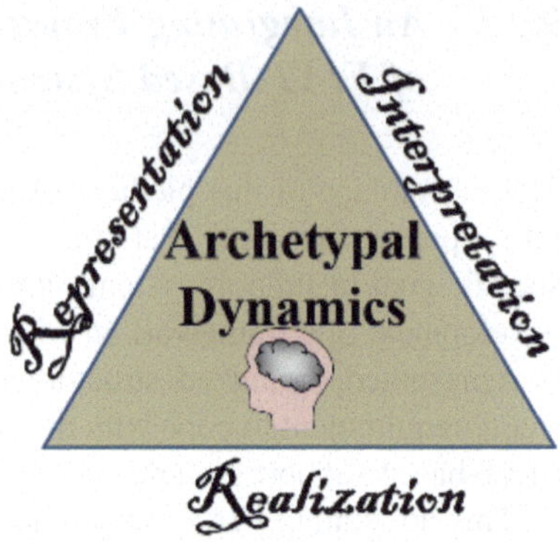

Fig. 9.9 Sulis' archetypal dynamic

The archetypal dynamics triad is being illustrated in Fig. 9.9. The semantic frame (defined later in the next section) provides partial answers to the six basic epistemic questions presented in Fig. 9.3 Full answers would suppose that one knows the ground truth. In archetypal dynamics, the way the information is understood is not in the sense of Shannon (i.e., the quantity of information), but in its active sense: "Information possesses content and elicits meaning."

A holonic computational model has been proposed in [16] using the ideas from archetypal dynamics and "holon" from complex systems theory to progress toward a potential FIAT computational model. Note that the three categories of FIAT in Fig. 9.2 follow the line of thought of the archetypal dynamics triad: representation (GIT)-interpretation (MS)-realization (distributed systems—MAS). Archetypal dynamics (AD) is almost unknown to the information fusion and analytics communities.

9.3 Awareness Quality and Decision Support

Situation awareness (SAW) quality can be related to the amount of information available to an individual: no information should result in poor SAW so very low decision quality as indicated on the lower left portion of Fig. 9.10. In such a case, a natural reaction would be to provide mechanisms to increase the amount of information available to the decision-makers to improve SAW quality. One could even claim that a good approach to reach better SAW and decision-making would be to provide as much information as possible: to provide "all information, everywhere, at all time," but more information does not automatically mean better SAW. First, all this information may exceed the human information processing capabilities, leading to cognitive overload (lower right portion of Fig. 9.10). Second, it is not all of the data and information available from the environment that is relevant and useful for reaching an optimal decision. In fact, in some situations, most of the data can be seen as distracters and noise for the decision-maker and may thus reduce his/her level of

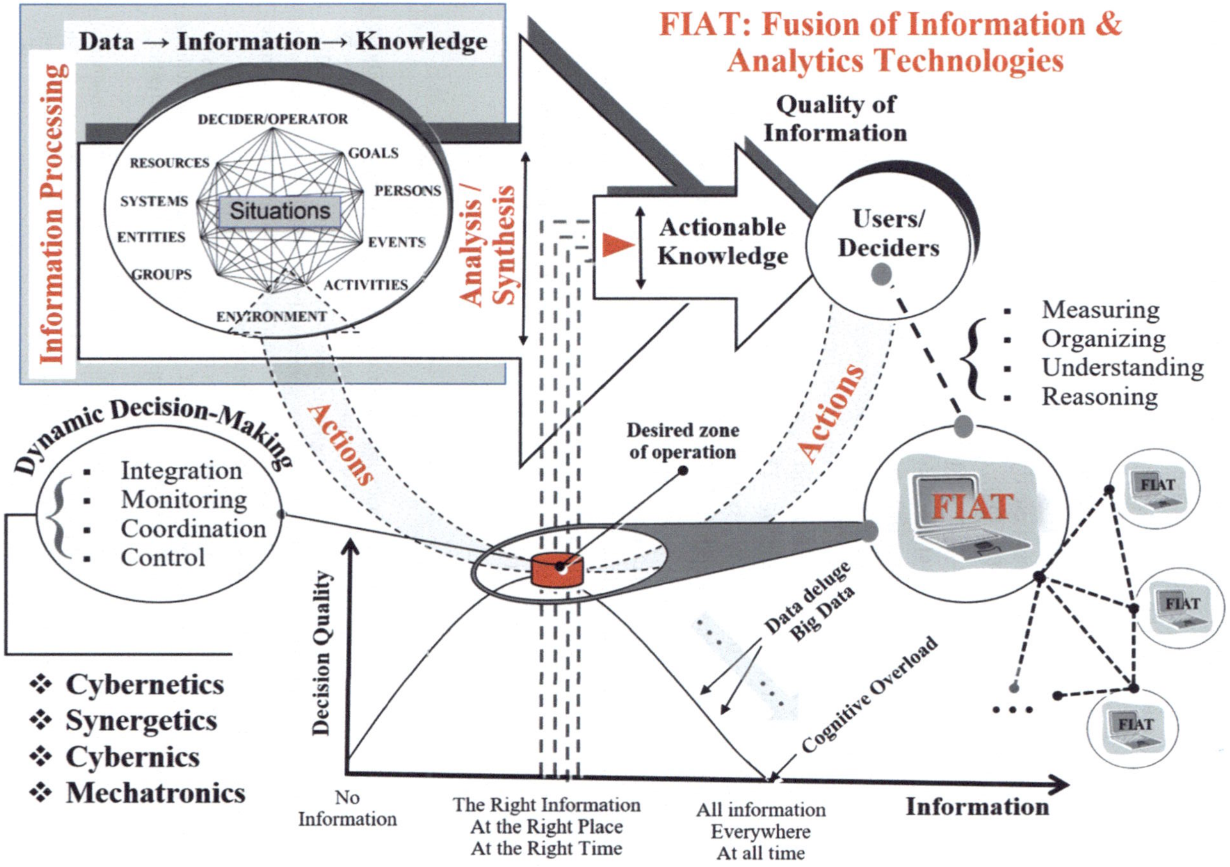

Fig. 9.10 FIAT supporting deciders to be in a desired zone of operation. (Adapted from [11])

SAW. The decision-maker must detect and use only a specific fraction, called "useful bandwidth" in Fig. 9.10, of this information to enhance his/her SAW and decision-making processes. Such considerations lead to the concept of "the right information, at the right place, at the right time," in turn leading to the notion of information relevance where the overall goal would be to provide only useful information (actionable knowledge) to deciders (desired zone of operation in Fig. 9.10).

A very crucial aspect of quality of information (QoI) is relevance of information: relevance shall be behind any intelligent filtering and context-aware processing concepts and impacts heavily on quality of decisions and the problem of volume of information in a Big Data environment. The analysis of information relevance is one very important prerequisite to build any FIAT-based support system. In Fig. 9.10, FIAT is represented by the analytics and information fusion processes to support measuring-organizing-understanding-reasoning with data-information-knowledge. Using an analogy borrowed from the signal processing community and pictured in Fig. 9.10, one can imagine a useful "information" bandwidth where the overall goal would be to provide only useful information to deciders, i.e., actionable knowledge. That conceptual multidimensional "useful bandwidth" could be defined by assembling appropriate FIAT such as in smart filtering and metadata-based approaches. In fact, in the presence of a deluge of data, that demands classification and prioritization of information required by a given task execution [19]: "As data become abundant, the main problem is no longer finding the information as such but laying one's hands on the relevant bits easily and quickly." New sciences or perhaps evolutions of cybernetics[1] are emerging to address issues related to networking and complexity such as synergetics,[2] cybernics,[3] and mechatronics[4] (Fig. 9.10).

[1]Cybernetics: Cybernetics is a transdisciplinary approach for exploring how the scientific study of how humans, animals, and machines control and communicate with each other. Cybernetics is very relevant to mechanical, physical, biological, cognitive, and social systems (e.g., CPSS). The essential goal of the broad field of cybernetics is to understand and define the functions and processes of systems that have goals and that participate in circular, causal chains that move from action to sensing to comparison with desired goal and again to action.

[2]Synergetics: Synergetics is the empirical study of systems in transformation, with an emphasis on total system behavior unpredicted by the behavior of any isolated components, including humanity's role as both participant and observer. (https://en.wikipedia.org/wiki/Synergetics_(Fuller)).

[3]Cybernics: Cybernics is a new interdisciplinary field study of technology that enhances, strengthens, and supports physical and cognitive functions of human beings, based on the fusion of human, machine, and information systems. The design of a seamless interface for interaction between the interior and exterior of the human body taking into account aspects such as the physical, neurophysiological, and cognitive levels [20].

[4]Mechatronics: Mechatronics is a multidisciplinary branch of engineering that focuses on the engineering of both electrical and mechanical systems and also includes a combination of robotics, electronics, computer, telecommunications, systems, control, and product engineering.

9.4 Where Does Possibility Theory Fit Within the Design of FIAT Systems?

As mentioned above, building a FIAT-based support system with only the tools provided by the possibility theory is not very wise since the characteristics of the information sources might be quite diverse. In fact, possibilistic mathematical tools of the previous chapters can be used in any of the generic functions of FIAT (Fig. 9.7): detection-association-merging-inference-prediction-truthfulness-partition. A possibilistic approach offers a great deal of performance when you need to compute these functions in the presence of vague or less complete information.

Figure 9.11 presents a prospective view of what the design of future DSS might require to cope with the complexity of the real-world applications. Table 9.1 gives some examples of real-world FIAT applications that have been designed using various kinds of theories (graphical-numerical-symbolic). So far, no integrating framework has been standardized for the design of FIAT DSS. The archetypal dynamics is still at its embryonic development through the proposed causal tapestries [16, 18, 37]. Considerable research effort still needs to be dedicated to materializing Fig. 9.11 into a practical design framework for FIAT DSS. In the meantime, we offer to use the mathematical tools, based on possibility theory, presented in the previous chapters that target only some of the FIAT core generic functions.

9.5 Conclusion

Fusion of Information and Analytics Technologies (FIAT) is a sine qua non enabler to cope with Cyber-Physical and Social Systems (CPSS) facing information overload and complexity. FIAT will support rethinking of cyber interface functions, such as coordination, integration, monitoring, and control, progress decision, and support concepts to be operational through these interface functions. The problem is tremendously complex, and the current book shows a promising theory (possibility theory) that can only solve a facet of the problem. The problem is multidisciplinary by nature, and several books and paper would be required to explore the vast solution space. However, the authors hope that this book would motivate the scientific and engineering communities to progress some of the approaches presented here and provide more implementable solutions to cope with that complexity. Possibility theory, eventually hybridized with other representations, can certainly contribute to the engineering of an efficient FIAT processing chain to transform data information to actionable knowledge in complex dynamic environments.

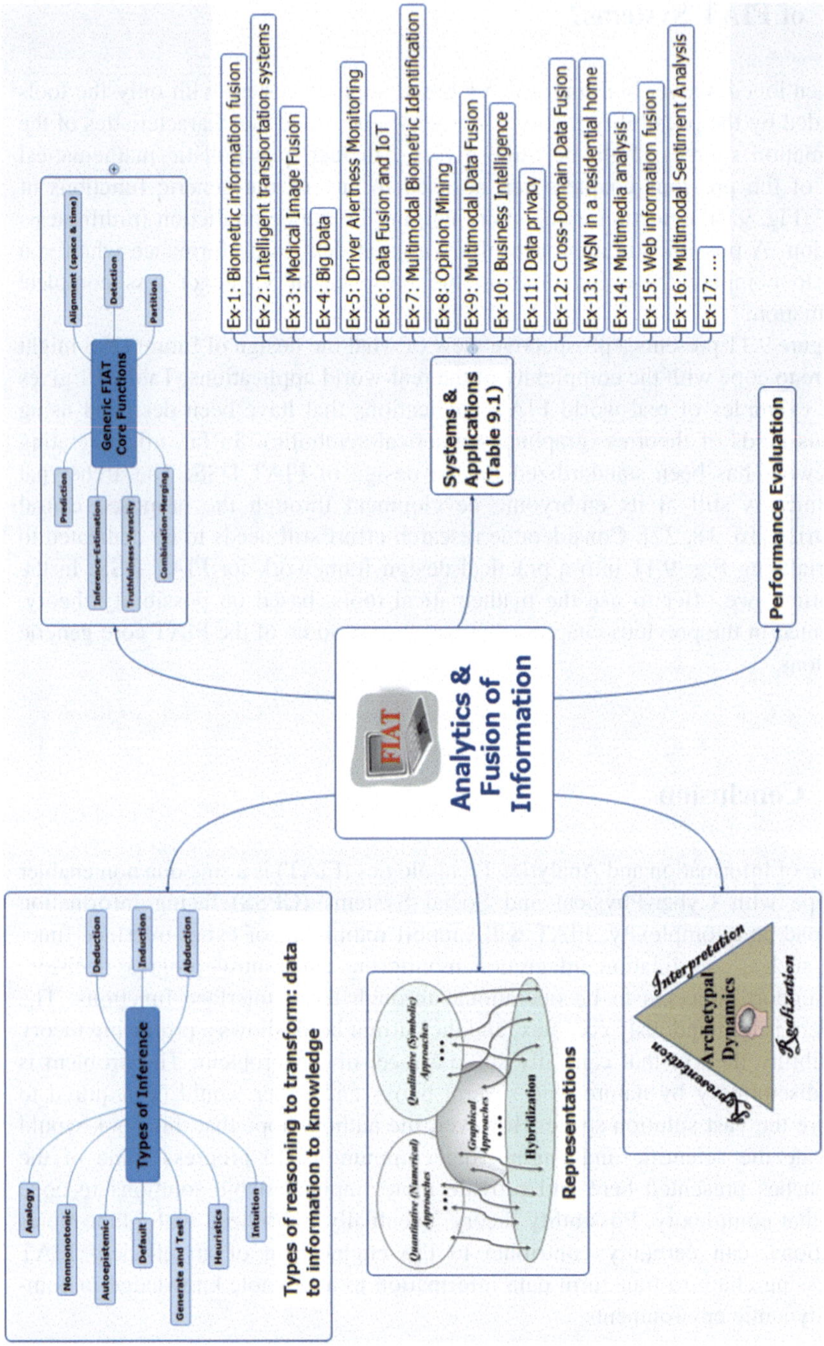

Fig. 9.11 A pictorial map of what is required for the design of a FIAT-based decision support system (DSS)

Table 9.1 Applications of FIAT in real-world problems

Examples of applications	References
Ex-1: Biometric information fusion	[21]
Ex-2: Intelligent transportation systems	[22]
Ex-3: Medical image fusion	[23]
Ex-4: Big Data	[24]
Ex-5: Driver alertness monitoring	[25]
Ex-6: Data fusion and IoT	[26]
Ex-7: Multimodal biometric identification	[27]
Ex-8: Opinion mining	[28]
Ex-9: Multimodal data fusion	[29]
Ex-10: Business intelligence	[30]
Ex-11: Data privacy	[31]
Ex-12: Cross-domain data fusion	[32]
Ex-13: WSN in a residential home	[33]
Ex-14: Multimedia analysis	[34]
Ex-15: Web information fusion	[35]
Ex-16: Multimodal sentiment analysis	[36]

References

1. L. Atzori, A. Iera, G. Morabito, The internet of things: A survey. Comput. Netw. **54**, 2787–2805 (2010)
2. M. Chen, S. Mao, Y. Liu, Big data: A survey. Mobile Netw. Appl. **19**, 171–209 (2014)
3. Z. Liu, D.-S. Yang, D. Wen, W.-M. Zhang, W. Mao, Cyber-physical-social systems for command and control. IEEE Intell. Syst. **26**, 92–96 (2011)
4. A. Sheth, P. Anantharam, C. Henson, Physical-cyber-social computing: An early 21st century approach. IEEE Intell. Syst. **28**, 78–82 (2013)
5. É. Bossé, B. Solaiman, *Information Fusion and Analytics for Big Data and IoT* (Artech House, Norwood, 2016)
6. M.R. Endsley, *Designing for Situation Awareness: An Approach to User-Centered Design* (CRC Press, Boca Raton, 2016)
7. E. Blasch, É. Bossé, D.A. Lambert, *High-Level Information Fusion Management and Systems Design* (Artech House, Boston, 2012)
8. M.R. Endsley, Automation and situation awareness. Autom. Hum. Perform. Theory Appl. **20**, 163–181 (1996)
9. J. Llinas, Reexamining information fusion-decision making inter-dependencies, in *2014 IEEE International Inter-Disciplinary Conference on Cognitive Methods in Situation Awareness and Decision Support (CogSIMA)* (2014), pp. 1–6
10. É. Bossé, J. Roy, S. Wark, *Concepts, Models, and Tools for Information Fusion* (Artech House, Boston, 2007)
11. E. Bossé, B. Solaiman, Fusion of information and analytics: A discussion on potential methods to cope with uncertainty in complex environments (big data and IoT). Int. J. Digital Signals Smart Syst. **2**, 279–316 (2018)
12. J. Esteban, A. Starr, R. Willetts, P. Hannah, P. Bryanston-Cross, A review of data fusion models and architectures: Towards engineering guidelines. Neural Comput. Appl. **14**, 273–281 (2005)
13. W.W. Eckerson, Predictive analytics, extending the value of your data warehousing investment, in *TDWI Best Practices Report*, vol. 1 (2007), pp. 1–36
14. N. Sheikh, *Implementing Analytics: A Blueprint for Design, Development, and Adoption* (Newnes, 2013)

15. S. Das, *Computational Business Analytics* (Chapman and Hall/CRC, Boca Raton, 2013)
16. B. Solaiman, É. Bossé, L. Pigeon, D. Gueriot, M.C. Florea, A conceptual definition of a holonic processing framework to support the design of information fusion systems. Inf. Fusion **21**, 85–99 (2015)
17. P. Zikopoulos, C. Eaton, *Understanding Big Data: Analytics for Enterprise Class Hadoop and Streaming Data* (McGraw-Hill Osborne Media, New York, 2011)
18. W. Sulis, Archetypal dynamics, emergent situations, and the reality game. Nonlinear Dyn. Psychol. Life Sci. **14**, 209–238 (2010)
19. K. Cukier, *Data, Data Everywhere: A Special Report on Managing Information* (Economist Newspaper, London, 2010)
20. Y. Sankai, *Cybernics: Fusion of Human, Machine and Information Systems* (Springer, Tokyo, 2014)
21. G. Slanzi, G. Pizarro, J.D. Velasquez, Biometric information fusion for web user navigation and preferences analysis: An overview. Inf. Fusion **38**, 12–21 (2017)
22. N.-E. El Faouzi, H. Leung, A. Kurian, Data fusion in intelligent transportation systems: Progress and challenges–a survey. Inf. Fusion **12**, 4–10 (2011)
23. A.P. James, B.V. Dasarathy, Medical image fusion: A survey of the state of the art. Inf. Fusion **19**, 4–19 (2014)
24. S. Ramírez-Gallego, A. Fernández, S. García, M. Chen, F. Herrera, Big data: Tutorial and guidelines on information and process fusion for analytics algorithms with MapReduce. Inf. Fusion **42**, 51–61 (2018)
25. B.-G. Lee, W.-Y. Chung, Driver alertness monitoring using fusion of facial features and bio-signals. IEEE Sens. J. **12**, 2416–2422 (2012)
26. F. Alam, R. Mehmood, I. Katib, N.N. Albogami, A. Albeshri, Data fusion and IoT for smart ubiquitous environments: a survey. IEEE Access **5**, 9533–9554 (2017)
27. S. Sree, N. Radha, A survey on fusion techniques for multimodal biometric identification. Int. J. Innovative Res. Comput. Commun. Eng. **2**, 7493–7497 (2014)
28. J.A. Balazs, J.D. Velásquez, Opinion mining and information fusion: A survey. Inf. Fusion **27**, 95–110 (2016)
29. D. Lahat, T. Adali, C. Jutten, Multimodal data fusion: An overview of methods, challenges, and prospects. Proc. IEEE **103**, 1449–1477 (2015)
30. G. Shroff, P. Agarwal, L. Dey, Enterprise information fusion for real-time business intelligence, in *14th International Conference on Information Fusion* (2011), pp. 1–8
31. G. Navarro-Arribas, V. Torra, Information fusion in data privacy: A survey. Inf. Fusion **13**, 235–244 (2012)
32. Y. Zheng, Methodologies for cross-domain data fusion: An overview. IEEE Trans. Big Data **1**, 16–34 (2015)
33. S. Rodríguez, J.F. De Paz, G. Villarrubia, C. Zato, J. Bajo, J.M. Corchado, Multi-agent information fusion system to manage data from a WSN in a residential home. Inf. Fusion **23**, 43–57 (2015)
34. P.K. Atrey, M.A. Hossain, A. El Saddik, M.S. Kankanhalli, Multimodal fusion for multimedia analysis: A survey. Multimedia Syst. **16**, 345–379 (2010)
35. J. Yao, V.V. Raghavan, Z. Wu, Web information fusion: A review of the state of the art. Inf. Fusion **9**, 446–449 (2008)
36. A. Zadeh, M. Chen, S. Poria, E. Cambria, L.-P. Morency, Tensor fusion network for multimodal sentiment analysis. *arXiv preprint arXiv:1707.07250* (2017)
37. W. Sulis, Causal tapestries. Bull. Am. Phys. Soc. **56**, 80–86 (2011)

Index

B. Solaiman, É. Bossé, *Possibility Theory for the Design of Information Fusion Systems*, Information Fusion and Data Science,
https://doi.org/10.1007/978-3-030-32853-5

The manufacturer's authorised representative in the EU is Springer Nature Customer Service Centre GmbH, Europaplatz 3, 69115 Heidelberg, Germany. If you have any concerns regarding our products, please contact ProductSafety@springernature.com

Printed and bound by CPI Group (UK) Ltd, Croydon, CR0 4YY
05/10/2026
02228145-0009